Properties of
GALLIUM ARSENIDE

INSPEC
The Institution of Electrical Engineers

EMIS Datareviews Series No. 2

Published by: INSPEC, The Institution of Electrical Engineers, London and New York

ISBN 0 85296 323 8

Printed in the United Kingdom by Unwin Brothers, The Gresham Press, Old Woking, Surrey, England.

Properties of
GALLIUM ARSENIDE

EMIS Datareviews Series

The series is bringing together materials property data, obtained in the course of many different and varied research programmes, into collections comprising short articles which have been called ´Datareviews´.

Each Datareview normally focuses on one property of a narrowly-defined material system. A Datareview contains both a summary of data extracted from selected articles which have been published in scientific journals or conference proceedings and bibliographic references to these articles and any other key sources. The Datareview is distinguished from the conventional review by its shorter length, high degree of specialisation and by its emphasis on hard data. However, like any scientific review it seeks to discuss, compare and illuminate research results, thus enabling the reader quickly to gain an insight and up-to-date awareness of its subject matter as well as to conduct further study by means of the references provided.

Datareviews are compiled by researchers from universities and industrial research laboratories and the EMIS Group at INSPEC (The Information Services Division of the Institution of Electrical Engineers). The selection of articles for the compilation of Datareviews, and the completed Datarviews themselves, are refereed by prominent researchers in the appropriate field of study. INSPEC is thus indebted to a large number of people who give of their time in the preparation of Datarviews for these volumes. They are gratefully acknowledged within the publications.

In addition to appearing in this and other publications of the series, Datareviews are published electronically on the EMIS (Electronic Materials Information Service) databank, a publicly available online service produced by INSPEC.

The EMIS databank was created and is being developed by INSPEC to satisfy the need for the latest and most authoritative data on the properties of electronic materials. Experts in the field, who pointed to this need, assist INSPEC staff in collecting, selecting and analysing the most recently published data.

This databank provides a collecting point for Datareviews prior to publication in the EMIS Datareviews Series, and also a facility for updating existing Datareviews as new data become available.

Individual publications in the EMIS Datareviews Series will be reprinted as and when new data added to the EMIS datbank represent a significant advance in the state-of-the-art.

Foreword

It was a pleasant surprise to be asked to write a foreword for this book, since I know that the contents will prove valuable to those such as myself involved with III-V compound semiconductor research, and also to members of the substantial - and growing - community working to exploit a wider range of applications for GaAs and related compounds. This is a book which few will read from cover to cover, though I do recommend that any reader - no matter how specialised his/her interests are - should take the time to become broadly familiar with the coverage provided in all of the 100+ topic sections before placing the volume on the reference shelf. From that time onwards, the individual Datareviews are available to be consulted whenever one needs an up-to-date critical distillate of what is and is not known about a specific property of GaAs; the collective effort of Mr Sears and the EMIS Group with Dr Brice and more than 30 other contributing authors.

When the III-V zincblende binaries were shown in the early 1950s to be semiconductors, there was a major research challenge for elucidation of the physics and chemistry of these and the subsequently created ternary and quaternary alloys. Acceptance of that challenge has resulted in a massive body of published - but widely dispersed - knowledge about GaAs. I became aware of just how much had been published about GaAs (with widely varying degrees of reliability), and which early measurements had never been reconfirmed, while assembling a personal 1980-era GaAs datafile, a project which evolved into a 1982 review article in 'Journal of Applied Physics'. It is thus pleasing to me that several numbers which worried me most then have since been remeasured, some in different and more direct ways. That information has been available to the authors of the various Datareviews in this book.

Popular today in science are compendium volumes, for which an editor collects chapters written by suitable experts. This book represents an unusually concentrated effort by the editing team and by the authorities recruited as contributing authors, to provide concise and critical evaluations of so many key physical, electronic, and processing aspects of GaAs. Moreover, as pointed out by Dr Brice in his Introduction, the EMIS databank has data stored on some less commonly wanted topics, and will have online availability for any Datareview in this book - with periodic updates as newly available data makes these necessary. I am old-fashioned enough to hope that the printed book will not become extinct, but do recognise and draw to the reader's attention the value of being able to summon electronically over the next several years an updated Datareview for any topic which appears not to be doubt-free as of press time. As attempts are made to use GaAs in electronic and optoelectronic applications of greater complexity and sophistication, the needs for precise information become even more pressing.

J.S Blakemore
Oregon Graduate Center
Beaverton, Oregon 97006, USA

Properties of Gallium Arsenide

Introduction

While it is useful to be able to describe phenomena in a qualitative manner, the best contributions to science and technology are usually quantitative. In order to produce quantitative models, it is often necessary to know the appropriate numerical values of the properties of the materials involved. These values will be those for the relevant conditions of temperature and pressure and, particularly when semiconductors are involved, will make allowances for the effects of the purity and perfection of the actual materials used. Finding the appropriate numerical value of a material property can be a major problem which may call for significant specialist knowledge. This book is one of a series aimed at simplifying the task of selection. Experts in various fields have looked at properties which they understand and have produced values recommended under various circumstances. Every entry has been checked by an independent referee. Thus we believe that this book contains the current best estimates for all the commonly needed data on gallim arsenide.

Less commonly wanted data are in the EMIS databank which can be consulted online. It is also worth mentioning that Datareviews (like those given here) are revised from time to time (usually about once per year) so that the online service should also be used when the latest recommendations are needed. A revised Datareview will have the same RN number as that given in its title in this volume and can be called by using that number.

When the demand provides justification, this book will be revised. EMIS will be happy to receive comments and suggestions. Offers to write Datareviews on topics not covered here will be particularly welcome.

J.C Brice
Philips Research Laboratories
Redhill
Surrey RH1 5HA
England

Contributing Authors

A R Adams	Surrey University, UK	5.11,5.12,6.3, 6.4,6.5,7.1
C I H Ashby	Sandia National Laboratories, USA	16.1 to 16.5
D E Aspnes	Bell Communications Research, USA	8.1 to 8.9
R S Bhattacharya	Universal Energy Systems, USA	3.1, 3.2
I W Boyd	University College, London, UK	10.2
J C Brice	Philips Research Laboratories, UK	1.1 to 1.10
M R Brozel	Trent Polytechnic, UK	10.1,13.1,13.2
J P R David	University of Sheffield, UK	4.1, 4.2, 4.3
P J Dobson	Philips Research Laboratories, UK	15.1
K Duncan	English Electric Valve Co, UK	5.13
L Eaves	University of Nottingham, UK	17.12
I H Goodridge	Plessey Research (Caswell), UK	3.3
C R M Grovenor	Oxford University, UK	17.3, 17.4
B Hamilton	UMIST, UK	12.8,12.9,12.10
H L Hartnagel	Institut fur Hochfrequenztechnik, German Federal Republic	15.2, 15.3
D Lancefield	Surrey University, UK	5.1 to 5.7,5.9
D Lee	RSRE, UK	2.1,18.5,18.6, 18.7
D C Look	Wright State University, USA	6.1,6.2,11.1
J D Medland	GEC Hirst Research Centre, UK	12.4 to 12.7
J Nicholas	UMIST, UK	13.8, 13.9
A R Peaker	UMIST, UK	13.6 to 13.10, 18.1, 18.2
W Potz	Arizona State University, USA	7.3, 13.5
D J Robbins	Plessey Research (Caswell), UK	9.1, 9.2
G P Schwartz	AT&T Bell Laboratories, USA	17.2
B J Sealy	Surrey University, UK	2.2, 5.10
S P Speakman	English Electric Valve Co, UK	18.8
R A Stradling	Imperial College, London, UK	11.2 to 11.7
B Tuck	University of Nottingham, UK	14.1 to 14.6
H H Weider	University of California, USA	17.11
D R Wight	RSRE, UK	5.13,5.14,6.6, 6.7
H V Winston	Hughes Research Laboratories, USA	13.4
S Yasuami	Toshiba R&D Center, Japan	13.3

Acknowledgements

It is a pleasure to acknowledge the work both of the contributing authors named on the previous page and of the following experts in the field who also have participated in the production of this book:

D.A Anderson	RSRE, UK
B Barry	STL, UK
J.S Blakemore	Oregon Graduate Center, USA
R.T Blunt	Plessey Research (Caswell), UK
J.B Clegg	Philips Research Laboratories, UK
H.G Drickamer	University of Illinois, USA
S Elliott	Cambridge University, UK
G.A Gledhill	Royal Holloway College, UK
P.D Greene	STL, UK
H.G Grimmeiss	University of Lund, Sweden
J.S Harris	Stanford University, USA
D.T.S Hurle	RSRE, UK
U Kaufmann	Fraunhofer Institut fur Angewandte Festkorperphysik, German Federal Republic
H Kuwamoto	Rockwell International, USA
I Lindau	Stanford University, USA
H Palfrey	Southampton University, UK
T.J Parker	Royal Holloway College, UK
R Pond	Liverpool University, UK
G.D Pitt	STL, UK
E.H Rhoderick	UMIST, UK
D Shaw	University of Hull, UK
M.S Skolnick	RSRE, UK
D Stirland	Plessey Research (Caswell), UK
C.W Turner	Kings College, London, UK
K Vanner	Plessey Research (Caswell), UK
J.R Waldrop	Rockwell International, USA
B.L Weiss	Surrey University, UK
A.F.W Willoughby	Southampton University, UK
D.R Wood	British Telecom Research Laboratories, UK

We are particularly grateful to Professor John S Blakemore of the Oregon Graduate Center, USA for writing the Foreword to this volume. Also to Dr John C Brice of Philips Research Laboratories, UK who, in addition to writing the Introduction and contributing several Datareviews, has been a constant source of advice and encouragement.

EMIS Group

Abbreviations

AC	alternating current
AES	Auger electron spectroscopy
BH	barrier height
CB	conduction band
CE	conductivity (electrical)
concn	concentration
CVD	chemical vapour deposition
DAP	donor acceptor pair
DC	direct current
diam	diameter
DLTS	deep level transient spectroscopy
Ec	conduction band edge
E(c)	conduction band edge
E(F)	Fermi level
Ev	valence band edge
E(v)	valence band edge
EPR	electron paramagnetic resonance
ESR	electron spin resonance
exp	exponential
FB	free to bound transition
FEC	fully encapsulated Czochralski
FET	field effect transistor
FWHM	full width at half maximum
HB	horizontal Bridgman
HF	high frequency
IBAE	ion-beam assisted etching
IR	infra-red
JFET	junction field effect transistor
LEC	liquid encapsulated Czochralski
LEED	low energy electron diffraction
LEK	liquid encapsulated Kyropoulos
LO	longitudinal optic
LPE	liquid phase epitaxy
LSI	large scale integration
LVM	localised vibrational modes

MBE	molecular beam epitaxy
ME	mobility of electrons
MESFET	metal-semiconductor field effect transistor
MIS	metal-insulator-semiconductor
MOCVD	metal-organic chemical vapour deposition
MOVPE	metal-organic vapour phase epitaxy
NMR	nuclear magnetic resonance
NTD	neutron transmutation doped
OTCS	optical transient current spectroscopy
PBN	pyrolytic boron nitride
PHC	photoconductivity
PL	photoluminescence
RBS	Rutherford backscattering
RF	radio frequency
RHEED	reflection high energy electron diffraction
RIBE	reactive ion-beam etching
RIE	reactive ion etching
SCLC	space charge limited current
sectn	section
SEM	scanning electron microscopy
SI	semi-insulating
SIMS	secondary ion mass spectroscopy
SQRT	square root
STEM	scanning transmission electron microscopy
SXPS	soft X-ray photoelectron spectroscopy
TD	thermal diffusivity
TEC	thermal expansion coefficient
TEM	transmission electron microscopy
temp	temperature
TO	transverse optic
TSC	thermally stimulated current
UHV	ultrahigh vacuum
UPS	ultra-violet photoelectron spectroscopy
UV	ultra-violet
VB	valence band
VPE	vapour phase epitaxy
wrt	with respect to
X	exciton
XPS	X-ray photoelectron spectroscopy

CONTENTS

Introduction to the Datareviews Series ii

Foreword to Properties of Gallium Arsenide iii

Introduction to Properties of Gallium Arsenide iv

Contributing Authors v

Acknowledgements vi

List of Abbreviations vii

1 PHYSICAL AND THERMAL PROPERTIES
J C Brice

1.1 Density of solid GaAs
1.2 Lattice parameter of GaAs
1.3 Lattice parameter of GaAs, doping dependence
1.4 Bulk modulus of GaAs
1.5 Stiffness of GaAs
1.6 Compliance of GaAs
1.7 Thermal expansion coefficient of GaAs, temperature dependence
1.8 Specific heat of GaAs
1.9 Thermal conductivity of GaAs
1.10 Thermal diffusivity of GaAs

2 RESISTIVITY
D Lee, B J Sealy

2.1 Resistivity of SI GaAs
2.2 Sheet resistivity of ion implanted GaAs

3 CARRIER CONCENTRATIONS
R S Bhattacharya, I H Goodridge

3.1 Carrier concentrations in ion implanted n-type GaAs
3.2 Carrier concentrations in ion implanted p-type GaAs
3.3 Carrier concentrations in doped and undoped epitaxial GaAs

4 CARRIER IONIZATION RATES
J P R David

4.1 Carrier ionization rates in GaAs, temperature dependence
4.2 Carrier ionization rates in GaAs, doping dependence
4.3 Carrier ionization rates in GaAs, electric field dependence

5 ELECTRON MOBILITY, DIFFUSION AND LIFETIME
D Lancefield, A R Adams, B J Sealy, D R Wight, K Duncan, EMIS Group

5.1 Electron mobility in bulk crystalline GaAs
5.2 Electron mobility in bulk crystalline GaAs, temperature dependence
5.3 Electron mobility in LPE GaAs
5.4 Electron mobility in VPE GaAs
5.5 Electron mobility in VPE GaAs, temperature dependence
5.6 Electron mobility in VPE GaAs, doping dependence
5.7 Electron mobility in MBE GaAs
5.8 Electron mobility in MBE GaAs, temperature dependence
5.9 Electron mobility in MBE GaAs, doping dependence
5.10 Electron mobility in ion implanted GaAs, doping dependence
5.11 Electron mobility in GaAs, pressure dependence
5.12 Electron effective mass in GaAs, pressure dependence
5.13 Electron diffusion length and coefficient in p-type GaAs
5.14 Electron lifetime in p-type GaAs

6 HOLE MOBILITY, DIFFUSION AND LIFETIME
D R Wight, D C Look, A R Adams

6.1 Hole mobility in undoped GaAs
6.2 Hole mobility in doped and ion implanted GaAs
6.3 Hole mobility in GaAs, temperature dependence
6.4 Hole mobility in GaAs, pressure dependence
6.5 Hole effective mass in GaAs, pressure dependence
6.6 Hole diffusion length and coefficient in n-type GaAs
6.7 Hole lifetime in n-type GaAs

7 ENERGY GAP
A R Adams, W Potz, EMIS Group

7.1 Direct band gap of GaAs, temperature dependence
7.2 Direct band gap of GaAs, pressure dependence
7.3 Optical deformation potential for holes at the GAMMA point in GaAs
7.4 Optical gap of amorphous GaAs

8 OPTICAL FUNCTIONS (Complex Refractive Index, Dielectric Function and Absorption Coefficient)
D E Aspnes

8.1 Optical functions of intrinsic GaAs: General remarks
8.2 Optical functions of intrinsic GaAs: Static and IR dielectric constants
8.3 Optical functions of intrinsic GaAs: Reststrahlen region (0-60 meV)
8.4 Optical functions of intrinsic GaAs: Transparent region (0.1-1.4 eV)
8.5 Optical functions of intrinsic GaAs: Direct bandgap region (1.3-1.5 eV)
8.6 Optical functions of intrinsic GaAs: Visible to near-UV
8.7 Optical functions of intrinsic GaAs: Vacuum UV (up to 155 eV)
8.8 Table of optical functions of intrinsic GaAs: Refractive index and absorption coefficient vs energy (0-155 eV)
8.9 Extrinsic effects on the optical functions of GaAs

9 ELECTRO-OPTIC PROPERTIES
D Robbins

9.1 Electro-optical coefficient of GaAs
9.2 Franz-Keldysh effect parameters of GaAs

10 INFRA-RED ABSORPTION
M R Brozel, I W Boyd

10.1 IR absorption bands of bulk GaAs
10.2 Two-photon IR absorption coefficient of GaAs

11 PHOTOCONDUCTIVITY SPECTRA
R A Stradling, D C Look

11.1 IR Photoconductivity spectra of bulk SI GaAs
11.2 Far-IR photoconductivity spectra of shallow donors in GaAs epilayers: General remarks
11.3 Far-IR photoconductivity spectra of shallow donors in LPE GaAs
11.4 Far-IR photoconductivity spectra of shallow donors in VPE GaAs
11.5 Far-IR photoconductivity spectra of shallow donors in MOCVD GaAs
11.6 Far-IR photoconductivity spectra of shallow donors in MBE GaAs
11.7 Far-IR photoconductivity spectra of shallow donors in GaAs doped by neutron transmutation

12 PHOTOLUMINESCENCE SPECTRA
B Hamilton, J D Medland, EMIS Group

12.1 Photoluminescence spectra of undoped SI GaAs
12.2 Photoluminescence spectra of Cr doped SI GaAs
12.3 Photoluminescence spectra of undoped LEC p-type GaAs
12.4 Photoluminescence spectra of LPE GaAs
12.5 Photoluminescence spectra of VPE GaAs
12.6 Photoluminescence spectra of MOCVD GaAs
12.7 Photoluminescence spectra of MBE GaAs
12.8 Photoluminescence spectra of Group II doped GaAs
12.9 Photoluminescence spectra of Group IV doped GaAs
12.10 Photoluminescence spectra of Group VI doped GaAs

13 DEFECTS AND IR IMAGING OF DEFECTS
M R Brozel, A R Peaker, J Nicholas, H V Winston, S Yasuami, W Potz

13.1 Analysis of defects in undoped SI LEC GaAs by near-IR imaging
13.2 Defect densities in SI LEC GaAs
13.3 Structure of native defects in SI LEC GaAs
13.4 Defect energy levels in melt-grown GaAs
13.5 Defect energy levels related to the anion antisite defect in GaAs
13.6 Defect energy levels in LPE GaAs
13.7 Defect energy levels in VPE GaAs
13.8 Defect energy levels in MOVPE GaAs
13.9 Defect energy levels in MBE GaAs
13.10 Defect energy levels in electron irradiated GaAs

14 DIFFUSION OF IMPURITIES
B Tuck

14.1 Diffusion of Cr in GaAs
14.2 Diffusion of Mn in GaAs
14.3 Diffusion of Zn in GaAs
14.4 Diffusion of Si in GaAs
14.5 Diffusion of Sn in GaAs
14.6 Diffusion of S in GaAs

15 SURFACE STRUCTURE AND OXIDATION
P J Dobson, H L Hartnagel

15.1 Surface structure of GaAs
15.2 Oxide layer structures on GaAs
15.3 Oxidation rates of GaAs

16 ETCHING RATES
C I H Ashby

16.1 Etching rates of wet etched GaAs
16.2 Etching rates of plasma etched GaAs
16.3 Etching rates of ion-beam milled and sputter etched GaAs
16.4 Etching rates of reactive ion and reactive ion-beam etched GaAs
16.5 Etching rates of GaAs etched by laser-assisted methods

17 INTERFACES AND CONTACTS
L Eaves, C R M Grovenor, G P Schwartz, H H Wieder, EMIS Group

17.1 Structure of the GaAs/Ag interface
17.2 Structure of the GaAs/Al interface
17.3 Structure of the GaAs/Au interface
17.4 Structure of the GaAs/Au-Ge interface
17.5 Structure of the GaAs/Cu interface
17.6 Barrier height at the GaAs/Ag interface
17.7 Barrier height at the GaAs/Al interface
17.8 Barrier height at the GaAs/Au interface
17.9 Barrier height at the GaAs/Pt interface
17.10 Barrier height at the GaAs/W interface
17.11 Band bending at GaAs/dielectric interfaces
17.12 Conduction band and valence band offset at the GaAs/AlGaAs heterostructure interface

18 AlGaAs: MISCELLANEOUS PROPERTIES
A R Peaker, D Lee, S Speakman, EMIS Group

18.1 Defect energy levels in AlGaAs
18.2 DX defect centre in AlGaAs
18.3 Photoluminescence spectra of AlGaAs
18.4 Electron mobilities in AlGaAs
18.5 Carrier concentrations in LPE AlGaAs
18.6 Carrier concentrations in MOCVD AlGaAs
18.7 Carrier concentrations in MBE AlGaAs
18.8 Etching rates of reactive ion and reactive ion-beam etched AlGaAs

INDEX

CHAPTER 1

PHYSICAL AND THERMAL PROPERTIES OF GALLIUM ARSENIDE

1.1 Density of solid GaAs

1.2 Lattice parameter of GaAs

1.3 Lattice parameter of GaAs, doping dependence

1.4 Bulk modulus of GaAs

1.5 Stiffness of GaAs

1.6 Compliance of GaAs

1.7 Thermal expansion coefficient of GaAs, temperature dependence

1.8 Specific heat of GaAs

1.9 Thermal conductivity of GaAs

1.10 Thermal diffusivity of GaAs

1.1 DENSITY OF SOLID GaAs

by J.C.Brice

September 1985
EMIS Datareview RN=15257

Estimates of the density of gallium arsenide crystals from X-ray and direct measurements have been given by many authors. Only two of these [1,2] fall outside the range 5.305 to 5.325 g/cm*3 and most of the values are clustered around 5.317 g/cm*3.

From the preferred values of the lattice constants of pure gallium arsenide [3] at 300 deg K, X-ray densities are found to be in the range 5.3158 to 5.3164 g/cm*3 with uncertainties of about 3 in the last decimal place. (A last place uncertainty of about 2 arises from the lack of precision in our knowledge of the atomic weights and Avogadro´s constant). At this level of accuracy, defect concentrations can be large enough to have a signficant effect on the density so that X-ray densities are only useful when it is known that the defect concentrations are small. In the case of gallium arsenide, it is believed, see references [4] to [6], that for near to stoichiometric crystals the defect concentrations are negligible and for these crystals the X-ray density is about 5.3163 g/cm*3 which can be compared with the directly measured value of 5.317 g/cm*3 [6].

There are two direct measurements of density with precisions comparable to the X-ray data. Using the preferred value of the expansion coefficient [7] to convert values to 300 deg K, these are:

(a) 5.3175 g/cm*3 found by Straumanis and Kim [5], and
(b) 5.3160 g/cm*3 found by Yarmolyu et al [8] who also show that the density does not change with dopant concentrations up to levels in excess of 2 X 10*18/cm*3.

Bublik et al [6] claim an order of magnitude greater precision and quote values of 5.31584 g/cm*3 (gallium rich samples) to 5.31691 g/cm*3 (arsenide rich samples) with precisions of about 2 in the last digit at 300 deg K. In the absence of other data with similar precision, it is impossible to assess the accuracy of these measurements.

However, it is suggested that at 300 deg K pure facet-free stoichiometric gallium arsenide has a density of 5.3165 g/cm*3 with a last place uncertainty of 2. Arsenic-rich samples could have densities up to about 5.3175 g/cm*3 and gallium-rich samples could have densities down to about 5.3158 g/cm*3 when facet-free. Material (particularly when doped) grown on a facet could perhaps have a density as low as about 5.3150 g/cm*3. (Facet growth has a large effect on the lattice constant [3]) . As noted above, impurities at levels up to 2 X 10*18/cm*3 have little effect on the density of facet-free crystals [6].

For most purposes we can assume at 300 deg K that gallium arsenide has a density of 5.3165 g/cm*3 with an uncertainty of +- 0.0015. Using the evaluated thermal expansion data [7] allows us to deduce the density as a function of temperature. It is then found that at T(deg K) the density (d, g/cm*3) has values:

T	0-100	150	200	250	300	400
d	5.332	5.329	5.326	5.322	5.3165	5.307
T	500	600	700	800	900	1000
d	5.298	5.289	5.279	5.269	5.259	5.249
T	1100	1200	1300	1400	1500	
d	5.239	5.228	5.217	5.207	5.196	

The apparent uncertainties (dd, g/cm*3) are:

T	0	300	600	900	1200	1500
dd	0.004	0.0015	0.005	0.010	0.015	0.020

The values of d and dd at intermediate temperatures can be found by linear interpolation.

Note that the tabulated density values are consistent with the dilatometric and X-ray data given by the authors quoted in [7] but ignore the data due to Glazov [1] which in the range 400 to 1400 deg K are consistently 0.028 g/cm*3 less than the tabulated versions. This large deviation makes Glazov's results completely inconsistent with those of other workers. If Glazov's results are incorporated, the values of d for T above 400 deg K decreased by about half the uncertainty (dd) quoted above.

REFERENCES
----

[1] V.Glazov et al [Liquid Semiconductors (Academic Press, New York 1969) p.325]

[2] R.S Feigelson, P.K.Route [Proc. Soc. Photo-Opt. Instrum. Eng. (USA) vol.214 (1979) p.76]

[3] J.C.Brice [EMIS Datareview RN=2003 (July 1985)]

[4] D.T.J.Hurle [J. Phys. & Chem. Solids (GB) vol.40 (1979) p.613]

[5] M.E.Straumanis, C.D.Kim [Acta Cryst. (Denmark) vol.19 (1965) p.256]

[6] V.T.Bublik et al [Sov. Phys. Cryst. (USA) vol.18 (1973) p.218]

[7] J.C.Brice [EMIS Datareview RN=2001 (July 1985) "Thermal expansion coefficient of GaAs, temperature dependence"]

[8] N.I.Yarmolyu et al [Sov. Phys. Semicond. (USA) vol. 14 (1980) p.773]

1.2 LATTICE PARAMETER OF GaAs

by J.C.Brice

July 1985
EMIS Datareview RN=2003

The following data assumes that Cu K(alpha) radiation has a wavelength of 1.537400 kXu = 1.540593 A with an uncertainty of 5 ppm. The data in refs [1-12] were compared on this basis.

For pure materials grown on faces other than (111) the following values at 25 deg C are recommended:

stoichiometric crystals	5.65359	(+- 6 X 10*-5) A
Gallium rich crystals	5.65367	(+- 4 X 10*-5) A
'rsenic rich crystals	5.65357	(+- 8 X 10*-5) A

For pure materials grown exactly on (111) (i.e. facet growth) add 3 X 10*-5 A. With doped crystals the facet effect is larger; add (7 +- 3) X 10*-5 A.

Ref [1] shows that dislocations broaden X-ray reflections but do not appreciably affect lattice constant.

Refs [1,4] show that mechanical damage increases the lattice constant near to the surface. Relative to an etched surface the following increases (in units of 10*-5 A) were noted:

sawn surface	+ 50
surface lapped with 600 grit	+ 30
mechanically polished surface	+ 20
lightly scratched surface	+ 30

At 25 deg C a one degree temperature rise increases the lattice constant by 3.2 X 10*-5 A. Note, however, that the thermal expansion coefficient changes significantly with temperature [13].

REFERENCES

[1] C.H.M.Driscoll et al [Inst. Phys. Conf. Ser. no.24 (1975) p.275]
[2] A.S.Cooper [Acta Cryst. vol.15 (1962) p.578]
[3] E.D.Pierron et al [Acta Cryst. vol.21 (1966) p.290]
[4] E.D.Pierron, J.B.McNeely [Advances in X-ray Analysis (New York, Plenum) vol.12 (1969) p.343]
[5] J.F.C.Baker et al [Solid State Electron. vol.19 (1976) p.331]
[6] R.Heckingbottom et al [Solid State Electron. vol.19 (1976) p.335]
[7] C.V.Ozolinsh et al [Sov. Phys. Cryst. vol.7 (1963) p.691]
[8] J.B.Mullin et al [J. Appl. Phys. (USA) vol.47 (1976) p.2584]
[9] P.F.Fewster, A.F.W.Willoughby [EMIS Original Data RN=500 (Sept 1980)]
[10] P.F.Fewster, A.F.W.Willoughby [EMIS Original Data RN=501 (Sept 1980)]
[11] M.E.Straumanis, C.D.Kim [Acta Cryst. vol.19 (1965) p.256]
[12] P.S.Dobson et al [Inst. Phys. Conf. Ser. no.45 (1979) p.163]

[13] J.C.Brice [EMIS Datareview RN=2001 (July 1985)]

1.3 LATTICE PARAMETER OF GaAs, DOPING DEPENDENCE

by J.C.Brice

July 1985
EMIS Datareview RN=2005

Dopants can have a large effect. Tentative values of the change in lattice parameter produced by 100 ppm of solute relative to either As or Ga (i.e. 2.2 X 10*18 cm*-3) are given below:

Impurity (100 ppm)	Lattice dilation 10*-5 A
Si (Ga rich)	-25 +- 5
Si (As rich)	+ 5 +- 10
Te	+12 +- 6
Ge	+10 +- 4
Sn	+20 +- 10
Pb	- 4 +- 2
O	+70 +- 50
S	+ 2 +- 2
Cr	-100 +- 70

Note that ref [1] shows that the anomalously large dilation due to tellurium at high concentrations (greater than 100 ppm) can be relieved by annealing. It is not known whether this effect occurs for other dopants. Likewise, silicon has an effect which depends on stoichiometry and it is not known if other dopants behave similarly.

REFERENCES
----

[1] P.S.Dobson et al [Inst. Phys. Conf. Ser. no.45 (1979) p.163]
[2] C.H.M.Driscoll, A.F.W.Willoughby, J.B.Mullin, B.W.Straughan [Inst. Phys. Conf. Ser. no.24 (1975) ch.5 p.275-91]
[3] J.F.C.Baker et al [Solid State Electron. vol.19 (1976) p.331]
[4] R.Heckingbottom et al [Solid State Electron. (GB) vol.19 (1976) p.335]
[5] J.B.Mullin et al [J. Appl. Phys. (USA) vol.147 (1976) p.2584]
[6] P.F.Fewster, A.F.W.Willoughby [EMIS Original Data RN=501 (Sept 1980)]

1.4 BULK MODULUS OF GaAs

by J.C.Brice

November 1985
EMIS Datareview RN=15414

ADIABATIC BULK MODULUS

Only two values of the adiabatic bulk modulus Bs at room temperature have been found [1,2]. These are respectively 7.540 and 7.53 X 10*11 dyne/cm*2. Ref [2] also gives values at 78 deg K and 0 deg K which are respectively 7.67 and 7.69 X 10*11 dyne/cm*2. However, using the relation

Bs = (C11 + 2 C12) / 3

where the Cij are the adiabatic values given in [3] it is trivial to calculate other values. In units of 10*11 dyne/cm*2 we obtain at 300 deg K recommended values of:

7.58 for heavily doped n-type samples,
7.55 for moderately doped n-type samples,
7.52 for heavily doped p-type samples.

At 77 deg K we find 7.75 and at 0 deg K we obtain 7.78. These last two values have uncertainties of about 2%. The 300 deg K values are probably in error by about 0.5%.

McSkimin et al [1] show that the rate of change of Bs with pressure is 4.49 with an error of 1% up to 250 kbar at 298 deg K. Note that Bs has units of pressure so that dB/dP is dimensionless.

Bs falls by 0.00085 X 10*11 dyne/cm*2 for each degree rise in temperature at 300 deg K.

ISOTHERMAL BULK MODULUS

Only one value of the isothermal bulk modulus has been found. Ref [1] gives 7.466 X 10*11 dyne/cm*2 at 298 deg K. We can calculate Bt from the relation

B = (C11 + 2C12)/3

where the Cij are the isothermal values given in ref [3]. Doping this yields

Bt = 7.67 - 0.000847 T

where Bt is in units of 10*11 dynes/cm*2 and T is the Kelvin temperature. This formula gives a value of 7.42 X 10*11 dyne/cm*2 at 298 deg K.

Ref [1] shows that the rate of change of Bt with pressure is 4.67 for pressures up to 250 kbar.

REFERENCES
----

[1] M.J.McSkimin, A.Jayaraman, P.Andreatch [J. Appl. Phys. (USA) vol.38 (1967) p.2362]

[2] R.I.Cottam, G.A.Saunders [J. Phys. C. (GB) vol.6 (1973) p.2105]

[3] J.C.Brice [EMIS Datareview RN=15412 (Nov 1985) "Stiffness of GaAs"]

1.5 STIFFNESS OF GaAs

by J.C.Brice

November 1985
EMIS Datareview RN=15412

ADIABATIC STIFFNESS CONSTANTS

Data relevant to the adiabatic stiffness constants of crystalline gallium arsenide can be found in refs [1 to 13]. There appear to be real variations between measured values on different samples The constants increase with the free electron concentration, so that heavily doped n-type samples (n about 10*19 cm*-3) have constants which are about 1% greater than those of heavily doped p-type samples. See refs [7,10] for discussion.

At 300 deg K the median values in units of 10*11 dyne/cm*2 are

C11 = 11.88   C12 = 5.38   C44 = 5.94

These values apply to n-type samples with melt electron concentrations in the range 10*16 to 10*18 cm*-3. Possible corrections for other samples are to add respectively 0.05, 0.02 and 0.08 for more heavily n-type doped samples and for p-type samples to subtract respectively 0.05,0.02 and 0.04.

McSkimin and Andreatch [5] give the rates of change of C11, C12 and C44 with pressure as 4.63, 4.42 and 1.10 at 300 deg K respectively. Note that dC/dP is dimensionless.

There are a few data relating to other temperatures. Suggested values in units of 10*11 dyne/cm*2 are:

T(K)	C11	C12	C44
77	12.15	5.56	6.02
0	12.17	5.58	6.04

These are subject to uncertainties of 1 to 2% and probably vary with doping levels in a way similar to the 300 deg K values. Use of the same corrections is suggested as a means of reducing the uncertainties. At room temperature and higher temperatures the rates of decrease with temperature are about 0.00139, 0.000576 and 0.000701 for C11, C12 and C44 respectively. These values could be used to extrapolate to other temperature using the 300 deg K data.

ISOTHERMAL STIFFNESS CONSTANTS

Burenkov et al [10] have determined the isothermal stiffness constants and Jordan [11] shows that for a temperature T(K) the experimental data are consistent with the following expressions in units of 10*11 dyne/cm*2 valid to 900 deg K:

C11 = 12.16 - 0.00139 T
C12 = 5.43 - 0.000576 T
C44 = 6.18 - 0.0000701 T

It is probably reasonable to assume uncertainties of 1 to 2% to allow for doping dependence.

Note that the theoretical expressions given in ref [10] have larger rates of decrease with temperature. These are 0.0025, 0.0021 and 0.00073 for C11, C12 and C44 respectively.

REFERENCES

[1] T.B.Bateman [H.J. McSkimin, J.M.Whelan [J Appl. Phys. (USA) vol.30 (1959) p.544]

[2] C.W.Garland, K.C.Parks [J. Appl. Phys. (USA) vol.33 (1962) p.759]

[3] G.Dolling, R.A.Cowley [Proc. Phys. Soc. vol.88 (1966) p.463]

[4] J.R.Drabble, A.J.Brammer [Solid State Comm. (USA) vol.4 (1966) p.467]

[5] H.J.McSkimin, P.Andreatch [J. Appl. Phys. (USA) vol.38 (1967) p.2610]

[6] H.J.McSkimin, A.Jayaraman, P.Andreatch [J. Appl. Phys. (USA) vol.38 (1967) p.2362]

[7] V.M.Beilin, Yu.Kh.Vekilov, O.M.Krasilnikov [Sov. Phys. Solid State (USA) vol.10 (1969) p.2443]

[8] B.A.Bobylev, A.F.Kravchenko [Sov. Phys. Acoustics (USA) vol.13 (1976) p.242]

[9] R.I.Cottam, G.A.Saunders [J. Phys. C(GB) vol.6 (1973) p.2105]

[10] Y.A.Burenkov, Yu.M.Burdokov, S.Yu.Davidov, S.P.Nikanorov [Sov. Phys. Solid State (USA) vol.15 (1973) p.1175]

[11] A.S.Jordan [J. Crystal Growth (Netherlands) vol.49 (1980) p.631]

[12] Landolt-Bornstein [Numerical Data and Functional Relationships in Science and Technology, Group III vol.11 (1979) p.27]

[13] Landolt-Bornstein [Numerical Data and Functional Relationships in Science and Technology Group III vol.17 subvolume a (1982) p.235-6]

1.6 COMPLIANCE OF GaAs

by J.C.Brice

November 1985
EMIS Datareview RN=15413

Data leading to adiabatic elastic compliances are given by [1] and [2]. See Landolt-Bornstein [3]. The results are scattered over ranges of about 3% and the numerical values for p-type samples generally exceed those for n-type. Median values at room temperature in units of 10*-12 cm*2/dyne are:

	S11	S12	S44
n-type	11.71	-16.71	3.61
p-type	11.80	-16.98	3.66

Note that S12 is negative. Uncertainties are likely to be about 1.5%.

REFERENCES

[1] B.A.Bobylev, A.F.Kravchenko [Sov. Phys. Acoustics (USA) vol.13 (1967) p.242]

[2] V.M.Beilin, Yu.Kh. Vekilov, O.M.Krasilnikov [Fiz Tverd. Tela (USSR) vol.10 (1968) p.3101]

[3] Landolt-Bornstein [Numerical Data and Functional Relationships in Science and Technology, Group III vol.11 (1979) p.27]

1.7 THERMAL EXPANSION COEFFICIENT OF GaAs, TEMPERATURE DEPENDENCE

by J.C.Brice

July 1985
EMIS Datareview RN=2001

TEC	-0.20	2.10	3.95	4.90	5.50	5.92	X 10*-6/K
Temperature	50	100	150	200	250	273	deg K
TEC	6.05	6.20	6.29	6.36	6.44	6.59	X 10*-6/K
Temperature	300	350	400	450	500	600	deg K
TEC	6.74	6.89	7.04	7.19	7.34	7.49	X 10*-6/K
Temperature	700	800	900	1000	1100	1200	deg K
TEC	7.64	7.80	7.95	X 10*-6/K			
Temperature	1300	1400	1500	deg K			

NOTES

1 Extrapolation to melting point (1513 deg K) gives 7.97 X 10*-6/K.

2 Table was obtained by drawing a smooth curve through the TEC values given or deduced from data in refs [1-9]. Average deviation of points from line is +- 0.3 X 10*-6/K but on a statistical basis the errors in the line position should be less than +- 0.1 X 10*-6/K.

3 Above Debye temperature (350 deg K) the tabulated values obey the Gruneisen relation in that the ratio of TEC to specific heat is 1.085 (+- 0.008) X 10*-6 gram atom/cal. Below 350 deg K the Gruneisen parameter varies with temperature. For further information and specific heat data see refs [10-12].

4 Numerical integration of the table values agree to about 2% with measured lattice dilations:

dilation	122	245	369	494	620	ppm
temperature	320	340	360	380	400	deg K

5 Above 400 deg K the expansion coefficient at T deg K is 5.69 X 10*-6 + 1.5 X 10*-9 X T per deg K and analytical integration can be used.

6 Jordan [13] using fewer sources gives an evaluated coefficient above 300 deg K of 4.68 X 10*-6 + 3.82 X 10*-9 X T per deg K.

7 There is no direct guidance that stoichiometry, purity or perfection affect thermal expansion but these factors might account for difference between values given by different authors.

REFERENCES
----

[1] L.Bernstein, R.J.Beales [J. Appl. Phys. (USA) vol.32 (1961) p.122]

[2] R.Feder, T.Light [J. Appl. Phys. (USA) vol.39 (1968) p.4870]

[3] S.I.Novikova [Sov. Phys. Solid State (USA) vol.3 (1961) p.129]

[4] M.E.Straumanis, C.D.Kim [Acta Crystall. (Denmark) vol.19 (1965) p.256]

[5] E.D.Pierron et al [Acta Crystall. (Denmark) vol.21 (1966) p.290]

[6] N.N.Sirota, Yu.I.Pashnitsev [Dokl. Akad. Nauk. SSSR (1959) p.609]
[7] I.Kidman, R.J.Paff [J. Appl. Phys. (USA) vol.43 (1972) p.3760]
[8] S.Nan, L.Yi-huan [Scientia Sinica vol.14 (1965) p.1582]
[9] V.Glazov et al [Liquid Semiconductor (Plenum, New York, 1969) p.325]
[10] U.Piesbergen [Semiconductors and Semimetals (Academic Press, New York) vol.2 (1966) p.49]
[11] S.Novikova [Semiconductors and Semimetals (Academic Press, New York) vol.2 (1966) p.33]
[12] B.Lichter, P.Sommelet [Trans. Met. Soc. AIME vol.245 (1969) p.1021]
[13] A.S.Jordan [J. Cryst. Growth (Netherlands) vol.49 (1980) p.631]

1.8 SPECIFIC HEAT OF GaAs

by J.C.Brice

October 1985
EMIS Datareview RN=15273

Data relevant to the specific heat at constant pressure of crystalline gallium arsenide are given in refs [1-6]. Up to 350 deg K the various authors agree to within about +-0.002 J/gK with the tabulated values of the specific heat and to within +- 2 deg with the tabulated values of the Debye temperature (D)

T(deg K)	C(J/gK)	D(deg K)
0	0	345
20	0.011	250 (note 1)
50	0.09	302
100	0.199	353
150	0.262	360 Note 2
200	0.294	358
250	0.314	352 Note 1
300	0.325	363
350	0.332	364

Note 1: the Debye temperature has minima at 20 and 250 deg K.
Note 2. the Debye temperature has a maximum at 150 deg K.

Above 350 deg K the specific heat is given by C(J/gK) = 0.307 + 7.25 X 10*-5 T with errors which increase to about +- 0.024 at 1500 deg K. The Debye temperature increases to about 375 deg K at the melting point (1513 deg K).

REFERENCES
----

[1] C.W.Garland, K.C.Parks [J. Appl. Phys. (USA) vol.33 (1962) p.759]
[2] B.D.Lichter, P.Sommeter [Trans. Met. Soc. AIME vol.245 (1969) p.1021]
[3] U.Piesbergen [Z. Naturforsch. (Germany) vol.18a (1963) p.141]
[4] U.Piesbergen [Semiconductors and Semimetals; Eds R.K.Willardson, A.C.Beer (New York, Academic Press) vol.2 (1966)]
[5] A.G.Dash et al [J. Chem. Eng. Data vol.19 (1974) p.113]
[6] H.R.Philipp, H.Ehrenreich [Phys. Rev. (USA) vol.129 (1963) p.1550]

1.9 THERMAL CONDUCTIVITY OF GaAs

by J.C.Brice

October 1985
EMIS Datareview RN=15275

Data on the thermal conductivity of gallium arsenide crystals is given by refs [1-8]. Below 77 deg K the values found go through a maximum and are extremely dependent on the sample. No data were found for temperatures above 1000 deg K. As Jordan [9] points out, if T is the Kelvin temperature, then the thermal conductivity for T above 77 is inversely proportional to T to the power n. Here we use n = 1.20 for the best fit, so that the thermal conductivity is

K = A / T*1.2 (watts/cm K)

where A depends on the doping level. The table gives A values for n-type samples with various donor concentrations. The same values of A may be used for p-type samples with acceptor concentrations which are three times larger (i.e. samples with 1 X 10*18 cm*-3 donors or 3 X 10*18 cm*-3 acceptors have very similar thermal conductivities).

Doping level (donors cm*-3)		A (watts K*0.2 cm*-1)
Less than about	5 X 10*16	544
About	1 X 10*17	488
About	3 X 10*17	404
About	1 X 10*18	376
Above about	7 X 10*18	357

Using these values of A gives a calculated value of K falling to within the 20% variation shown by experimental results.

REFERENCES

[1] A.Amith, I.Kndman, E.F.Steigmeier, [Phys. Rev. (USA) vol.138 (1956) p.A1270]
[2] E.F.Steigmeier, I.Kudman, [Phys. Rev. (USA) vol.132 (1963) p.508]
[3] M.G.Holland [Semicond. and Semimet. (USA), Eds R.K.Willardson & A.C.Beer (New York, Academic Press) vol.2 (1966)
[4] H.Wagini [Z. Naturforsch. a (Germany) vol.20a (1965) p.494]
[5] A.D.Stuckes [Brit. J. Appl. Phys. (GB) vol.12 (1961) p.675]
[6] R.O.Carlson, G.A.Slack, S.J.Silverman [J. Appl. Phys. (USA) vol.36 (1965) p.505]
[7] M.G.Holland [Phys. Rev. (USA) vol.134 (1963) p.A471]
[8] M.G.Holland [Proc. 7th Int. Conf. Semicond. Paris (1964)]
[9] A.S.Jordan [J. Crystal Growth (Netherlands) vol.49 (1980) p.631]

1.10 THERMAL DIFFUSIVITY OF GaAs

by J.C.Brice

October 1985
EMIS Datareview RN=15274

From the data given in refs [1-3] it is possible to calculate the thermal diffusivity (TD) of a sample. The value obtained depends strongly on the thermal conductivity. The values given in the table are for an average sample (one containing 2 X 10*17 cm*-3 donors or 6 X 10*17 cm*-3 acceptors). Purer samples could have diffusivities a factor of up to 1.24 larger. Less pure samples could have values less by a factor of up to 1.24.

T(deg K)	100	150	200	250	300	350
TD(cm*2/s)	1.65	0.77	0.49	0.35	0.27	0.22

T(deg K)	400	500	600	700	800	900
TD(cm*2/s)	0.186	0.140	0.110	0.090	0.075	0.0064

REFERENCES

[1] J.C.Brice [EMIS Datareview RN=15273 (Oct 1985) "Specific heat of GaAs"]
[2] J.C.Brice [EMIS Datareview RN=15275 (Oct 1985) "Thermal conductivity of GaAs"]
[3] J.C.Brice [EMIS Datareview RN=15257 (Sept 1985) "Density of solid GaAs"]

CHAPTER 2

RESISTIVITY OF GALLIUM ARSENIDE

2.1 Resistivity of SI GaAs

2.2 Sheet resistivity of ion implanted GaAs

2.1 RESISTIVITY OF SEMI-INSULATING GaAs

by D.Lee

January 1986
EMIS Datareview RN=14405

Depending upon growth conditions, semi-insulating (SI) GaAs can exhibit resistivities close to that of intrinsic perfect material (3.3 X 10*8 ohm cm with intrinsic carrier density of 2.1 X 10*6 cm*-3 at 300 deg K [1,2]). Typical room temperature values of resistivity are in the range 10*7 to 10*9 ohm cm. Measured resistivities are very sensitive to surface preparation and improper surface treatment has been reported to lead to order of magnitude reductions in apparent resistivity [3]. Recent reviews of SI GaAs can be found in ref [4].

SI GaAs is available by horizontal Bridgman and gradient freeze techniques (boat growth) or liquid encapsulated Czochralski (LEC). Boat grown material contains a large concentration (up to approx. 10*16 cm*-3) of silicon acceptors from decomposition of the silica boat [5] which must be compensated by doping with chromium or minimised by growth from an oxygen-containing melt [5]. A further disadvantage of boat grown material is that ingots take the shape of the boat and wafers are much smaller than those from large diameter cylindrical LEC crystals [6].

Most commercial interest, therefore, lies in LEC growth using boric oxide as encapsulant. Whilst LEC GaAs grown from silica crucibles can contain similar concentrations of Si acceptors to boat grown material, this can be reduced to less than 10*15 cm*-3 by use of pyrolytic boron nitride (pBN) crucibles [5,7] or wet boric oxide encapsulant [7].

SI GaAs can be produced either by compensation of residual shallow donors by chromium (deep acceptors) or by growth of undoped material under carefully controlled conditions. Cr doped GaAs generally has a higher resistivity than undoped material grown by the same technique [8]. Betko and Herinsky [3] report resistivities of Cr doped SI GaAs from several sources (after careful surface preparation) to be in very close agreement over the temperature range 296 to 430 deg K, with resistivities in the range 8.3 X 10*8 to 9.5 X 10*8 ohm cm at 296 deg K, falling to about 5.5 X 10*4 ohm cm at 430 deg K.

The electrical properties of Cr doped material are thermally unstable and wafer surfaces can become conducting to a depth of up to several microns on annealing [5,9,10]. Both n- and p-type conversion have been observed, n-type resulting from outdiffusion of Cr, leaving behind uncompensated shallow donors [11,12] and p-type conversion (more common [10]) attributed to the incorporation of Mn from the ambient gas into Cr depleted sites [12]. (Kutt et al [13] report that V has a factor of 10 lower diffusion coefficient than Cr and hence may be a superior dopant for SI GaAs). Also Cr doped substrates can contain more than 10*17 cm*-3 ionized impurities that severely degrade the electron mobility [5].

Current interest has therefore concentrated on undoped LEC GaAs, in which SI properties depend upon compensation of background donor impurities (mainly Si) by shallow acceptors (carbon) which in turn are compensated by the native deep donor EL2 [14] (believed to be associated with the As antisite defect) [5,7,9,15]. Oliver et al [7] note that the Si impurity concentration must be kept to below 10*15 cm*-3 to ensure SI behaviour, hence it is now standard practice to grow from pBN crucibles.

The resistivity of undoped LEC GaAs is strongly dependent upon melt stoichiometry (because of the variation of the EL2 concentration) [5,9,16]. SI material is obtained from melts with [Ga]/[As] ratios below about 1.1 with room temperature peak resistivities typically about 10*8 ohm cm and mobilities up to 6500 cm*2/V.s whilst material grown from more Ga-rich melts is p-type with resistivity of the order of 1 ohm cm and mobility about 300 cm*2/V.s [5,9,16,17]. The resistivity also decreases very gradually as the As fraction increases from the critical composition [9].

Crystals grown from slightly As-rich melts remain SI over their entire length whereas those grown from stoichiometric or slightly Ga-rich melts undergo a transition from SI to low resistivity p-type at some point in the crystal, due to As depletion of the melt during growth [5,9,17]. Thomas et al [5] report uniform n-type high resistivity over the full crystal length only for [Ga]/[As] = 0.98. SI GaAs wafers grown from near stoichiometric melts undergo p-type thermal conversion to a depth of more than 8 microns, due to a reduction of the EL2 concentration at the surface during annealing at 860 deg C, whilst As-rich material exhibits excellent thermal stability [17]. Nakamura et al [18] report that the degree of thermal conversion is strongly dependent on the carbon concentration. Hughes and Li [19] observed thermal conversion on annealing at 850 deg C in a hydrogen atmosphere whilst the same material remained semi-insulating on annealing in nitrogen.

The free carrier concentration of undoped LEC GaAs is dependent upon the carbon concentration [18,20]. Hunter et al [20] report carrier concentrations always below 5 X 10*7 cm*-3 for carbon concentrations in excess of 5 X 10*14 cm*-3 rising to 10*11 cm*-3 for samples containing less carbon. Carbon content is reported to decrease with increasing water content of the boric oxide encapsulant from 8 X 10*15 cm*-3 for dry boric oxide (100-150 ppm water) to 3-5 X 10*14 cm*-3 for boric oxide containing 500 ppm water [20]. Emori et al [16] report that addition of 2000 ppm of water results in an increase in the critical [Ga]/[As] ratio from 1.1 to 1.4 so that SI crystals with resistivities of 10*7 to 10*8 ohm cm over their entire length could be produced from highly Ga-rich melts. The critical [Ga]/[As] ratio is also reported to increase with the pressure of argon above the boric oxide encapsulant from 1.10 at 2.0 MPa to 1.20 at 3.1 MPa [21].

Several workers have studied the radial variation of resistivity across undoped SI LEC GaAs wafers. This has been found to exhibit a characteristic `M´ shaped profile [22-25] which has been correlated with corresponding `W´ shaped profiles of both dislocation density and EL2 concentration. Kikuta et al [26] have measured the 2 dimensional microscopic variation of resistivity (at a resolution of 70 microns) and report a detailed correlation between areas of reduced resistivity (by a factor of approx 10) and high EL2 concentration in regions of high dislocation density (at dislocation cell boundaries).

The macroscopic non-uniformity of resistivity is reported to be reduced, or even eliminated, and mobility to be increased (typically from 4000 to 6000 cm*2/V.s) on annealing at temperatures between 650 and 950 deg C [24,27,28]. Reduction of the thermal gradients in the growing crystal is reported to lead to a low and homogeneous dislocation density and a reduction of the radial non-uniformity of resistivity [29,30]. SI LEC GaAs can be produced with zero dislocation density by doping with indium [31-33]. Such material has a uniform EL2 distribution [32] and is reported to exhibit thermally stable high resistivity in the 10*7 to 10*8 ohm cm range [33].

Terashima et al [34,35] investigated the effects of a magnetic field during LEC growth. The resistivity was observed to fall from 10*8 to 10 ohm cm as H increased from zero to 1300 Oe, attributed to a decrease in the EL2 concentration [34]. Temperature fluctuations in the melt were reduced from 18 to 0.1 deg C in an applied field of 1300 Oe [35]. Background concentrations of boron, silicon and carbon (segregation coefficient, k, above 1) were found to be reduced, whereas the chromium (k less than 1) doping concentration was increased (attributed to the suppression of melt convection by the magnetic field) [35].

Emori et al [36] studied various gas ambients during the growth of SI GaAs from stoichiometric melts and found the resistivity (295 deg K) to be increased in the order krypton greater than argon greater than nitrogen with a corresponding decrease in dislocation density.

Several authors have reported theoretical analyses of the transport properties of SI GaAs [37-40]. Winter et al [37] study mixed conduction effects, Lee and Park [38] employ a three band model, Pistoulet et al [39] study the effects of potential fluctuations on the transport properties and Martin et al [40] discuss the compensation mechanism in SI GaAs.

REFERENCES

[1] J.S.Blakemore [J. Appl. Phys. (USA) vol.53 no.1 (1982) p.520-31]

[2] J.S.Blakemore [J. Appl. Phys. (USA) vol.53 no.10 (1982) p.R123-81]

[3] J.Betko, K.Merinsky [J. Appl. Phys. (USA) vol.50 no.6 (1979) p.4212-6]

[4] Semiconductors and Semimetals vol.20: Semi-Insulating GaAs Ed R.K.Willardson & A.C.Beer (Academic Press, 1984)

[5] R.N.Thomas, H.M.Hobgood, G.W.Eldridge, D.L.Barrett, T.T.Braggins, L.B.Ta, S.K.Wang [in ref [4] p.1-87]

[6] C.A.Stolte [in ref [4] p.89-158]

[7] J.R.Oliver, R.D.Fairman, R.T.Chen [Electron. Lett. (GB) vol.171 no.22 (1981) p.839-41]

[8] I.Grant, D.Rumsby, R.M.Ware, M.R.Brozel, B.Tuck [Proc. Semi-Insulating III-V Materials, Evian 1982, Ed B.Makram-Ebeid & B.Tuck (Shiva, 1982) p.98-106]

[9] C.G.Kirkpatrick, R.T.Chen, D.E.Holmes, P.M.Asbeck, K.R.Elliott R.D.Fairman, J.R.Oliver [in ref [4] p.159-231]

[10] B.Tuck [Proc. Semi-Insulating III-V Materials, Kah-nee-ta 1984, Ed D.C.Look & J.S.Blakemore (Shiva, 1984) p.2-18]

[11] P.K.Vasudev, R.G.Wilson, C.A.Evans, V.R.Deline [Solid State Electron. (GB) vol.26 no.6 (1983) p.565-7]
[12] M.Oshima, K.Watanabe, S.Miyazawa [J. Electrochem. Soc. (USA) vol.131 no.1 (1984) p.130-6]
[13] W.Kutt, D.Bimberg, M.Maier, H.Krautle, F.Kohl, E.Bauser [Appl. Phys. Lett. (USA) vol.44 no.11 (1984) p.1078-80]
[14] S.Makram-Ebeid, P.Langlade, G.M.Martin [Proc. Semi-Insulating III-V Materials, Kah-nee-ta, 1984, Ed D.C.Look & J.S.Blakemore (Shiva, 1984) p.184-203]
[15] D.E.Holmes, K.R.Elliott, R.T.Chen, C.G.Kirkpatrick [Proc. Semi-Insulating III-V Materials, Evian 1982, Ed S.Makram-Ebeid & B.Tuck (Shiva, 1982) p.19-27]
[16] H.Emori, T.Kikuta, T.Inada, T.Obokata, T.Fukuda [Jpn. J. Appl. Phys. (Japan) vol.24 no.5 (1985) p.L291-3]
[17] L.B.Ta, H.M.Hobgood, A.Rohatgi, R.N.Thomas [J. Appl. Phys. (USA) vol.53 no.8 (1982) p.5771-5]
[18] Y.Nakamura, Y.Ohtsuki, Y.Ihoh, J.Kikawa, Y.Kashawayanagi [Proc. GaAs & Related Compounds, Karauizawa 1985 (Inst. Phys. Conf. Ser. no.79) p.49-54]
[19] B.Hughes, C.Li [Proc. GaAs & Related Compounds, Albuquerque 1982 (Inst. Phys. Conf. Ser. no.65) p.57-64]
[20] A.T.Hunter, H.Kimura, J.P.Baukus, H.V.Winston, O.J.Marsh [Appl. Phys. Lett. (USA) vol.44 no.1 (1984) p.74-6]
[21] Y.Kadota, M.Kamimoto, Y.Masa, A.Tanaka, T.Yokohama [Proc. Semi-Insulating III-V Materials, Kah-nee-ta 1984, Ed D.C.Look & J.S.Blakemore (Shiva, 1984) p.157-9]
[22] T.Honda, Y.Ishii, S.Miyazawa, H.Yamazaki, Y.Nanishi [Jpn. J. Appl. Phys. (Japan) vol.22 no.5 (1983) p.L270-2]
[23] T.Matsumura, H.Emori, K.Terashima, T.Fukuda [Jpn. J. Appl. Phys. (Japan) vol.22 no.3 (1983) p.L154-6]
[24] D.Rumsby, I.Grant, M.R.Brozel, E.J.Foulkes, R.M.Ware [Proc. Semi-Insulating III-V Materials, Kah-nee-ta 1984, Ed D.C.Look & J.S.Blakemore (Shiva, 1984) p.165-70]
[25] M.R.Brozel, I.Grant, D.J.Stirland, M.S.Skolnick [J. Appl. Phys. (USA) vol.56 no.4 (1984) p.1109-18]
[26] T.Kikuta, T.Katsumata, T.Obokata, K.Ishida [Proc. GaAs & Related Compounds, Biarritz 1984 (Inst. Phys. Conf. Ser. no.74) p.47-52]
[27] T.Obokata, T.Matsumura, K.Terashima, F.Orito, T.Kikuta, T.Fukuda [Jpn. J. Appl. Phys. (Japan) vol.23 no.8 (1984) p.L602-5]
[28] M.Yokogawa, S.Nishine, K.Matsumoto, H.Morishita, K.Fujita, S.Akai [Proc. GaAs & Related Compounds, Biarritz 1984 (Inst. Phys. Conf. Ser. no.74) p.29-34]
[29] T.Shimada, K.Terashima, H.Nakajima, T.Fukuda [Jpn. J. Appl. Phys. (Japan) vol.23 no.1 (1984) p.L23-5]
[30] T.Shimada, T.Obokata, T.Fukuda [Jpn. J. Appl. Phys. (Japan) vol.23 no.7 (1984) p.L441-4]
[31] M.Duseaux, S.Martin [Proc. Semi-Insulating III-V Materials, Kah-nee-ta 1984, Ed D.C.Look & J.S.Blakemore (Shiva, 1984) p.118-25]
[32] S.Martin, M.Duseaux, E.Erman [Proc. GaAs & Related Compounds, Biarritz 1984 (Inst. Phys. Conf. Ser. no.74 p.53-8]
[33] H.M.Hobgood, R.N.Thomas, D.L.Barrett, G.W.Eldridge, M.M.Sopira, M.C.Driver [Proc. Semi-Insulating III-V Materials, Kah-nee-ta 1984, Ed D.C.Look & J.S.Blakemore (Shiva, 1984) p.149-56]
[34] K.Terashima, T.Katsumata, F.Orito, T.Kikuta, T.Fukuda [Jpn. J. Appl. Phys. (Japan) vol.22 no.6 (1983) p.L325-7]
[35] K.Terashima, F.Orito, T.Katsumata, T.Fukuda [Jpn. J. Appl. Phys.

(Japan) vol.23 no.7 (1984) p.L485-7]
[36] H.Emori, K.Terashima, F.Orito, T.Kikuta, T.Fukuda [Proc. Semi-Insulating III-V Materials, Kah-nee-ta 1984, Ed D.C.Look & J.S.Blakemore (Shiva, 1984) p.111-7]
[37] J.J.Winter, H.A.Leupold, R.L.Ross, A.Ballato [J. Appl. Phys. (USA) vol.54 no.9 (1983) p.5176-82]
[38] H.J.Lee, Y.S.Park [J. Appl. Phys. (USA) vol.55 no.2 (1984) p.402-5]
[39] B.Pistoulet, P.Girard, G.Hamamdjian [J. Appl. Phys. (USA) vol.56 no.8 (1984) p.2268-74]
[40] G.M.Martin, J.P.Farges, G.Jacob, J.P.Hallais [J. Appl. Phys. (USA) vol.51 no.5 (1980) p.2840-52]

2.2 SHEET RESISTIVITY OF ION-IMPLANTED GaAs

by B.J.Sealy

January 1986
EMIS Datareview RN=15470

It is well known that the act of ion implantation changes the electrical properties of GaAs. For example, conducting layers can be made semi-insulating and semi-insulating layers can become partially conducting. The percentage change in electrical conductivity depends on ion mass, atomic number and dose and, to a lesser extent, on ion energy and implant temperature. As an example, Table 1 illustrates the effect of varying the dose of silicon ions on the sheet resistivity of semi-insulating GaAs [1]. The resistivity saturates at high doses to a value of 10*4 to 10*5 ohms/sq.

TABLE 1: variation of sheet resistivity, rho(s), with dose of 50 keV Si+ ions.

dose (cm*-2)	0	10*12	10*13	10*14	10*15
rho(s) (ohms/sq.)	10*9	2 X 10*8	10*6	10*5	8 X 10*4

These characteristics are true regardless of the energy and mass of the implanted ions, which suggest that the resistivity is a consequence of the lattice disorder produced by the implantation process [2]. Some of the ions that have been observed to behave in this way are Kr, Ar, Zn, S, Ne, N, He [2]. Starting with doped GaAs, it is possible to increase the resistivity by performing low dose implants. For example, a dose of 1.5 X 10*11 cm*-2 of 400 keV O+ ions was found to increase the resistivity of an epitaxial layer of GaAs from 150 ohms/sq. to 5 X 10*8 ohms/sq. [3]. As the dose increased from 10*13 O+ cm*-2, the resistivity decreased to a value of about 10*4 ohms/sq. A similar dependence of resistivity on dose has been observed for protons and deuterons in GaAs [4,5]. Thus for light mass ions, the resistivity of doped GaAs increases with increasing dose, reaches a peak value and finally decreases at high doses as described above for heavy mass atoms. The conduction mechanism of the disordered GaAs has been studied by various groups [2,6,7] and at low temperatures (below 180 deg K) is characterised by a T*-0.25 dependence, where T is the temperature of measurement, that is

rho(s) is proportional to exp (B/T*0.25)

This behaviour is similar to that for amorphous semiconductors, i.e. a thermally activated hopping conduction mechanism is responsible.

For material where a saturation resistivity has been attained following implantation, annealing produces an increase in resistivity with increasing temperature. A maximum value of the resistivity of 10*8-10*9 ohms/sq. is obtained at some intermediate temperature, and at higher temperatures and for implants of dopant ions, the resistivity decreases as the temperature is increased further. Electrical activity of the implanted dopant ions occurs generally at temperatures above about 600 deg C. This phenomenon is well known and the peak in the sheet resistivity occurs at a temperature in the range 400 to 700 deg K, depending on the ion species and dose [2,8]. Thus the high resistivity layers produced by radiation damage can be removed following an anneal at a sufficiently high temperature. Thus, in the case of protons which are often used to produce high resistivity layers in doped GaAs, the effect is mostly annealed out at a temperature of about 400 deg C [4]. However, a multi-energy implant at a dose sufficient to produce a saturation resistivity (10*5-10*6 ohms/sq.) improves the thermal stability once the sample has been annealed at about 500 deg C. It is as if the resistivity follows the dose dependence effect in reverse. Thus, as the temperature is raised the resistivity increases, reaches a maximum at 500 deg C, and above this temperature the resistivity decreases sharply [4]. In contrast, high dose oxygen implants in GaAs produce increasingly resistive material so that at temperatures in excess of 700 deg C the resistivity is maximised and thermally stable at a value of about 10*8 ohms/sq. [3].

The aim of dopant implants is to change the electrical properties of the GaAs in a controlled and reproducible way. However, this is still difficult to achieve due to variations in the properties of the substrate material (normally undoped semi-insulating GaAs) and to the variability of the encapsulation and annealing process. The standard industrial technique is to implant Si ions followed by an anneal at 850 to 900 deg C for about 15 minutes. For low dose implants suitable for the channel region of a MESFET doses of about 5 X 10*12 cm*-2 produce layers with resistivity of about 500 ohms/sq. [9,10]. Normally Si ions are used for the doping of the channel of FETs and the resistivity of such implants is a minimum at about 950 deg C. For example, a dose of 5 X 10*12 Si+ cm*-2 at 150 keV produced a minimum resistivity of 440 ohms/sq. at 1000 deg C for 5s [9]. The reason for the minumum is that at higher temperature compensation occurs due to the amphoteric nature of Si in GaAs. The higher doping levels required for ohmic contact regions produce resistivities down to a minimum of about 20 ohms/sq., but annealing temperatures above 900 deg C are generally required to achieve this. This lowest sheet resistivity recorded for furnace annealed GaAs was for a room temperature implant of 10*15 Si+ cm*-2 at 400 keV. The cap was a CVD Si3N4 layer and the anneal was at 900 deg C for 15 minutes [11]. The same authors also studied samples implanted with 400 keV Se ions to a dose of 10*14 cm*-2. Following an anneal at 900 deg C for 15 minutes, the resistivity was found to decrease from 252 to 54 ohms/sq. as the implant temperature was raised from room temperature to 500 deg C. Thus, the implantation temperature can have a significant effect on the electrical properties of GaAs, in particular for heavy mass ions such as Se, Te and Sn. However, if the annealing temperature is limited to 900 deg C, then, for low mass ions such as Si and also for low doses of the heavy mass ions, which do not create an amorphous layer, it is unnecessary to implant above room temperature. The implication is that high doses of heavy mass ions implanted at room temperature require very high temperatures to creat significant electrical activity. This has been studied recently for doses of 10*14 and 10*15 cm*-2 of Se and Sn ions [12,13]. It was found that annealing at 1000 deg C for 10-50s produced sheet resistivties of about 110 ohms/sq. (10*14 Se+ cm*-2), 80 ohms/sq. (10*15 Se+ cm*-2 and 10*14 Sn+ cm*-2) and 60 ohms/sq. (10*15 Sn+ cm*-2). Thus, the lowest resistivities were obtained for the highest doses. By raising the temperature to 1050 deg C for 10s, the 10*15 Se+ cm*-2 dose gave a sheet resistivity of 37 ohms/sq. The 10*15

Se+ cm*-2 dose produced a value of 28 ohms/sq. when annealed at 1090 deg C for 5s. Implants of 400 keV, 10*15 Se+ cm*-2 into GaAs at 200 deg C have produced resistivities as low as 22 ohms/sq. when annealed at 1075 deg C for 5s [13]. A more detailed study of a hot implant of 400 keV, 10*15 Se+ cm*-2 at 300 deg C has shown that the resistivity decreases with increasing annealing temperature for an anneal time of 10s (Table 2) [14].

TABLE 2: sheet resistivity as a function of annealing temperature for 10s anneals of 10*15 Se+ cm*-2, 400 keV implanted at 300 deg C [14]

rho(s) ohms/sq.	100	70	33	25
annealing temperature deg C	900	1000	1100	1140

Similarly low values, 24 ohms/sq., have been obtained for 4 X 10*14 Si+ cm*-2 annealed at a temperature in excess of 1000 deg C [15]. In addition there have been a number of reports of sheet resistivities in the range 30-60 ohms/sq. for Si [16,17], Sn [18] and S [19] following rapid thermal annealing at temperatures in the range 1000 to 1100 deg C.

In contrast to the above very low resistivities, a typical implant used for the contact regions of MESFET´s that is, 3 X 10*13 Si cm*-2 at 100 keV, can produce resistivities of 210 and 120 ohms/sq. following a 10s anneal at 800 deg C and 1000 deg C respectively [20]. Thus, it is clear that to obtain significant reductions in sheet resistivity for device applications, it is necessary to use higher doses and higher temperatures. Values of sheet resistivity in the range 20-49 ohms/sq. will then be possible which should be advantageous for device fabrication.

So far, the electrical properties of donor implants have been considered. Acceptors, however, are also important, mostly for creating highly doped regions. For example, implants of Zn and Mg have been used for forming thin p+ regions for JFETs [21]. Compared with donors, it is much easier to activate acceptor implants in GaAs. Thus, for high doses of the important ions Be, Mg, Zn and Cd implanted at room temperature, it is possible to obtain high electrical activities at temperatures of 700 to 800 deg C. As with donors, for a given annealing condition, the resistivity decreases with increasing dose (see Table 3).

TABLE 3: sheet resistivity as a function of dose for 135 keV Cd annealed at 900 deg C for 10 minutes [22]

dose (ions/cm*-2)	10*13	10*14	10*15	10*16
approx. rho(s) (ohms/sq.)	10*4	2 X 10*3	700	800

Also, as the annealing temperature increases the resistivity decreases. Thus, for 150 keV, 10*15 Zn+ cm*-2, the sheet resistivity falls from 440 to 76 ohms/sq. as the temperature is raised from 650 to 900 deg C [23]. The resistivity may also depend on the cap [19,24,25]. For example, for 10*15 Zn+ cm*-2 at 60 keV the resistivity after annealing at 900 deg C for 25s was 67 and 191 ohms/sq. for Si3N4 and AlN capped samples respectively [24]. Samples annealed in the same way also showed a weak energy dependence, the resistivity being 67, 69 and 76 ohms/sq. for 60, 150 and 300 keV implants of Zn respectively [24]. These resistivities are the lowest recorded for acceptor implants in GaAs. However, they are obtained at the expense of control over the elctrical profile which is broadened significantly due to diffusion of zinc during annealing. Very short time anneals of about 1s can reduce profile broadening but the resistivity is higher and typically in excess of 100 ohms/sq. [17,26,27].

In summary, radiation damage in GaAs converts semi-insulating material to partially conducting with a saturation resistivity of 10*4-10*5 ohms/sq. At a dose below the value required to produce this saturation effect, the resistivity is maximised to a value of about 10*7-10*8 ohms/sq. Heat-treatment of such layers removes radiation damage. The resistivity decreases either to a value corresponding to the activation of the implanted dopant or to the value prior to implantation. However, oxygen ions are an exception, since they can produce thermally stable, high resistivity layers in GaAs. To obtain the lowest sheet resistivities, it is necessary to anneal at temperatures in excess of 1000 deg C for donors, but low temperatures may be sufficient for acceptor implants.

REFERENCES
----

[1] M.Tamura, T.Miyazaki, N.Natsuaki [Suppl. J. Jap. Soc. Appl. Phys. (Japan) vol.43 (1972) p.504-7]

[2] Y.Kato, T.Shimada, Y.Shiraki, K.F.Komatsubara [J. Appl. Phys. (USA) vol.45 no.3 (Mar 1974) p.1044-9]

[3] S.Gecim, B.J.Sealy, K.g.Stephens [Electron. Lett. (GB) vol.14 no.10 (May 1978) p.306-8]

[4] J.P.Donnelly, C.O.Bozler, R.A.Murphy [Circuit Manuf. (USA) (April 1978) p.45-50]

[5] K.Steeples, I.J.Saunders, J.G.Smith [IEEE Electron. Dev. Lett. (USA) vol.EDL-1 no.5 (May 1980) p.72-4]

[6] B.K.Shin [J. Appl. Phys. (USA) vol.47 no.8 (Aug 1976) p.3612]

[7] V.N.Brudnyi, M.A.Kriov, A.I.Potapov [Sov. Phys. J. (USA) vol.25 no.1 (Jan 1982) p.37-40]

[8] B.J.Sealy [J. Mater. Sci. (GB) vol.10 (1975) p.683-91]

[9] M.Kuyuhara, H.Kohzu, Y.Takayama [Appl. Phys. Lett. (USA) vol.41 no.8 (Oct 1982) p.755-8]

[10] S.A.Kitching, M.H.Badawi, S.W.Bland, J.Mun [Mat. Res. Soc. Conf. Proc. (USA) vol.35 (9185) p.503-8]

[11] J.P.Donnelly, W.T.Lindley, C.E.Hurwitz [Appl. Phys. Lett. (USA) vol.27 no.1 (July 1975) p.41-3]

[12] K.K.Patel, R.Bensalem, M.A.Shahid, B.J.Sealy [Nucl. Instr. Meth. B (Netherlands) vol.7/8 (1975) p.418-22]

[13] B.J.Sealy, R.Bensalem, K.K.Patel [Nucl. Instr. Meth. B (Netherlands) vol.6 (1985) p.325-9]

[14] R.L.Chapman, J.C.C.Fan, J.P.Donnelly, B.Y.Tsaur [Appl. Phys. Lett. (USA) vo.40 no.9 (May 1982) p.805-7]

[15] D.E.Davies, P.J.McNally, J.P.Lorenzo, M.Julian [IEEE Electron. Dev. Lett. (USA) vol.EDL-3, no.4 (April 1982) p.102-3]

[16] S.K.Tiku, W.M.Duncan [J. Electrochem. Soc. (USA) vol.12 no.9 (Sept 1985) p.2237-9]

[17] K.Tabatabail-Alavi, A.N.M.Masum Choudhury, C.G.Fonstad, J.C.Gelpey [Appl. Phys. Lett. (USA) vol.43 no.3 (Sept 1983)

p.505-7]

[18] R.Gwilliam, R.Bensalem, B.J.Sealy, K.G.Stephens [Physica B (Netherlands) vol.129 (1985) p.440-4]

[19] M.Kuzuhara, H.Kohzu, Y.Takayama [Mat. Res. Soc. Conf. Proc. (USA) vol.23 (1984) p.651-61]

[20] T.Hiramoto, T.Saito, T.Ikoma [Jpn. J. Appl. Phys. (Japan) vol.24 no.3 (March 1985) p.L193-5]

[21] R.Zuleeg [Jpn. J. Appl. Phys. (Japan) vol.19 suppl. 19-1 (1980) p.315-8]

[22] B.K.Shin, D.C.Look, Y.S.Park, J.E.Ehret [J. Appl. Phys. (USA) vol.47 no.4 (April 1976) p.1574-9]

[23] S.S.Kular, B.J.Sealy, K.G.Stephens, D.Sadana, G.R.Booker [Solid State Electron. (GB) vol.23 (1980) p.831-8]

[24] N.J.Barrett, J.D.Grange, B.J.Sealy, K.G.Stpehens [J. Appl. Phys. (USA) vol.57 no.12 (June 1985) p.5470-6]

[25] D.E.Davies, J.K.Kennedy, C.E.Ludington [J. Electrochem. Soc. (USA) vol.122 no.10 (Oct 1975) p.1374-7]

[26] D.E.Davies, P.J.McNally [Appl. Phys. Lett. (USA) vol.44 no.3 (Feb 1984) p.304-6]

[27] N.J.Barrett, B.J.Sealy [Electron. Lett. (GB) vol.20 no.4 (Feb 1984) p.175-7]

CHAPTER 3

CARRIER CONCENTRATIONS IN GALLIUM ARSENIDE

3.1 Carrier concentrations in ion implanted n-type GaAs

3.2 Carrier concentrations in ion implanted p-type GaAs

3.3 Carrier concentrations in doped and undoped epitaxial GaAs

3.1 CARRIER CONCENTRATIONS IN ION-IMPLANTED n-TYPE GaAs

by R.S.Bhattacharya

June 1985
EMIS Datareview RN=15130

Doping by ion implantation is an established production technology in the case of GaAs for which a diffusion doping technology has not been developed. The production of n-type layers in GaAs by ion implantation doping can be achieved by either implanting column VI impurities (Se, Te, S) or column IV impurities (Si, Sn). The doping efficiency of n-type impurities is found to be a function of not only the substrate temperature during implantation and the subsequent anneal temperature but also the energy and dose of ions. Low doses can lead to a high proportion of the implanted impurities becoming electrically active, whereas increasing the dose results in a general decrease in doping efficiency which is attributed to the compensating residual damage after high temperature annealing. The carrier concentration profile after thermal annealing is also strongly dependent on the diffusion behavior of implanted ions. Sulphur undergoes considerable diffusion in GaAs whereas Si+, Se+, and Te+ are found to be relatively stable. Most of the published results are discussed in two recent reviews [1,2].

Kwor et al [3] have reported carrier concentration and its depth distribution for S+ implanted GaAs in the energy range 40-600 keV and a dose range 4 X 10*12 - 4 X 10*15/cm*2. Hall measurements were combined with chemical etching techniques for profiles. Annealing was done between 700 and 900 deg C with a Si3N4 cap in a furnace for 15 minutes. Indiffusion of S is seen for all doses, becoming significant at higher doses. A maximum carrier concentration of 2.5 X 10*18/cm*3 was obtained for the sample implanted with 10*14/cm*2 and annealed at 900 deg C with a plasma deposited cap. Kuzuhara et al [4] have used infrared rapid thermal annealing between 750-1100 deg C for 2 secs. Electrical activation was found to increase with increasing annealing temperature up to 1100 deg C. A maximum electrical activation of 78% was obtained for a dose of 5 X 10*13/cm*-2. Also, more than 5 X 10*18/cm*3 peak carrier concentration was obtained for a dose of 1 X 10*14/cm*2, indicating two times higher peak concentration than that obtained after conventional furnace annealing. Bhattacharya et al [5] have measured carrier concentrations of 120 keV Se+ implanted GaAs after furnace annealing between 850 to 950 deg C for 15 min. At 900 deg C activation efficiency for 1 X 10*13/cm*2 has been reported to be 30% which goes down to 6% and 4% for 1 X 10*14 and 3 X 10*14/cm*2 respectively. Maximum carrier concentrations of 1 X 10*19/cm*3 has been found near the surface at both doses. Inada et al [6] have performed orientation dependent study using (111) and (100) GaAs crystals. For 100 keV Se+, 5 X 10*13cm*2, after annealing at 900 deg C, 15 min. they have obtained peak carrier concentrations of 3 X 10*18/cm*-3 at implant temperature of 400 deg C and 9 X 10*17/cm*3 at room temperature for (111) GaAs. The carrier concentrations were about the same for (100) crystals.

A number of publications dealing with the electrical properties of Si implants in GaAs have appeared in the literature [1-2]. There are some inconsistent reports about depth profiles of carrier concentrations and diffusion of Si during annealing. Yeo [7] has done a comprehensive study on the electrical properties of Si implanted GaAs - both Cr doped and undoped. For a dose of 1 X 10*13/cm*2 or below, the maximum electrical activation occurs at the 850 deg C anneal, while for doses of 3 X 10*13 and 1 X 10*14/cm*2, the maximum occurs at 900 deg C. The activation efficiencies increase with ion dose up to about 6 - 8 X 10*12/cm*2 and then they decrease at higher doses. Carrier concentrations in Cr-doped substrates were found to be higher than those in the undoped substrates. Peak carrier concentration of the order of 1 X 10*18/cm*3 was obtained for a dose of 1 X 10*13/cm*2, 100 keV Si+ after annealing at 850 deg C for 15 min. Dhiman and Wang [8] have developed empirical models for simulating the implantation and annealing steps used for doping GaAs with Si and Se. The universal activation efficiency plots were made from published experimental results that provide percentage activation efficiency as a function of atomic concentration (/cm*3) at different annealing temperatures. Kohzu et al [9] have shown that infrared rapid thermal annealing provided LSS tight carrier concentration profile for both n and n+ type Si dopant due to negligible diffusion. However the peak carrier concentration is about the same as that obtained by furnance annealing.

Davies et al [10] have studied incoherent light annealed (1000 deg C) 200 keV, 4 X 10*14/cm*2 Si+ implanted GaAs. A maximum carrier concentration of 6 X 10*18/cm*3 was obtained after annealing for 2.5 secs. Nojima [11] has studied the annealing effects of Si implanted GaAs by a Q-switch pulsed ruby laser. No electrical activation was observed for Q-switch pulsed ruby laser. No electrical activation was observed for doses below 10*14/cm*2. However, a carrier concentration 3 X 10*18/cm*3 was reported for 1 X 10*15/cm*2, 200 keV Si after laser annealing. Sealy [12] has obtained a peak carrier concentration 1 X 10*19/cm*3 after laser annealing 10*15/cm*2 of Se, Te and Sn.

A single implant of Ge in GaAs has shown amphoteric behavior [7]. Low doses favour p-type whereas high doses (more than 1 X 2-*14) favour n-type conductivity. Dual implantation of Ge and Ga produced p-type conductivity for all doses below 3 X 10*15/cm*3 whereas dual implantation of Ge and As has produced a significant enhancement of n-type activity for ion doses exceeding 10*15/cm*2 [13].

Not much work has been done on Te and Sn implanted GaAs. Eisen et al [14] have implanted 400 keV Te+ at doses of 10*13 and 10*14 cm*-2 at elevated substrate temperature (350 deg C). A peak carrier concentration of 2 X 1*18/cm*3 was reported after annealing with an aluminum oxy-nitride cap. Studies [15] on the Sn has also been reported for elevated temperature (200 deg C) implantation. A maximum n-type carrier concentration of 1.5 X 10*18/cm*3 has been reported for 400 keV Sn+, 1 X 10*14/cm*2, annealed by thermal pulse at 900 deg C for 30 sec.

REFERENCES
----

[1] D.V.Morgan, F.H.Eisen, A.Ezis [IEEE Proc. (USA) vol.128 (1981) p.109]

[2] K.G.Stephens [Nucl. Instr. & Meth. (Netherlands) vol.209/210 (1983) p.589]

[3] P.Kwor, Y.K.Yeo, Y.S.Park [J. Appl. Phys. (USA) vol.53 (1982) p.4786]

[4] M.Kuzuhara, H.Kohzu, Y.Takayama [J. Appl. Phys. (USA) vol.54

(1983) p.3121]
[5] R.S.Bhattacharya, P.P.Pronko, Y.K.Yeo, A.K.Rai, Y.S.Park, J.Narayan [J. Appl. Phys. (USA) vol.53 (1982) p.4821]
[6] T.Inada, K.Tokunaga, S.Taka, Y.Yuge, H.Kohzu [Nucl. Intstr. & Meth. (Netherlands) vol.182/183 (1981) p.709]
[7] Y.K.Yeo ["Ion Implantation in III-V Compound Semiconductors", Final report no. AFWAL-TR-84-1122, Sept. 1984]
[8] J.K.Dhiman, K.L.Wang [J. Electrochem. Soc. (USA) vol.131 (1984) 295]
[9] H.Kohzu, M.Kuzuhara, Y.Takayama [J. Appl. Phys. (USA) vol.54 (1983) p.4998]
[10] D.E.Davies, J.P.Lorenzo, M.Julian, P.J.McNally [IEEE Electron Device Lett. (USA) vol.3 (1982) p.102]
[11] S.Nojima [J. Appl. Phys. (USA) vol.53 (1982) p.5028]
[12] B.J.Sealy [Microelectron. J. (GB) vol.13 (1982) p.21]
[13] Y.S.Park, Y.K.Yeo, F.O.Pedrotti [Nucl. Instr. & Meth. (Netherlands) vol.182/183 (1981) p.617]
[14] F.H.Eise, B.M.Welch, H.Muller, K.Gammo, T.Inada, J.W.Mayer [Solid State Electron. vol.20 (1977) p.219]
[15] B.J.Sealy, R.K.Surridge, E.C.Bell, A.D.E.D´Cruz [Inst. of Phys. Conf. Ser. (GB) vol.28 (1976) p.69]

3.2 CARRIER CONCENTRATIONS IN ION-IMPLANTED p-TYPE GaAs

by R.S.Bhattacharya

June 1985
EMIS Datareview RN=15425

In general, p-type doping of GaAs by ion implantation was found to be successful with about 100% doping efficiency reported for doses up to 10*14 ions cm*-2. A saturation effect was observed with a decrease in doping efficiency at larger dose which is attributed to solubility limits, and in some cases outdiffusion. The species that have been successfully activated are: Zn, Be, Cd, Mg, and C. The optimum temperature for annealing was found to be about 800 deg C.

Reviews [1,2] of Zn implantation in the dose range 1 X 10*12 - 1 X 10*16 cm*-2 and energy range 20-450 keV reveal the activation efficiency to be in the range 6-100%. Doses below 10*14 cm*2 could be activated to 100% upon annealing at 800 deg C. No significant improvement in activity was observed by implanting at elevated temperature (up to 400 deg C). Cd shows generally the same properties at those of Zn. Very little work has been reported regarding the concentration profiles of Cd. However, some work has been pubished on the concentration profiles of Zn after conventional and rapid thermal annealing. Zolch et al [3] have shown that 150 keV Zn+ at a dose of 10*15 cm*-2 extend beyond the anticipated 0.2 micron and penetrate as deep as 1 micron below the surface upon conventional furnace annealing as 950 deg C for 30 min. Kular et al [4] have shown that for 150 keV Zn+ at a dose of 1 X 10*15 cm*-2, annealing up to 750 deg C for 15 min, does not produce significant broadening of the hole concentration profile; however, significant broadening is observed at higher temperatures. Rapid annealing at high temperature (approx. 800-1200 deg C) for times up to 10 s can achieve superior results to conventional thermal annealing in terms of the doping profiles, peak carrier concentrations, activation efficiencies and mobilities [5-8]. Typical peak carrier concentration about 2 X 10*19 cm*-3 has been obtained by furnace annealing whereas rapid annealing has been found to provide carrier levels up to about 8 X 10*19 cm*-3.

Be implant in the dose range 1 X 10*13 - 1 X 10*16 cm*-2 and energy range 40 - 400 keV shows activation in the range 10-100% at anneal temperature between 500-900 deg C [1]. Nojima and Kawasaki [9] have shown that for 350 keV Be+, 4 X 10*15 cm*-2 a maximum hole concentration approx. 2.5 X 10*19 cm*-3 was obtained by annealing at 700 deg C, 20 min with SiO2 encapsulant. Significant diffusion was observed after annealing at higher temperatures. Considerably less diffusion has been observed after rapid annealing using arc lamp [7].

Yeo et al [10] have studied the correlation of the electrical carrier concentration and the atomic profiles of Mg implants for doses between 10*13 and 3 X 10*15 cm*-2 at 120 keV. For doses below 10*14 cm*-2 and anneal temperature of 750 deg C no significant diffusion was measured. However the amount of outdiffusion increased with implanted dose during annealing for doses of 10*14 cm*-2 or more: it was 13%, 62% and 81% of the implanted dose for 10*14, 10*15 and 3 X 10*15 cm*-2 respectively.

Thus electrical activations show a strong dependence on the implanted dose. Very little work has been done on C implants. Implantation of 10-50 keV C at doses 4 X 10*13 - 4 X 10*14 cm*-2 and followed by annealing at above 600 deg C has been found to provide p-type activity [1]. Somewhat more detailed investigations by Danilov et al [11] indicate that the highest activation efficiency of about 5% can be obtained at low doses. They also found that additional bombardment with rare gas ions increased the doping efficiency drastically.

The concept of dual implantation in GaAs, as proposed by Heckingbottom and Ambridge [12] has been proved by several workers to improve the electrical activity and inhibit the diffusion of certain implant species. For example, the improvement of p-type activity has been observed as a result of the dual implantation of C and Ga into GaAs [13]. Dual implants of Zn or Cd and As into GaAs have been found to considerably decrease the diffusion of the implanted Zn or Cd [14,15].

Implantation of Ge up to doses below 10*14 cm*-2 produces p-type conductivity at all annealing temperatures from 700-950 deg C since more Ge ions occupy As sites. A maximum efficiency of 38% was obtained at a dose of 10*13 cm*-2 and annealing temperature of 950 deg C. At higher doses n-type activity results because of the Ga site occupancy. Dual implantation of Ge and Ga produces p-type conductivity at all doses (below 3 X 10*15 cm*-2) for an annealing temperature of up to 900 deg C. The p-type electrical activity of Ge in the dual implantation has been found to be improved significantly over that of the single implantation for all doses less than 3 X 10*14 cm*-2 while the original n-type behaviour of the Ge single implant has been found to change to p-type for doses above 10*-15 cm*-2 except at anneal temperatures higher than 900 deg C [17,18.]

Only a few reports are available in the literature on laser annealing of GaAs. Inada et al [19] have used Nd:YAG, wavelength 1.06 micron laser at a pulse length of 200 ns to anneal 100 keV Zn+ implanted GaAs. A peak carrier concentration of about 4 X 10*19 cm*-3 was obtained at a dose of 1 X 10*15 cm*-2 and laser energy density of 1.5 J/cm*2. Kular and Sealy [20] have also measured carrier concentration of Zn implanted GaAs by using a ruby laser with wavelength 0.69 micron and pulse length 0.8 ms. A peak hole concentration of 1 X 10*20 cm*-3 following irradiation with four pulses of 3 J/cm*2 was obtained in the case of 60 keV, 1 X 10*15 cm*-2 Zn+ implanted GaAs.

REFERENCES

[1] D.V.Morgan, F.H.Eisen, A.Ezis [IEE Proc. I (GB) vol.128 (1981) p.109]

[2] K.G.Stephens [Nucl. Instrum. & Methods. (Netherlands) vol.209/210 (1983) p.589]

[3] R.Zolch, H.Ryssel, H.Krapz, H.Reichl, I.Ruge ["Ion Implantation of Semiconductors and Other Materials" (Plenum Press, New York, 1977) p.593]

[4] S.S.Kular, B.J.Sealy, Y.Ono, K.G.Stephens [Solid State Electron. (GB) vol.27 (1984) p.83]

[5] D.E.Davies, P.J.McNally [IEEE Electron Device Lett.(USA) vol.ELD -4 (1983) p.356]

[6] T.Suzuki, H.Sakurai, M.Arai [Appl. Phys. Lett. (USA) vol.43

(1983) p.951]
[7] K.Tabatabaie-Alavi, A.N.M.Masum Choudhury, C.G.Fonstad, J.C.Gelpey [Appl. Phys. Lett. (USA) vol.43 (1983) p.505]
[8] S.Liu, J.Narayan [J. Electron. Mat. (USA) vol.43 (1983) p.505]
[9] S.Nojima, Y.Kawasaki [Jpn. J. Appl. Phys. Japan vol.17 (1978) p.1897]
[10] Y.K.Yeo, Y.S.Park, F.L.Pedrotti, B.D.Choe [J. Appl. Phys. (USA) vol.53 (1982) p.6148]
[11] Y.A.Danilov, V.T.Karmanov, P.V.Pavlov, E.A.Pitrimova [Sov. Phys. Semicond. (USA) vol.15 (1981) p.971]
[12] R.Heckingbottom, T.Ambridge [Radiation Eff. (GB) vol.17 (1973) p.31]
[13] B.K.Shin, J.E.Ehret, Y.S.Park, M.Stenfiniw [J. Appl. Phys. (USA) vol.49 (1978) p.2988]
[14] A.A.Gavrilou, G.A.Kachurin, N.B.Pridachin, C.S.Smirov [Sov. Phys. Semicond. (USA) vol.8 (1975) p.1455]
[15] K.Aoki, K.Gamo, K.Masuda, S.Namba ["Ion Implantation in Semiconductors" ed F.Chernow, J.A.Borders, D.K.Brice (Plenum, New York, 1977) p.123]
[16] Y.K.Yeo, J.E.Ehret, F.L.Pedrotti, Y.S.Park, W.M.Theis [Appl. Phys. Lett. (USA) vol.35 (1979) p.197]
[17] Y.S.Park, Y.K.Yeo, F.L.Pedrotti [Nucl. Instrum. & Methods. (Netherlands) vol.182/183 (1981) p.617]
[18] P.P.Pronko, Y.K.Yeo, R.S.Bhattacharya, A.K.Rai ["Ion Implantation in III-V Compound Semiconductors", Final report No.AFWAL-TR-84-112 (Sept 1984) p.66-82]
[19] T.Inada, S.Kato, Y.Maeda, K.Tokunaga [J. Appl. Phys. (USA) vol.50 (1979) p.6000]
[20] S.S.Kular, B.J.Sealy [Solid State Electron. (GB) vol.23 (1983) p.875]

3.3 CARRIER CONCENTRATIONS IN UNDOPED AND DOPED EPITAXIAL GaAs

by I.H.Goodridge

December 1985
EMIS Datareview RN=15490

It is usual for epitaxial GaAs to contain both acceptor and donor impurities simultaneously. The difference between the two concentrations, the net carrier concentration, (Nd-Na), determines whether the GaAs is p-type or n-type.

Donors, present in epitaxially grown GaAs as either residual impurities or deliberately added as dopants, can include Sn, Ge, Si, C, Te, Se and S. The Group VI elements listed are always donors in GaAs. All the donors listed have similar low ionisation energies of approx. 5.8 - 5.9 meV below the conduction band, which means that absolute identification of these residual shallow donors is difficult. Analysis techniques used, include far infrared photoconductivity, photoluminescence in a magnetic field and more recently photothermal ionisation spectroscopy.

The shallow acceptors which are often seen in undoped GaAs, of which some may be used as doping elements, include Ge, Cd, Si, Zn, Mg, Be, C. The Group II elements listed are always acceptors. Residual shallow acceptors can be determined using photoluminescence techniques. The ionisation energies, in ranging from 26 meV for C up to 40 meV for Ge, i.e. above the valence band, simplifies the determination.

The usual techniques used for growing epitaxial GaAs include MBE, LPE, VPE and MOCVD. Although all techniques can produce high purity GaAs, conditions for the minimisation of (Nd-Na) are not necessarily optimum for the production of multilayer structures as required by the device engineer. Thus the reactor parameters are not always adjusted to minimise (Nd-Na) per se and so the term residual background concentration should be treated with caution.

From the above listings, the elements Ge, Si and C are seen to be amphoteric in nature and can therefore substitute preferentially either for arsenic or gallium during epitaxy. Specific site occupancy is determined by the V/III ratio. In MOCVD this is the AsH_3/TMG ratio and in VPE, the As/Ga ratio. Interestingly one or more of these amphoteric impurities is invariably observed as the dominant impurity in normally undoped epitaxial GaAs, independent of the method used.

In MOCVD and VPE, where the V/III much greater 1, typically (Nd-Na) approx. 10^{15} cm^{-3} with Si seen as the dominant residual impurity. In VPE, the Si is widely considered to originate from in-situ side reactions between the hot quartz wall of the reactor and HCl. However in chloride VPE, when N_2 replaces [1] H_2 as the carrier gas, Si is not seen in VPE layers. Such layers still have (Nd-Na) approx. 10^{15} cm^{-3} but with S as the dominant residual impurity. Presumably in the $N_2/AsCl_3$/Ga system, since HCl cannot be formed, then there is no mechanism for Si transport. In MOCVD, a similar situation exists. Since the reactor walls are cold (and chlorides absent), contamination by Si from the reactor is not possible. However, Si is known to be the major contaminant in TMG (trimethylgallium) and is often seen at the ppm level. It is considered that the present purity of the alkyl sources used in MOCVD limits the (Nd-Na) to approx. 10*15 cm*-3 with mu(77) up to 100,000 cm*2/V.s (c.f. VPE where mu(77) above 200,000 cm*2/V.s. However using highly refined TMG, it is possible to better this [2], e.g. 3.7 X 10*14 cm*-3 with mu(77)degK = 139,000 cm*2/V.s.

If the V/III ratio is reduced, IN MOCVD, layers become p-type approx. 10*15 cm-3 with C seen as the dominant residual impurity. The (C) in MOCVD GaAs can be reduced by using TEG at low reactor pressure with a low V/III ratio, or more simply, by using a higher V/III ratio with the standard TMG/AsH3/H2 system at atmospheric pressure. In chloride VPE, when the AsCl3 molar fraction [3] is increased i.e. V/III ratio decreased (Nd-Na) approx. 10*13 - 10*14 cm*-3 are obtained. If further increased, a higher molar fraction results in highly resistive, (10*4-10*6 ohm cm) GaAs being formed. The molar fraction effect is considered as suppressing the parasitic wall reactions yielding Si chlorosilanes rather than a true V/III effect.

In the LPE system, the reverse situation exists, i.e. V/III below or equal to 1. The effect of this is that Si can be either a donor or an acceptor dependent on temperature. In LPE, not only is the reactor more simple, but also the chemistry involved. This can result in the attainment of very low backgrounds at the low 10*12 cm*-3 level. This requires boats and long bake-out times of the Ga melt. High purity layers where mu(77) as high as 244,000 cm*2/V.s have been obtained [4] with (Nd-Na) = 3.7 X 10*12 cm*-3. The advantage of the LPE system is that once volatile impurities have been removed by baking, a low growth rate can be selected which minimises the segregation coefficients of the remaining impurities. This results in the achievement of low backgrounds.

In MBE, using a solid As source, p-type GaAs at approx. 10*14/cm*3 is usually obtained with C seen as the dominant residual impurity. If AsH3 is used as the As source then (Nd-Na) = approx. 2 X 10*14/cm*3 with mu(77) = 110,000 cm*2/V.s can result [5]. Semi-insulating layers have also been grown [6], again using a solid As source, which could then be back-doped with Si to give (Nd-Na) = 10*14 cm*-3 and mu(55) = 140,000 cm*2/V.s.

Controlled intentional doping of GaAs occurs from about an order of magnitude above the residual background carrier concentration and can continue to saturation. This can be as high as (Nd-Na) = 2.3 X 10*20 cm*-3 using MOCVD and Be from the precursor DMBe [7]. There is some evidence that impurity uptake continues beyond saturation but with no effect on carrier concentration.

In MOCVD and VPE, good doping control is achieved by selecting both the optimum dopant and precusor to suit the special requirements. For p-type doing, Zn as either DEZn or DMZn is favoured although Be, Cd and Mg have been used. (Nd-Na) is found [8] to depend linearly on P(DMZn) up to the onset of saturation at approx. 3 X 10*19 cm*-3. Mg appears to be a better dopant at lower concentrations. This perhaps reflects its lower vapour pressure and the use [9] of the precursor, bis cyclopentadienyl magnesium, which may well dissociate differently from other precursors.

Again, in MOCVD and VPE, the choice of n-type dopant is wide. In VPE, S from H2S is favoured, perhaps for historical reasons. Again a linear dependence of (Nd-Na) on P(H2S) is found, with, in this case, saturation seen at 3 X 10*18 cm*-3. In MOCVD, Se is favoured since higher doping concentrations are possible. Linear doping behaviour is again observed up to saturation at approx. 10*19 cm*-3. It is found that surface kinetics control the uptake of those elements with high vapour pressures, e.g. Cd, Zn, Se, S, whereas those elements with lower vapour pressures, e.g. Sn, Si, Ge, are incorporated at levels independent of deposition temperature which is indicative of a mass transport limited process. Thus, using Si2H6 as a precursor, Si has been used to produce very uniformly doped n-type layers in MOCVD.

In MBE, vapour pressure of the dopant element is also critical. Only those with relatively low vapour pressure have sticking coefficients of unity or near unity, e.g. Sn, Si, Ge, Mg, Be. Even though Sn, Si and Ge are useful n-type dopants in MBE, levels higher than approx. 5 X 10*18 cm*-3 are difficult. However, higher levels can be achieved by using precursors [10] of the type SnS2, PbS, PbSe, Tel, to yield (Nd-Na) above or equal to 10*19 cm*-3.

In LPE, obviously only these elements with low vapour pressures can be used. Si in high temperature melts yields n-type conductivity whereas at lower temperatures this changes to p-type. Te doping reaches saturation at approx. 4 X 10*19 cm*-3 and Ge is used as an acceptor in doped LPE GaAs up to approx. 3 X 10*18 cm*-3.

Intentional doping using deep acceptors has successfully been used in LPE, VPE and MOCVD to produce highly resistive material with Cr and Fe being preferred. In VPE and MOCVD volatile precursors such as FeCl2, CrO2Cl2, (CO)6Cr and VO(OC2H2)3 have been used.

REFERENCES

[1] Lin Lan-Ying, Lin Yao-Wang, Zhong Xung-Ru, Zhang Yanji-Yun, Li Hsiu-Lan [J. Crystal Growth (Netherlands) vol.56 (1982) p.344-439]

[2] T.Nakanisi, T.Udagawa, K.Kamei [J. Crystal Growth (Netherlands) vol.55 (1981) p.255-62]

[3] J.V.Dilorenzo, G.E.Moore [J. Electrochem. Soc. (USA) vol.118 (1971) p.1823-30]

[4] H.Miki, M.Otsubo [Jpn. J. Appl. Phys.(Japan) vol.10 (1971) p. 509]

[5] A.R.Calawa [Appl. Phys. Lett. (USA) vol.38 (1981) p.701-3]

[6] J.C.M.Hwang, H.Temkin, T.M.Brenan, R.E.Frahm [Appl. Phys. Lett. (USA) vol.42 (1983) p.66-8]

[7] R.Mellet, R.Azoulay, L.Dugrand, E.V.K.Rao, A.Mircea [Inst. Phys Conf. Ser. (GB) vol.63 (1982) p.583-4]

[8] I.H.Goodridge, P.M.Edwardson [Physica b (Netherlands) vol.129 (1985) p.408-12]

[9] C.R.Lewis, W.T.Dietze, M.J.Ludowise [J. Electron. Mater. (USA) vol.12 (1983) p.507-24]

[10] J. De-Sheng, Y.Makita, K.Ploog, H.J.Queisser [J. Appl. Phys. (USA) vol.53 (1981) p.999-1006]
vol.47 (1976) p.1574]

CHAPTER 4

CARRIER IONIZATION RATES IN GALLIUM ARSENIDE

4.1 Carrier ionization rates in GaAs, temperature dependence

4.2 Carrier ionization rates in GaAs, doping dependence

4.3 Carrier ionization rates in GaAs, electric field dependence

4.1 CARRIER IONIZATION RATES IN GaAs, TEMPERATURE DEPENDENCE

by J.P.R.David

September 1984
EMIS Datareview RN=12287

Pearsall et al [1] considered the effects of the ionization threshold energy and its location in the Brillouin zone in determining ionization rates. Temperature and orientation affect the band structure and changes in a(E) and b(E) are found, where a and b are the electron and hole ionization coeffs respectively. LPE grown p+n (N = 2 X 10*16 cm*-3) junctions grown on (100), (110) and (111) orientation substrates were used. Over the field range 3 X 10*5 - 5.5 X 10*5 Lee et al [2] fitted the a,b curves of Pearsall et al [1] to the expression:

a,b = C exp [-(x/E)*m]

giving:

a,b	Orientation	(100)	(110)	(111)	
a	C	9.12 X 10*4	2.19 X 10*7	7.76 X 10*4	(cm*-1)
	x	4.77 X 10*5	2.95 X 10*6	4.45 X 10*5	(V/cm)
	m	3.48	1	6.91	
b	C	3.47 X 10*6	3.47 X 10*6	6.31 X 10*6	(cm*-1)
	x	2.18 X 10*6	2.27 X 10*6	2.31 X 10*6	(V/cm)
	m	1	1	1	
Field range		3.13-4.76	3.13-4.76	3.33-5.56	(10*5 V/cm)

Conditions for hole ionization are similar in all three directions hence the similarity in b(E). Differences in a(E) are explained by the thresholds varying in different orientations. In GaAs the bandgap decreases with temperature and Pearsall et al [1] investigated the effect on K (a/b) for the (100) orientation device over the range 20-250 deg C. For T less than 200 deg C, K is less than 1 but for T greater than 200 deg C, K is greater than 1 over the entire field range 4-5 X 10*5 V/cm. Capasso et al [3] further investigated temperature effects on (110) and (111) orientation junctions. The (111) orientation had similar trends to the (100) orientation while the (110) orientation had K greater than or equal to 1 from 20-200 deg C. In all orientations a,b decreased with temperature, this being attributed to phonon scattering. These results are explained in terms of the band structure and threshold energy.

REFERENCES

[1] T.P.Pearsall F.Capasso, R.E.Nahory, M.A.Pollack, J.R.Chelikowsky [Solid-State Electron. (GB) vol.21 (1978) p.297-302]

[2] M.H.Lee, S.M.Sze [Solid-State Electron. (GB) vol.23 (1980) p.1007-9]

[3] F.Capasso, R.E.Nahory, M.A.Pollack [Electron. Lett. (GB) vol.15 no.4 (1979) p.117-8]

4.2 CARRIER IONIZATION RATES IN GaAs, DOPING DEPENDENCE

by J.P.R. David

September 1984
EMIS Datareview RN=12289

There is not as much published on the effect of doping on electron (a) and hole (b) ionization coefficients as of electric field. Law et al [1] are the only investigations to notice a doping dependence. The doping range of 2.94 X 10*15 - 5.5 X 10*16 cm*-3 with corresponding field values of 2.2 X 10*5 - 4.7 X 10*5 V/cm was covered by Pt-Schottky GaAs diodes. b(E) was fitted to be continuous over the entire doping range with the form:

b,a = C exp [-(d/E)*2] cm*-1

while a(E) was fitted for each doping range. Their results gave:

	C	d(V/cm)
Low doping a (2.94 X 10*15 cm*-3)	1.97 X 10*5	5.58 X 10*5
Medium doping a (1.2 X 10*16 cm*-3)	4.54 X 10*5	6.28 X 10*5
High doping a (5 X 10*16 cm*-3)	1.83 X 10*5	5.79 X 10*5
b (2.94 X 10*15 - 5 X 10*16 cm*-3)	4.10 X 10*5	6.32 X 10*5

Thus at low doping a/b is greater than 1 and at high doping a/b is less than 1.

This behaviour is explained by the transit time of electrons in the high field region being similar to the band to band scattering time. Ando et al [2] carried out noise measurements on p+n mesa type (100) GaAs avalanche photodiodes with a doping range 6 X 10*15 - 9 X 10*16 cm*-3 corresponding to a field variation of 2.4 X 10*5 - 5.6 X 10*5 V/cm. The measured excess noise factor gave a K(eff) (or b/a) = 0.5 over the entire field and doping range. a and b were found to be exponential functions of field and independent of the doping density.

a(E) = 1.1 X 10*7 exp (-2.2 X 10*6/E) cm*-1

b(E) = 0.5 a(E) cm*-1

Recently Bulman et al [3] have used specially fabricated (100) GaAs pn junction diodes covering the doping range 2.2 X 10*15 - 1.1 X 10*17 cm*-3 corresponding to a field range of 2.2 X 10*5 - 6.25 X 10*5 V/cm. Their structures eliminated errors due to Franz-Keldysh electroabsoprtion and the analysis took account of the dead-space where carriers gain the threshold energy without ionizing This dead-space correction enabled a(E) and b(E) to line up over the entire field range with the form:

a(E) = 2.994 X 10*5 exp [-(6.84 X 10*5/E)*1.6] cm*-1

b(E) = 2.215 X 10*5 exp [-(6.570 X 10*5/E)*1.75] cm*-1

and no effect of doping is seen.

The results of [1] may be in part due to the lack of this dead space correction.

REFERENCES

[1] H.D.Law, C.A.Lee [Solid-State Electron. (GB) vol.21 (1978) p.331-40]

[2] H.Ando, H.Kanbe [Solid-State Electron. (GB) vol.24 (1981) p.629-34]

[3] G.E.Bulman, V.M.Robbins, K.F.Brennan, K.Hess, G.E.Stillman [IEEE Electron Device Lett. (USA) vol.EDL-4 no.6 (1983) p.181-5]

4.3 CARRIER IONIZATION RATES IN GaAs, ELECTRIC FIELD DEPENDENCE

by J.P.R. David

September 1984
EMIS Datareview RN=12288

A detailed review of work on ionization measurements in GaAs, including theoretical considerations, until 1975 is given by Stillman et al [1] and only a brief recapitulation is given here.

Logan et al [2] were the earliest to measure ionization coefficients in GaAs. They assumed $a = b$, where a and b are the electron and hole ionization coefficients respectively, and used narrow diffused pn junctions to obtain a field dependent rate.

Stillman et al [3] were the first to show that a is not equal to b by using a Pt Schottky barrier avalanche photodiode. By shining light of different wavelengths, nearly pure injections of holes and electrons were achieved. Over the electric field range $2.5-3.3 \times 10^5$ V/cm they found that b/a is greater than 1.

Law et al [4] covered a wider electric field range of $2.2-4.7 \times 10^5$ V/cm with different doped layers. Pure electrons and holes were injected by shining light on either side of a Pt Schottky diode. They found that a/b is greater than 1 at low fields and a/b is less than 1 at high fields with a crossover occurring at about 3×10^5 V/cm . b was curve fitted over the entire doping range to give:

$$b(E) = 4.1 \times 10^5 \exp [(-6.34 \times 10^5/E)^2] \text{ cm}^{-1}$$

At low doping range (2.94×10^{15} cm^{-3}),

$$a(E) = 1.97 \times 10^5 \exp [-(5.58 \times 10^5/E)^2] \text{ cm}^{-1}$$

At medium doping range (1.2×10^{16} cm^{-3})

$$a(E) = 4.54 \times 10^5 \exp [-(6.28 \times 10^5/E)^2] \text{ cm}^{-1}$$

At high doping range (5×10^{16} cm^{-3})

$$a(E) = 1.83 \times 10^5 \exp [-(5.79 \times 10^5/E)^2] \text{ cm}^{-1}$$

Pearsall et al [5] used LPE grown (100) GaAs pn junctions with $N(D) = 2 \times 10^{16}$ cm^{-3}. Over the electric field range $3-5.5 \times 10^5$ V/cm they found b/a is greater than 1 at the low fields with a comparable to b at the higher fields.

Ito et al [6] measured a and b of a (100) GaAs pn diffused diode with $n = 1.2 \times 10^{17}$ cm^{-3} using a novel crater mesa structure. Different wavelengths of light focused at the top and bottom of the mesa ensured pure carrier injection. Over the field range $4-6 \times 10^5$ V/cm the results obtained were:

a(E) = 5.6 X 10*6 exp (-2.41 X 10*6/E)cm*-1

b(E) = 1.5 X 10*6 exp (-1.57 X 10*6/E)cm*-1

b is greater than a for E less than or equal to 5 X 10*5 V/cm with b/a tending to unity for E greater than 5 X 10*5 V/cm. Ando et al [7] carried out noise measurements on p+n mesa type (100) GaAs avalanche photodiodes over the electric field range 2.4-5.6 X 10*5 V/cm. The measured excess noise factor gave an effective K (K = b/a) of 0.5. Ionization coefficient measurements using the constant K gave:

a(E) = 1.1 X 10*7 exp (-2.2 X 10*6/E) cm*-1

b(E) = 0.5 a(E) cm*-1

Thus this agrees with [4] over the low field region only and disagrees with [6] about K although b is similar. Recently Bulman et al [8] have used specially fabricated (100) GaAs pn junction diodes covering the electric field range 2.2-6.25 X 10*5 V/cm. This structure uses an etch-top to enable reproducible fabrication of wells in the substrate, which eliminates the Franz-Keldysh electroabsorption under back illumination. Several devices with different dopings were used and a and b were corrected to account for the dead space where carriers need to gain the threshold energy before ionization. The ratio a/b decreased from 2.5 to 1.3 as the electric field increased from 2.22-6.25 X 10*5 V/cm. Excess noise measurements also showed K(eff) of 0.6 confirming that a is greater than b in GaAs and agreeing with [7]. Their data give:

a(E) = 2.994 X 10*5 exp [-(6.848 X 10*5/E)*1.6] cm*-1

b(E) = 2.215 X 10*5 exp [-(6.57 X 10*5/E)*1.75] cm*-1

REFERENCES

[1] G.E.Stillman, C.M.Wolfe [Semiconductors. and Semimetals, Ed. R.K.Willardson, A.C.Beer, Academic Press (1977) vol.12 ch.5]

[2] R.A.Logan, A.G.Chynoweth, B.G.Cohen [Phys. Rev. (USA) vol.128 no.6 (Dec 1962) p.2518-23]

[3] G.E.Stillman, C.M.Wolfe, J.A.Rossi, A.G.Foyt [Appl. Phys. Lett. (USA) vol.24 no.10 (1974) p.471-4]

[4] H.D.Law, C.A.Lee [Solid-State Electron. (GB) vol.21 (1978) p.331-40]

[5] T.P.Pearsall, F.C.Capasso, R.E.Nahory, M.A.Pollack, J.R.Chelikowksy [Solid-State Electron. (GB) vol.21 (1978) p.297-302]

[6] M.Ito, S.Kagawa, T.Kaneda, T.Yamaoka [J. Appl. Phys. (USA) vol.49 no.8 (1976) p.4607-8]

[7] H.Ando, H.Kanbe [Solid-State Electron. (GB) vol.24 (1981) p.629-34]

[8] G.E.Bulman, V.M.Robbins, K.F.Brennan, K.Hess, G.E.Stillman [IEEE Electron. Dev. Lett. (USA) vol. EDL-4 no.4 (1983) p.181-4]

CHAPTER 5

ELECTRON MOBILITY, DIFFUSION AND LIFETIME IN GALLIUM ARSENIDE

5.1 Electron mobility in bulk crystalline GaAs

5.2 Electron mobility in bulk crystalline GaAs, temperature dependence

5.3 Electron mobility in LPE GaAs

5.4 Electron mobility in VPE GaAs

5.5 Electron mobility in VPE GaAs, temperature dependence

5.6 Electron mobility in VPE GaAs, doping dependence

5.7 Electron mobility in MBE GaAs

5.8 Electron mobility in MBE GaAs, temperature dependence

5.9 Electron mobility in MBE GaAs, doping dependence

5.10 Electron mobility in ion implanted GaAs, doping dependence

5.11 Electron mobility in GaAs, pressure dependence

5.12 Electron effective mass in GaAs, pressure dependence

5.13 Electron diffusion length and coefficient in p-type GaAs

5.14 Electron lifetime in p-type GaAs

5.1 ELECTRON MOBILITY OF BULK CRYSTALLINE GaAs

by D.Lancefield

July 1985
EMIS Datareview RN=14024

Parsey et al [1] have shown that low dislocation electron-trap-free GaAs can be grown reproducibly by the Bridgman method. Carrier mobilities of 2400 and 3300 cm*2/V.s were obtained at room temperature for crystals grown from Ga2O3 doped and undoped melts respectively.

Bonnafe et al [2] reported that the Hall mobilities of HB GaAs were in the range 1600 to 4200 cm*2/V.s at 350 deg K and 2400 to 3350 cm*2/V.s at 470 deg K. For LEC GaAs mobilities were 890 to 4050 cm*2/V.s at 350 deg K and 1940 to 3400 cm*2/V.s at 470 deg K.

Lent et al [3] found that 400 deg K Hall mobilities for undoped LEC GaAs were in the range 1375-4270 cm*2/V.s depending on crucible material, dopant concn, growth rate, water content of B2O3 encapsulant, post-growth annealing schedule and whether the sample was from the top, the middle or bottom of the crystal. Workers at Thomson CSF [4] have obtained mobilities of 3500-4800 cm*2/V.s at 400 deg K. Automatic Hall effect measurements on 50 mm diameter wafers indicate variations of 25% in mobility along crystal axis as well as variations across the wafer.

Mizutani et al [5] reported that mobilities for various HB GaAs:Cr samples after Si implantation and annealing were 3600-4000 cm*2/V.s at room temperature and 4000-5000 cm*2/V.s at 77 deg K. LEC GaAs:Cr showed similar results but undoped LEC GaAs showed higher mobilities (4000-4400 cm*2/V.s at room temperature and 5500-6500 cm*2/V.s at 77 deg K). As HB GaAs cannot be prepared without Cr dopant it is likely to be inferior to undoped LEC GaAs for fabrication of direct-implanted GaAs ICs.

Thomas et al [6] studied Si-29 implanted LEC semi-insulating GaAs grown from Si-free pyrolytic boron nitride crucible and found that at peak donor implant concentrations of about 10*17 cm*-3 mobilities of 4800-5100 cm*2/V.s were achieved. This material was likely to be more suitable than Si doped LEC or boat grown GaAs:Cr (mobility 3700-4500 cm*2/V.s) for application to devices such as power FETS. Barrett et al [7] find that In-doping of LEC GaAs significantly reduces the dislocation densities in large diameter wafers while maintaining 300 deg K mobilities of 5000 cm*2/V.s. Initial studies of implanted FET's fabricated on these wafers indicate a much tighter distribution of threshold voltage and source drain currents.

Nakai et al [8] have obtained mobilities of up to 6100 cm*2/V.s at room temperature using a LEC system incorporating specially designed heaters and shields, which allows good quality 3 inch diameter wafers to be produced.

Wood et al [9] reported calculations showing that the drift mobility in crystalline GaAs was almost unaffected by dislocation density at

300 deg K and above (operating conditions of GaAs microwave devices).

REFERENCES

[1] J.M.Parsey,Jr., Y.Nanishi, J.Lagowski, H.C.Gatos [J. Electrochem. Soc. (USA) vol.128 no.4 (1981) p.936-8]

[2] J.Bonnafe et al [Mater. Res. Bull. (USA) vol.16 no.10 (Oct 1981) p.1193-212]

[3] B.Lent, M.Bonnett, N.Visentin, J.P.Duchemin [Microelectron. J. (GB) vol.13 no.1 (1982) p.5-9]

[4] News item [`Distribution of Electrical Parameters in GaAs Monocrystals´, in Semicond. Int. (USA) vol.7 no.12 (Dec 1984) p.22]

[5] T.Mizutani, T.Honda, S.Ishida, Y.Kawasaki [Solid State Electron. (GB) vol.25 no.9 (1982) p.885-91]

[6] R.N.Thomas, H.M.Hobgood, D.L.Barrett, C.W.Eldridge [Semi-Insulating III-V Materials (Conf.) Nottingham (1980) Shiva Publishing Ltd ISBN 0-906812-05-4]

[7] D.L.Barrett, S.McGuigan, H.M.Hobgood, G.W.Eldridge, R.N.Thomas [J. Cryst. Growth (Netherlands) vol.70 (1984) p.179]

[8] R.Nakai et al [Sumimoto Electr. Tech. Rev. (Japan) vol.23 (1984) p.159-67]

[9] J.Wood, M.J.Howes, D.V.Morgan [Phys. Status Solidi a (Germany) vol.74 no.2 (1982) p.493-6]

5.2 ELECTRON MOBILITY IN BULK CRYSTALLINE GaAs, TEMPERATURE DEPENDENCE

by D.Lancefield

July 1985
EMIS Datareview RN=14027

Ciupina [1] has reported that the ratio of magnetoresistance mobility to Hall mobility for bulk GaAs is very close to unity over a wide temperature range (77-300 deg K). Hall mobility decreases with increasing temperature from 19000 cm*2/V.s at 77 deg K to 6000 cm*2/V.s at 296 deg K. Magnetoresistance mobility behaved similarly with values of 19500 cm*2/V.s at 77 deg K and 6100 cm*2/V.s at 296 deg K for the same sample.

Fritzche [2] reported Hall mobilities of undoped GaAs and GaAs:Cr subjected to nuclear transmutation doping. Mobilities were higher than in non-irradiated samples with similar carrier concentrations. The presence of Cr doping led to a reduction in mobility. For doped samples irradiated with 9.4 X 10*17 neutrons/cm*2 and annealed at 800 deg C for 4.5 hours mobilities at 77 deg K were 5000 cm*2/V.s for samples with and without Cr. Mobilities decreased with decreasing temperature, reaching a constant value. For neutron transmutation doped GaAs:Cr mobility at 2.5 deg K and below was about 500 cm*2/V.s. For samples without Cr, mobility at 10 deg K and below was about 1700 cm*2/V.s. Kolin et al [3] reported that for NTD LEC GaAs mobilities of 4000 cm*2/V.s were obtained in samples irradiated with 1.1 X 10*18 neutrons/cm*2, while the electron mobilities obtained generally depend in a complex manner on the neutron fluence, and on the anneal temperature and duration.

Jervis [4] measured low field mobility by the magnetoresistance method for n-GaAs in the range 90-300 deg K and found that mobility increased with decreasing temperature. Values at 300 deg K and 90 deg K were 7500 and 45000 cm*2/V.s respectively.

Theoretical mobilities have been reported by Jastrzebski et al [5] who found that for carrier concentration 4 X 10*17 cm*-3 mobility was 3100 cm*2/V.s at 300 deg K decreasing with increasing temperature to 2800 cm*2/V.s at 600 deg K.

Chattopadhyay et al [6] calculated Hall mobilities of compensated GaAs which shows a mobility degraded by a contribution varying as temperature raised to the power -0.5.

REFERENCES

[1] V.Ciupina [Rev. Roum. Phys. (Rumania) vol.26 no.2 (1981) p.211-9]
[2] H.Fritzsche [Report no. AD A093217 (Oct 1980)]
[3] N.G.Kolin, L.V.Kulikova, V.B.Osvenskii, S.P.Solov´ev, V.A.Kharchenko [Sov. Phys. Semicond. vol.18 (1984) p.1364]
[4] T.R.Jervis [Phys. Status Solidi a (Germany) vol.53 (1979) p.K199-202]
[5] L.Jastrzebski, J.Lagowski, H.C.Gatos, W.Walukiewicz [Inst.

Phys. Conf. Ser. (USA) no.45 ch.5 (1979) p.437-44]
[6] D.Chattopadhyay, H.J.Queisser [Proc. 15th Internat. Conf. on Phys. of Semiconductors, Kyoto, 1980, J. Phys. Soc. Jpn (Japan) vol.49 suppl. A (1980) p.293-6]

5.3 ELECTRON MOBILITY IN LPE GaAs

D.Lancefield

January 1986
EMIS Datareview RN=15600

In 1968 LPE GaAs with carrier concentrations about 2 X 10*14 cm*-3 had 300 deg K mobilities of 8000 cm*2/V.s [1], while at 77 deg K Solomon [2] had obtained mobilities of 110000 cm*2/V.s for a carrier concentration about 5 X 10*12 cm*-3, which he ascribed to residual oxygen impurities incorporated during growth. Since that time the carrier concentration dependence of the electron mobility has been measured by a number of authors including Vilms and Garrett [3], Panish [4], and Rosztoczy and Kinoshita [5]. Vilms and Garrett [3] measured mobilities in Sn doped GaAs at 300 deg K varying from 9000 cm*2/V.s for 5 X 10*13 cm*-3 to about 1000 cm*2/V.s at 10*19 cm*-3. At 77 deg K the mobilities decreased from about 130000 cm*2/V.s at 5 X 10*13 cm*-3 to about 8000 cm*2/V.s at 4 X 10*17 cm*-3.

Abrokwah et al [6] have shown that the simultaneous bake out of both melt and substrate at 775 deg C results in layers with carrier concentrations about 2 X 10*14 cm*-3 and 77 deg K mobilities in excess of 10*5 cm*2/V.s. Morkoc and Eastman [7] obtained mobilities at 77 deg K of 163000 cm*2/V.s with carrier concentrations of 9 X 10*13 cm*-3 using a 24 hour bake out and a growth temperature of 700 deg C. Bake out is considered to remove volatile impurities but C contamination from the graphite boat is the main remaining impurity.

Hicks and Manley [8] using a Spectrosil dipping boat (rather than graphite), reported on a high purity layer with liquid nitrogen mobilities in excess of 1.7 X 10*5 cm*2/V.s. Using a Spectrosil quartz reaction tube and a glassy carbon boat Miki and Otsubo [9] obtained mobilities of up to 244000 cm*2/V.s for samples with carrier concentrations down to 3 X 10*12 cm*-3.

Otsubo and Miki [10] obtained 77 deg K mobilities of between 1.7 X 10*5 cm*2/V.s and 2 X 10*5 cm*2/V.s reproducibly by LPE under a mixed gas flow of purified H2 and arsenic vapour which was found to reduce contamination due to oxygen, Si and Cu.

REFERENCES
----

[1] A.R.Goodwin, C.D.Dobson, J.Franks [Proc. Int. Symp., GaAs (1968) Dallas (Inst. Phys. Conf. Ser. no.7) p.36]
[2] R.Solomon [Proc. Int. Symp., GaAs (1968) Dallas (Inst Phys. Conf. Ser. no.7) p.11]
[3] J.Vilms, J.P.Garrett [Solid-State Electron. (GB) vol.15 (1972) p.443]
[4] M.B.Panish [J. Appl. Phys. (USA) vol.44 (1973) p.2667]
[5] F.E.Rosztoczy, J.Kinoshita [J. Electrochem. Soc. (USA) vol.121 (1974) p.439]
[6] J.K.Abrokwah, M.L.Hitchell, J.E.Borell, D.R.Schulze [J. Electron. Mater. (USA) vol.10 no.4 (1981) p.723]
[7] H.Morkoc, L.F.Eastman [J. Cryst. Growth (Netherlands) vol.36

(1976) p.109]
[8] H.G.B.Hicks, D.F.Manley [Solid State Commun. (USA) vol.7 (1969) p.1463]
[9] H.Miki, M.Otsubo [Jpn. J. Appl. Phys. (Japan) vol.10 (1971) p.509]
[10] M.Otsubo, H.Miki [Jpn. J. Appl. Phys. (Japan) vol.14 (1975) p.621]

5.4 ELECTRON MOBILITY IN VPE GaAs

by D.Lancefield

November 1985
EMIS Datareview RN=14023

Electron Hall mobilities of up to 8900 cm*2/V.s at room temperature and up to 200000 cm*2/V.s at 77 deg K have been observed in VPE GaAs [1].

Enstrom and Appert [2] showed that VPE layers grown using the hydride method with cooled HCl had superior mobilities to those grown using uncooled HCl. The HCl cooling increased room temperature mobilities by about 20% and 77 deg K mobilities by up to 75%. Look and Colter [1] obtained high mobility and very low compensations by the hydride method by prebaking the Ga source at 820 deg C for 16 hours immediately prior to growth.

Chang et al [3] have reported that films with room temperature Hall mobilities above 7000 cm*2/V.s can be grown by MOVPE using triethylgallium (TEG) as the Ga source under optimum growth conditions (As/Ga mole ratio = 13; growth pressure = 20 torr; growth temperature = 660 deg C; AsH3, TEG, and H2 flow rates 250, 10 and 1000 cm*3/min respectively). Kobayashi et al [4] showed that the mobility of GaAs grown from AsH3 + TMG by MOCVD at reduced pressure (76 torr) was not significantly different from that of layers grown at atmospheric pressure. Escobosa et al [5] reported mobility values decreasing with decreasing growth temperature for MOCVD layers grown from TMG and AsH3 at atmospheric pressure. Sealy et al [6] showed that at pressures of 76 torr the best mobility was obtained for As:Ga ratio of about 70, giving mobilities at 77 deg K of 93000 cm*2/V.s.

However, Smith has obtained 77 deg K mobilities of above 120000 cm*2/V.s at pressures of 1.5 torr with As:Ga ratios in excess of 300 [7]. Fraas [8] found that GaAs layers deposited from TEG and AsH3 using a vacuum MOCVD system at growth pressures of 10*-2 torr had room temperature mobilities of 4990 cm*2/V.s for unintentionally doped layer with carrier concentration of 1 X 10*16 cm*-3. Kuech and Potemski [9] using a similar system obtained mobilities of 97000 cm*2/V.s at 77 deg K with an As:Ga ratio of 20, and a growth pressure and temperature of 0.1 atm and 650 deg C respectively. Blaauw et al [10] obtained 80 deg K mobilities of 13000 cm*2/V.s at carrier concentrations of 2 X 10*16 cm*-3 by MOCVD using a Si susceptor, and found that it had no effect on the layers grown.

Yoshida et al [11] have grown GaAs using TMG and AsCl3. This has the advantage of being free from the source instability of chloride VPE and avoids the use of arsine. Mobilities of 36000 cm*2/V.s have been obtained at 77 deg K for carrier concentrations of 1.4 X 10*14 cm*-3.

S doped films grown from AsCl3-GaAs-H2 on single crystal GaAs using the cleft process had 77 deg K mobility of 20000 cm*2/V.s (Bozler et al [12]).

Hariu et al [13] studied GaAs grown by plasma-assisted epitaxy.

Undoped GaAs deposited in H2 and Sn-doped GaAs deposited in Ar were n-type with electron mobilities up to 2350 cm*2/V.s after annealing.

REFERENCES
----

[1] D.C.Look, P.C.Colter [Phys. Rev. B (USA) vol.28 (1983) p.1151]

[2] R.E.Enstrom, J.R.Appert [J. Electrochem. Soc. (USA) vol.129 no.11 (1982) p.2566-9]

[3] C.Y.Chang, Y.K.Su, M.K.Lee, L.G.Chen, M.P.Houng [J. Cryst. Growth (Netherlands) vol.55 no.1 (1981) p.24-9]

[4] N.Kobayashi, T.Fukui, Y.Horikoshi [Jpn. J. Appl. Phys. (Japan) vol.21 no.11 (1982) p.L705-6]

[5] A.Escobosa, H.Krautle, H.Beneking [J. Cryst. Growth (Netherlands) vol.57 no.3 (1982) p.605-6]

[6] J.R.Sealy et al [Inst. Phys. Conf. Ser. (GB) no.65 (1983) GaAs & Related Compounds, Albuquerque (1982) p.109-16]

[7] F.T.J.Smith [J. Cryst. Growth (Netherlands) vol.67 (1984) p.573-8]

[8] L.M.Fraas [J. Appl. Phys. (USA) vol.52 no.11 (1981) p.6939-43]

[9] T.F.Kuech, R.Potemski [Appl. Phys. Lett. vol.47 no.8 (1985) p.821]

[10] C.Blaauw, A.J.Springthorpe, B.Emmerstorfer, C.Miner [J. Vac. Sci. & Technol. A vol.3 no.2 (1985) p.438]

[11] M.Yoshida, H.Terao, H.Wantanabe [J. Electrochem. Soc. vol.132 no.4 (1985) p.930]

[12] C.O.Bozler, R.W.McClelland, J.P.Salerno, J.C.C.Fan [J. Vac. Sci. & Technol. vol.20 no.3 (1982) p.720-5]

[13] T.Hariu, K.Matsushita, Y.Komatsu, S.Shibuya, S.Igarashi, Y.Shibata [Inst. Phys. Conf. Ser.no.65 (1983) GaAs & Related Compounds, Albuquerque, 1982, p.141-8]

5.5 ELECTRON MOBILITY IN VPE GaAs, TEMPERATURE DEPENDENCE

by D.Lancefield

July 1985
EMIS Datareview RN=14026

Gudz et al [1] have shown that for high purity VPE GaAs deposited onto semi-insulating GaAs substrates etching away of the surface layer led to a marked increase in mobility at higher temperatures (100-300 deg K). For both unetched and etched samples there was a maximum in the mobility near 50 deg K. At 50 deg K values of log(mobility/cm*2/V.s) were 5.07 and 5.17 for unetched samples and a sample with 5 microns etched off, respectively.

Look and Colter [2] studied low compensation VPE GaAs grown in an AsCl3-Ga-H2 reactor with Ga source bakeout prior to growth. The mobility of this material also showed a maximum (about 2 X 10*5 cm*2/V.s) at 50 deg K. An additional maximum, thought to be associated with extremely low compensation, occurred near 9 deg K. Yao-Wang et al [3] using an AsCl3-Ga-Ar system in which Si contamination is suppressed have obtained 77 deg K mobilities of 205000 cm*2/V.s with carrier concentrations of 7.4 X 10*12 cm*-3, and peak mobilities at 35 deg K of 378000 cm*2/V.s.

Walsh et al [4] reported Hall mobility vs temperature data for compensated MOVPE n-GaAs on (100) oriented semi-insulating GaAs. Their results suggested the existence of a metallic impurity donor band and there was a maximum in the mobility at about 105 deg K. Mobilities at 105 deg K were in the range 2000-3000 cm*2/V.s for samples with various donor and acceptor concentrations.

REFERENCES

[1] E.S.Gudz et al [Inorg. Mater. (USA) vol.16 no.2 (1980) p.128-32]

[2] D.C.Look, P.C.Colter [Phys. Rev. B (USA) vol.28 no.2 (1982) p.1151-3]

[3] Lin Yao-Wang, Zhang Yan-Yun, Li Hsiu-Lan, Liang Jun-Wu, Lin Lan-Ying [J. Cryst. Growth (Netherlands) vol.74 (1984) p.108]

[4] D.Walsh, A.Poirier, N.Christoforou, J.Auclair [J. Phys. C (GB) vol.15 no.34 (1982) p.7043-8]

5.6 MOBILITY OF ELECTRONS IN VPE GaAs, DOPING DEPENDENCE

by D.Lancefield

November 1985
EMIS Datareview RN=14028

Since the work of Poth et al [1] who showed that Hall mobility of VPE S-, Se-, or Si-doped GaAs films was strongly dependent on free carrier concentration but was independent of growth parameters and produced a universal mobility vs carrier concentration curve, several authors have presented data on decrease of mobility of VPE GaAs with increasing carrier concentration.

Hollan et al [2] reported that mobility of S doped VPE GaAs was less than that of Se doped layers for carrier concentration above 4 X 10*17 cm*-3. For carrier concentrations 10*18 cm*-3 Hall mobilities were 2900 cm*2/V.s for GaAs:Se and 2600 cm*2/V.s for GaAs:S. For carrier concentrations 4 X 10*18 cm*-3 mobilities of GaAs:Se and GaAs:S were 1900 and 1500 cm*2/V.s respectively.

Glew [3] also studied MOCVD epitaxial S and Se doped GaAs grown using H2S and H2Se as dopant sources. Differences in behaviour were ascribed to differing S and Se incorporation mechanisms. For dopant concentration 3 X 10*16 cm*-3 mobilities of GaAs:Se and GaAs:S were 4815 and 5500 cm*2/V.s respectively. For dopant concentration 1 X 10*18 cm*-3 respective mobilities were 3060 and 2720 cm*2/V.s.

Veuhoff et al [4] studied VPE GaAs grown in the HCl-Ga-AsH3-H2 system on GaAs of four different orientations and found that Hall mobility (77 deg K) was similar in all four orientations for carrier concentrations up to 10*18 cm*-3 decreasing with increasing electron concentration before decreasing rapidly near saturation. For layers on (111)As surface mobilities as low as 100 cm*2/V.s were observed (for carrier concentrations about 4 X 10*18 cm*-3).

Beneking et al [5] found that Hall mobility of unintentionally doped GaAs grown in a MOCVD reactor in which the susceptor is radiatively heated fell from about 16000 cm*2/V.s at carrier concentration 10*16 cm*-3 to 4000 cm*2/V.s at carrier concentrations of 10*17 cm*-3. Se doped material behaved similarly with carrier concentrations up to 10*19 cm*-3 (corresponding mobility 390 cm*2/V.s) being achieved. The behaviour of both undoped and doped material was in agreement with the data of Poth et al.

Maier et al [6] studied VPE GaAs:Sn and found 77 deg K Hall mobilities ranging from 2680 cm*2/V.s at carrier concentrations of 3 X 10*17 cm*-3 to 1240 cm*2/V.s at 3.4 X 10*19 cm*-3, in agreement with Poth's curve.

Parsons and Krajenbrink [7] found that 300 deg K Hall mobilities of GaAs:Sn grown by MOCVD, with carrier concentrations in the range about 7 X 10*14 to 10*19 cm*-3, had corresponding mobilities of 8000 cm*2/V.s and 1300 cm*2/V.s respectively. Roth et al [8] found that GaAs grown by low pressure MOCVD gave mobilities at 77 deg K that lay significantly above Poth's curve.

Kuech et al [9] have shown that for GaAs:Si grown by MOCVD using disilane as the dopant source 300 deg K mobilities at about 7 X 10*18 cm*-3 are about 1200 cm*2/V.s and increase to about 5500 cm*2/V.s as the carrier concentration decreases to about 3 X 10*16 cm*-3. Krautle et al [10] have demonstrated that sheet carrier concentrations and mobilities of 5 X 10*14/cm*2 and 5000 cm*2/V.s, respectively, can be obtained in 20 nm widths by laser pulse initiated diffusion of surface adsorbed Se.

REFERENCES

[1] H.Poth, H.Bruch, M.Heyen, P.Balk [J. Appl. Phys. (USA) vol.49 no.1 (Jan 1978) p.285-8]

[2] L.Hollan, M.Boulou, J.P.Chane [J. Electron Mater. (USA) vol.10 no.1 (Jan 1981) p.193-212]

[3] R.W.Glew [Inst. Phys. Conf. Ser. (GB) no.63 ch.12 (1982) p.581-2]

[4] E.Veuhoff, M.Maier, K.-H.Bachem, P.Balk [J. Cryst. Growth (Netherlands) vol.53 no.3 (June 1981) p.598-604]

[5] H.Beneking, A.Escobosa, H.Krautle [J. Electron Mater. (USA) vol.10 no.3 (May 1981) p.473-80]

[6] M.Maier, B.Hanel, P.Balk [J. Appl. Phys. (USA) vol.52 no.1 (Jan 1981) p.342-3]

[7] J.D.Parsons, F.G.Krajenbrink [J. Crystal Growth (Netherlands) vol.68 (1984) p.60]

[8] A.P.Roth, R.Yakimova, V.S.Sundaram [J. Crystal Growth (Netherlands) vol.68 (1984) p.65]

[9] T.F.Kuech, B.S.Meyerson, E.Veuhoff [Appl. Phys. Lett. (USA) vol.44 (1984) p.986]

[10] H.Krautle, W.Roth, A.Krings, H.Beneking [Mat. Res. Soc. Symp. Proc. (USA) vol.29 (1984) p.353]

5.7 ELECTRON MOBILITY IN MBE GaAs

by D.Lancefield

November 1985
EMIS Datareview RN=14022

Electron Hall mobilities of up to 8100 cm*2/V.s at room temperature [1] and up to 126000 cm*2/V.s at 77 deg K [2] have been observed for MBE GaAs. This required outgassing of the effusion cells and a reduction of the water vapour in the growth chamber. Mobilities of Sn doped samples [1] were above 10*5 cm*2/V.s at 77 deg K for sample thicknesses of 27 microns. Comparable samples 3 microns thick had 77 deg K mobilities of 60000 cm*2/V.s.

Ballingall and Wood [3] found that for n-GaAs:Si layers grown on Cr doped substrates with various orientations and substrate temperatures 500-600 deg K, Hall mobilities were in the range 2840-5090 cm*2/V.s at 300 deg K.

Metze and coworkers [4,5] have demonstrated that GaAs:Si films grown at low substrate temperatures and low growth rates show electrical properties comparable with those of films grown under conventional conditions (substrate temperature 570 deg C, growth rate 1 micron/h). Substrate temperatures as low as 380 deg C could be used if growth rate was reduced to 0.02 micron/h. 77 deg K mobilities for layers grown at 380 deg C, 0.02 micron/h and at 450 deg C, 0.17 micron/h were about 800 and 10000 cm*2/V.s respectively.

Erickson et al [6] showed that, using As2, mobilities of 7370 cm*2/V.s at room temperature and 60000 cm*2/V.s at 77 deg K can be obtained for Si doped GaAs at growth rates which are relatively insensitive to substrate temperature.

Davies and Andrews have shown [7] that oil-diffusion pumped MBE systems, the most appropriate for production oriented systems, can produce S doped samples with 77 deg K mobilities of 50000 cm*2/V.s.

Kudou et al [8] have shown that metalorganic MBE (MOMBE) using triethylgallium (TEG) as the Ga source can produce n-type GaAs at growth temperatures below 580 deg C. At 300 deg K mobilities of 1600 cm*2/V.s were obtained which increased to 2700 cm*2/V.s when a partial pressure of hydrogen equal to that of TEG was introduced into the growth chamber. Erickson et al [9] have obtained mobilities as high as 110000 cm*2/V.s at 77 deg K and 7900 cm*2/V.s at 300 deg K for samples with carrier concentrations in the mid 10*14 cm*-3 range, using a direct substrate heating technique.

REFERENCES
----

[1] H.Markoc, A.Y.Cho [J. Appl. Phys. (USA) vol.50 (1979) p.6413]
[2] J.C.M.Hwang, H.Temkin, T.M.Brennan, R.E.Frahm [Appl. Phys. Lett vol.42 no.1 (1983) p.66]
[3] J.M.Ballingall, C.E.C.Wood [Appl. Phys. Lett. vol.41 no.10

(1982) p.947-9]
[4] G.M.Metze, A.R.Calawa [Appl. Phys. Lett. vol.42 no.9 (1983) p.818-20]
[5] G.M.Metze, A.R.Calawa, J.G.Mavroides [J. Vac. Sci. & Technol. vol.1 no.2 (1983) p.166-9]
[6] L.P.Erickson et al [J. Appl. Phys. vol.56 (1984) p.2231]
[7] G.J.Davies, D.A.Andrews [Vacuum vol.34 no.5 (1984) p.543-5]
[8] Y.Kudou et al [in Extended Abstracts of the 16th Conf. of Solid State Devices and Materials, Kobe (1984) p.687]
[9] L.P.Erickson et al [J. Vac. Sci. & Technol. B vol.3 no.2 (1985) p.536]

5.8 ELECTRON MOBLILTY IN MBE GaAs, TEMPERATURE DEPENDENCE

by EMIS Group

January 1985
EMIS Datareview RN=14025

Peak mobilities of above 10*5 cm*2/V.s have been reported for MBE GaAs.

Calawa [1] found that use of cracked AsH3 as the As source resulted in high quality MBE layers with peak Hall mobility above 130000 cm*2/V.s and 77 deg K mobility above 110000 cm*2/V.s for carrier concn 2.4 X 10*14/cm*3 and thickness 5.4 microns.

Dingle et al [2] measured the Hall mobility of Si doped GaAs layers grown by high purity MBE and found peak mobilities above 10*5 cm*2/V.s (at about 60 deg K) for films of thickness 27 microns.

Hwang et al [3] studied high purity Si doped GaAs layers grown by MBE using carefully controlled growth conditions. For film of thickness 6 microns the peak mobility was 140000 cm*2/V.s (at 55 deg K), the highest value yet reported for MBE GaAs.

REFERENCES
----

[1] A.R.Calawa [Appl. Phys. Lett. vol.38 no.9 (1981) p.701-3]
[2] R.Dingle, C.Weisbuch, A.Y.Cho, H.L.Stormer, H.Morkoc [Appl. Phys. Lett. vol.40 no.6 (1982) p.507-10]
[3] J.C.M.Hwang, H.Temkin, T.M.Brennan, R.E.Frahm [Appl. Phys. Lett. vol.42 no.1 (1983) p.66-8]

5.9 ELECTRON MOBILITY IN MBE GaAs, DOPING DEPENDENCE

by D.Lancefield

January 1985
EMIS Datareview

Several workers have reported data on the decrease in electron mobility of MBE GaAs with increasing dopant concentration.

Hwang et al [1] studied high purity Si-doped GaAs layers grown by MBE using carefully controlled growth conditions. For film thicknesses of 6 microns the Hall mobility decreased from 126000 to 66000 cm*2/V.s as carrier concentration increased from 1 X 10*14 cm*-3 to 6.5 X 10*14 cm*-3 at 77 deg K. Li and Xin [2] found that for Ge doped MBE GaAs measured room temperature mobilities were lower than theoretical values. Experimental Hall mobilities (300 deg K) were 5300 and 1000 cm*2/V.s for electron concentrations 6.7 X 10*15 cm*-3 and 2 X 10*18 cm*-3 respectively.

Smith et al [3] reported that SnSe2 was a suitable n-type dopant source for MBE GaAs. Hall mobilities (77 deg K) of the resultant GaAs:Se layers were 2.4 X 10*4 cm*2/V.s at free electron concentration 3 X 10*15 cm*-3 and 1.6 X 10*3 cm*2/V.s at free electron concentration 6 X 10*19 cm*-3.

Stringfellow et al [4] studied C doped MBE GaAs and found that use of a relatively high CO partial pressure during deposition (i.e. introduction of a high C content) greatly reduced the room temperature mobility. For film with high C content grown at substrate temperature 535 deg K carrier concentration and mobility at 300 deg K were 2.6 X 10*15 cm*-3 and 5800 cm*2/V.s respectively. For film with low C content grown at 600 deg K carrier concentration was 2.0 X 10*15 cm*-3 and the mobility was 7500 cm*2/V.s.

REFERENCES

[1] J.C.M.Hwang, H.Temkin, T.M.Brennan, R.E.Frahm [Appl. Phys. Lett. (USA) vol.42 no.1 (1983) p.66-8]
[2] A.-Z. Li, S.-H. Xin [J. Electron. Mater. (USA) vol.12 no.1 (1983) p.71-91]
[3] R.S.Smith, P.M.Ganser, H.Ennen [J. Appl. Phys. (USA) vol.53 no.12 (1982) p.9210-3]
[4] G.B.Stringfellow, R.Stall, W.Koschel [Appl. Phys. Lett. (USA) vol.38 no.3 (1981) p.156-7]

5.10 ELECTRON MOBILITY IN ANNEALED ION IMPLANTED GaAs, DOPING DEPENDENCE

by B.J.Sealy

January 1985
EMIS Datareview

The electrical effects of many different ions have been studied following implantation into GaAs. Most recent work has been concerned with Si ions implanted into undoped semi-insulating substrates, since these appear to produce more reliable results than Cr-doped layers do. Also, recently many groups have concentrated on an assessment of rapid thermal annealing techniques which produce similar and sometimes better results than those obtained following an anneal in a furnace. In general, as the ion dose is increased from about 3 X 10*12 cm*-2 to 10*15 cm*-2 the sheet mobility after annealing decreases from about 4000 cm*2/V.s to about 1500 cm*2/V.s [1,2,3]. However, the decrease in mobility is linked more closely to an increase in the sheet carrier concentration, than to the dose, and the sheet carrier concentration varies with dose, ion energy and annealing conditions.

For high doses (10*14 ions cm*-2 and above) the mobility tends to decrease with increasing annealing temperature due to the simultaneous and often large improvement in electrical activity [2,3,4,5]. For example, the sheet mobility of a sample implanted with 1.5 X 10*14 Si+ cm*-2 at 200 keV was reduced from about 1750 to 1600 cm*2/V.s as the annealing temperature was raised from 920 deg C to 1050 deg C for 1 s [4]. Similar results were found also for 200 keV Se+ at a dose of 2 X 10*14 cm*-2 [4], for 4 X 10*14 Si+ cm*-2 at 200 keV [6] and for 10*15 S+ cm*-2 at 120 keV [5]. For a 10*15 cm*-2 dose, it has been found that Sn implants produce higher mobilities (1400 cm*2/V.s) than do Se implants (900 cm*2/V.s) following an anneal at 1000 deg C for 10 s [2].

Low dose implants (below 10*13 ions cm*-2) produce the highest mobilities since implant damage is less and can be removed more easily than for high doses, and the resultant carrier concentration is low, perhaps 10*17 cm*-3, which should produce mobilities of the order of 5000 cm*2/V.s. In practice, doses of 3-5 X 10*12 Si+ cm*-2 have mobilities in the range 4000 to 4500 cm*2/V.s [7,8,9]. It has been claimed for low doses of Si, that a two stage anneal, 900 deg C / 5 s plus 850 deg C / 30 s, can produce improvements in both mobility and electrical activity compared with single stage annealing [8]. This treatment involved a rapid thermal anneal using a GaAs proximity cap. Similarly high mobilities have been recorded for low dose Se implants [10,11].

In general, annealing in arsine and face to face capless anneals are not very successful due to surface degradation at temperatures above about 950 deg C. To anneal above 1000 deg C it seems necessary to employ a cap such as AlN [2], Si3N4 + AlN [2], Si3N4 [4,12], SiO2 [6,13] and SiOxNy [13].

REFERENCES

REFERENCES

[1] W.H.Haydl [IEEE Electron Device Lett. (USA) vol EDL-5 no.3 (Mar 1984) p.78-81]

[2] K.K.Patel, R.Bensalem, M.A.Shahid, B.J.Sealy [Nucl. Instrum. & Methods Phys. Res. Sect. B vol B7/8 pt.1 (Mar 1985) p.418-22]

[3] Y.K.Yeo, R.Kwor, Y.S.Park [J. Appl. Phys. (USA) vol.53 no.3 (Mar 1982) p.1812-4]

[4] D.E.Davies, P.J.McNally, T.G.Ryan, K.J.Soda, J.J.Comer [Inst. Phys. Conf. Ser No.65 (1983) p.619-25]

[5] R.Kwor, Y.K.Yeo, Y.S.Park [J. Appl. Phys. (USA) vol.53 no.7 (July 1982) p.4786-92]

[6] K.Tabatabaie-Alavi, A.N.M.Choudhary, C.G.Fonstad,J.C.Gelpey [Appl. Phys. Lett. (USA) vol.43 no.5 (Sept 1983) p.505-7]

[7] C.P.Stewart, R.T.Blunt, G.R.Booker, I.R.Sanders [Physica (Netherlands) vol.116B nos.1-3 (Feb 1983) p.635-40]

[8] K.S.Seo, S.Dhar, P.K.Bhattacharya [Appl. Phys. Lett. (USA) vol.47 no.5 (Sept 1985) p.500-2]

[9] M.H.Badawi, J.Mun [Electron. Lett. (GB) vol.20 no.3 (2 Feb 1984) p.125-6]

[10] P.A.Leigh [IEE Proc. I (GB) vol.130 no.6 (Dec 1983) p.265-74]

[11] N.J.Barrett, J.D.Grange, B.J.Sealy, K.G.Stephens [J. Appl. Phys. (USA) vol.56 no.12 (Dec 1984) p.3503-7]

[12] S.A.Kitching, M.H.Badawi, S.W.Bland, J.Mun [Mater. Res. Soc. Symp. Proc. (USA) vol.35 (1985) p.503-8]

[13] M.Kuzuhara, H.Kohzu, Y.Takayama, [Mater. Res. Soc. Symp. Proc. (USA) vol.23 (1984) p.651-62]

5.11 ELECTRON MOBILITY IN GaAs, PRESSURE DEPENDENCE

by A.R.Adams

December 1985
EMIS Datareview RN=15458

Lancefield et al [1] showed that at room temperature the electron mobility in ultra pure GaAs decreases with pressure at a rate of 0.73%/kbar in the range 0-8 kbar. This is due predominantly to the increasing electron effective mass with pressure. The value was determined from Hall effect and resistivity measurements on samples grown by LPE on semi-insulating substrates. Lancefield et al [1] compared their results with the theoretical pressure variation of the Hall mobility calculated by iterative solution of the Boltzmann equation including piezoelectric, deformation potential, and polar optical phonon scattering. They were able to obtain good agreement between theory and experiment.

In the LPE samples studied by Lancefield et al [1] it was observed that the pressure dependence of the electron mobility increased as the mobility was decreased by ionized impurity scattering. In samples with a mobility of 1600 $cm^2/V.s$ the pressure dependence of the mobility,

$$\frac{d(\ln \mu)}{dP} = 1.63 \times 10^{-2}/\mathrm{kbar}$$

Correlated impurity scattering as described by Yanchev et al [2] appears to partially explain the results. Similar results in pure and heavily doped InP support the validity of the experimental approach.

Above about 33 kbar at room temperature Pitt and Lees [3] observed a swift decrease in mobility to 375 $cm^2/V.s$. This can be attributed to electron transfer to the X minima where the mobility is lower due to the higher effective mass and due to intervalley scattering. No measurements of the pressure dependence of the electron mobility in the X valleys have yet been made, but by comparison with GaP it can be expected to increase with pressure at approximately 0.25%/kbar.

REFERENCES

[1] D.Lancefield, A.R.Adams, B.J.Gunney [Appl. Phys. Lett. (USA) vol.45 (1984) p.1121]
[2] I.Y.Yanchev, S.K.Evtimova [J. Phys. C (GB) vol.18 (1985) p.L377]
[3] G.D.Pitt, J.Lees [Phys. Rev. B (USA) vol.2 (1970) p.4144]

5.12 ELECTRON EFFECTIVE MASS IN GaAs, PRESSURE DEPENDENCE

by A.R.Adams

December 1985
EMIS Datareview RN=15459

The electron effective mass in GaAs is almost directly proportional to the direct energy gap, Eo. This increases with pressure as described in ref [2] leading to an increase in effective mass of 0.75%/kbar.

Shantharama et al [1] studied high purity GaAs grown by LPE on semi-insulating substrates. They determined the effective mass from magnetophonon oscillations in magnetic fields up to 9 T while applying hydrostatic pressure up to 20 kbar. By also measuring the photoconductive response of the crystal they were able to determine Eo and m*(e) simultaneously as pressure was applied. Thus they were able to compare their results directly with k.p theory. After correcting for non-parabolicity and polaron enhancement effects best fit to the effective mass was obtained by

$$\frac{d[m^*(e)]}{dE_o} = 0.045\ m(o)/eV$$

Taking an average value of the pressure dependence of Eo of 11.2 meV/kbar [2] and the atmospheric value of

$$m^*(e) = 0.0667\ m(o)$$

leads to the pressure dependence of m*(e) of 0.75%/kbar.

REFERENCES

[1] L.G.Shantharama, A.R.Adams, C.N.Ahmad, R.J.Nicholas [J. Phys. C (GB) vol.17 (1984) p.4429]

[2] EMIS Group [EMIS Datareview RN=15500 (Dec 1985), "Direct band gap of GaAs, pressure dependence"]

5.13 ELECTRON DIFFUSION LENGTH AND COEFFICIENT IN p-TYPE GaAs

by D.R.Wight and K.Duncan

January 1986
EMIS Datareview RN=15612

The diffusion length, L, of a minority carrier is the average distance travelled by the carrier before recombination terminates its existence. Its most basic definition for an extrinsic semiconductor is a `bulk´ value which pertains to a uniform low density of minority carriers in an infinite and uniform crystal, with recombination proceeding via `Shockley-Read´ point defects, as is common in silicon. Under basic diffusion theory it is then simply related to two more fundamental parameters, the minority carrier lifetime, tau, and the minority carrier diffusion coefficient, D, i.e.

$$L = \sqrt{D \tau} \qquad (1)$$

D is further related to the minority carrier mobility, mu(min), by the Einstein relationship

$$D = \frac{\mu(min)\, kT}{q} \qquad (2)$$

In p-type GaAs a fairly large number of measurements of L have been reported [1-6], using typically the analysis of collected current in p-n junctions or surface barrier contacts, excited by electron beams or photons. However many of these results were compromised (either knowingly or unknowingly) by the presence of surface or interface recombination processes such that only where values of around 3 microns or shorter are obtained should these be considered valid.

In ingot materials [1] it would seem that L takes values around 2 microns or less and is dominated at low hole concentrations by non-radiative point defect recombination, but in epitaxial materials of higher purity and crystallographic quality a near ideal situation can be reached. In the ideal crystal at carrier concentrations below 10^{19} cm^{-3} the lifetime is controlled by the band to band radiative process which is resonant with the edge absorption process in a direct band gap material such as GaAs. This means that the recombination event does not terminate the minority carrier´s existence because the generated photon can be re-absorbed to re-create the minority carrier at a different point in the crystal. This `photon recycling´ is, of course, a physical process not encompassed by diffusion theory, so the basic concept of diffusion length breaks down and, for example, the migration distance can be controlled by the optical geometry of the sample. By eliminating the effects of interface/surface recombination and measuring luminescence intensity profiles this new process has been accounted for and very long "true" diffusion lengths have been deduced [6] in liquid phase epitaxial materials. These values and values obtained for MOCVD grown [5] and other LPE p-type GaAs [4] are shown in the table below.

To show that these values are near ideal, values L´ obtained by using Eqn (1), and predicting tau from absorption edge data and D from majority electron mobility at the same carrier concentration (D´) are included in the table. The agreement is satisfactory over the doping range 10*15 - 10*20 cm*-3 when the uncertainty in the data and prediction of D´ is considered.

Systematic direct measurements of the diffusion constants of minority carriers are difficult to measure using the Shockley-Haynes technique because of the short diffusion lengths usually encountered. However, systematic indirect measurements have been made by measuring L and tau for the same set of samples and using Eqn (1). These values are also listed in the table and are again in tolerable agreement when the uncertainty in the data and the estimated D´ is considered.

TABLE: minority carrier diffusion length and diffusion constant for p-GaAs at 300 deg K

P(cm*-3)	L (10*-4 cm)	L´	D (cm*2/s)	D´
1.9 X 10*15	140	300	130	180
5 X 10*15	90	170	110	170
3 X 10*16	45	70	110	135
8 X 10*16	28	40	75	115
1.6 X 10*17	15	20	45	100
9 X 10*17	8	10	48	75
2 X 10*18	10(a)(b)	7	100(a)(c)	65
2 X 10*18	7(a)	7	80(a)	65
8 X 10*18	3.5(a)(b)	4.0		40
1 X 10*19	2.5(a)	3.8		35
4 X 10*19	0.8(a)(b)	0.7		25

(a) photon recycling not accounted for; (b) MOCVD material; (c) MOCVD, direct measurement.

REFERENCES

[1] T.S.Rao-Sahib, D.B.Witty [J. Appl. Phys. (USA) vol.40 (1969) p.3745]
[2] H.C.Casey, B.I.Miller, E.Pinkas [J. Appl. Phys. (USA) vol.44 (1973) p.1281]
[3] M.Ettenburg, C.J.Nuese [J. Appl. Phys. (USA) vol.46 (1975) p.3500]
[4] G.A.Ackett, W.Nijman, H.t´Lam [J. Appl. Phys. (USA) vol.45 (1974) p.3033]
[5] D.R.Wight, P.E.Oliver, T.Prentice, V.W.Steward [J. Cryst. Growth (Netherlands) vol.55 (1981) p.183]
[6] R.J.Nelson [`GaAs and Related Compounds´, Inst. Phys. Conf. Ser. no.45 (1979) p.256]

5.14 ELECTRON LIFETIME IN p-TYPE GaAs

by D.R.Wight

January 1986
EMIS Datareview RN=15613

The electron lifetime in p-type GaAs at 300 deg K can be controlled by recombination at point defects, surfaces or interfaces, dislocations [1] or bulk radiative recombination processes. Most direct experimental measurements have involved the study of small signal decay times for band to band luminescence, excited using electron or laser beams, when the presence of strong surface recombination complicates the measurement if bulk lifetimes are to be accurately determined. The most simple and basic definition of minority carrier lifetime is that which applies to a bulk infinite sample with a uniform low density of excess carriers recombining via a uniform density of `Shockley-Read´ defects or band to band radiative processes. It can be shown that the transient value is then identical to the steady-state lifetime. When dislocations or surface/interface recombination are present the steady state and transient values can be different [3]. The fundamental limit to the lifetime is the band to band recombination process which at hole concentrations below 10*19 cm*-3 is radiative and can be deduced from absorption coefficient data [2] providing the `photon-recycling´ process is not significant [6]. The most thorough and systematic study of minority electron lifetime in near ideal material has been reported by Nelson et al [4] and additional careful measurements at high hole densities have been reported by Ackett et al [5]. Both these studies were carried out on liquid phase epitaxial material, and the results are shown in the table together with the predictions for the limiting values for the band to band radiative process. Agreement is rather satisfactory even for very low doping if the effects of interfacial recombination are accounted for.

The situation in bulk crystals is usually less perfect in that the values only approach the ideal limit at hole concentrations around 3 X 10*18 cm*-3 indicating that recombination at intrinsic or impurity point defects controls the lifetime at lower doping where the radiative lifetime becomes relatively much longer.

Table: electron lifetime in p-type GaAs (300 deg K)

p (cm*-3)	tau (ns)	predicted tau (ns)
1.9 X 10*15	2000	3000
5 X 10*15	1400	1300
3 X 10*16	260	280
1.6 X 10*17	60	60
9 X 10*17	16	10
3.5 X 10*18	3	1.8
1 X 10*19	1.2	0.8

REFERENCES
----

[1] D.R.Wight, P.E.Oliver, T.P.Prentice, V.W.Steward [J. Cryst. Growth

(Netherlands) vol.55 (1981) p.183]
[2] D.R.Wight [`Gallium Arsenide: Materials, Devices and Circuits' Ed. M.J.Howes & D.V.Morgan, Wiley (1985), ch.1 p.1-37]
[3] D.R.Wight, I.Blenkinsop, W.Harding, B.Hamilton [Phys. Rev. B (USA) vol.23 (1981) p.5495]
[4] R.J.Nelson, R.G.Sobers [J. Appl. Phys. (USA) vol.49 (1978) p.6103]
[5] G.A.Ackett, W.Nijman, H.t'Lam [J. Appl. Phys. (USA) vol.45 (1974) p.3033]
[6] D.R.Wight [EMIS Datareview RN=15612 (Jan 1986) `Electron diffusion length and coefficient in p-type GaAs']

CHAPTER 6

HOLE MOBILITY, DIFFUSION AND LIFETIME IN GALLIUM ARSENIDE

6.1 Hole mobility in undoped GaAs

6.2 Hole mobility in doped and ion implanted GaAs

6.3 Hole mobility in GaAs, temperature dependence

6.4 Hole mobility in GaAs, pressure dependence

6.5 Hole effective mass in GaAs, pressure dependence

6.6 Hole diffusion length and coefficient in n-type GaAs

6.7 Hole lifetime in n-type GaAs

6.1 HOLE MOBILITY IN UNDOPED GaAs

by D.C.Look

December 1985
EMIS Datareview RN=15450

Mobilities in GaAs are generally measured by means of the Hall effect method. Typical contacting materials for p-type GaAs include AuZn, AuIn, AgZn, and AgIn [1-3]. The lowest contact resistivity so far, 2 X 10*-7 ohm cm*2, has been achieved by first diffusing Zn and In, and then depositing a refractory metal, such as Ti/Pt/Au [4]. However, for Hall-effect measurements, it is usually sufficient to simply solder pure In onto the surface, and then anneal at 425 deg C for 2 minutes in a flowing inert gas [5].

The hole concentration is determined from the Hall coefficient by p = r(p)/R(p), where r(p), the "Hall factor", is the ratio between the Hall and conductivity mobilities. Unfortunately, the Hall factor for holes is difficult to measure, and unlike the situation for electrons, is even more difficult to calculate [6]. The reason lies in a degenerate valence band, with holes in the two bands exhibiting different effective masses ("light" and "heavy" holes). Two recent calculations suggest that r(p) is approx. 1.8 at 300 deg K for "pure" GaAs and decreases at lower temperatures [7,8]. However, it should be recognized that these calculations are not as accurate as those for n-type material [9] and should be treated as preliminary.

The dominant scattering mechanisms for pure p-type GaAs at 300 deg K include both acoustic-mode (deformation-potential and piezoelectric), and optic-mode (non-polar deformation-potential and polar) lattice processes. The theoreticl results of ref [8] show that polar optic-mode scattering is dominant at 300 deg K, and acoustic-mode at 77 deg K. However, it should be noted that ionized-impurity scattering [16], will normally be the dominant mechanism at 77 deg K. Wiley [6] suggests that the of-quoted experimental results of three different groups [10-12] can be reasonably well fitted, in the lattice-scattering regime (say, 100-400 deg K), by

mu(Hp) = approx. 400 (300/T)*2.3 cm*2/V.s

where mu(Hp) is the lattice-limited Hall mobility for holes. This formula should be sufficient for estimates of the lattice-scattering contribution over the temperature range, 100-400 deg K.

One final note of caution as regards measurements in material which is inhomogeneous in the plane of current flow. The apparent mobility in such cases is normally lower than expected [13,14], but can also be higher [15]. Inhomogeneous samples will often exhibit non-ohmic conduction.

REFERENCES
----

[1] O.Ishihara, K.Nishitani, H.Sawano, S.Mitsui [Jpn. J. Appl. Phys. (Japan) vol.15 (1976) p.1411]

[2] V L.Rideout [Solid-State Electron. (GB) vol.18 (1975) p.541]
[3] A.Piotrowska, A.Guivarch, G.Pelous [Solid-State Electron. (GB) vol.26 (1983) p.979]
[4] J.R.Shealy, S.R.Chinn [Appl. Phys. Lett. (USA) vol.47 (1985) p.410]
[5] B.K.Shin, D.C.Look, Y.S.Park, J.E.Ehret [J. Appl. Phys. (USA) vol.47 (1976) p.1574]
[6] J.D.Wiley ["in Semiconductors and Semimetals", (USA) vol.10 Ed. R.K.Willardson & A.C.Beer (Academic, New York 1975), p.91]
[7] H.J.Lee, D.C.Look [J. Appl. Phys. (USA) vol.54 (1983) p.4446]
[8] K.Takeda, N.Matsumoto, A.Taguchi, H.Taki, E.Ohta, M.Sakata [Phys. Rev. B (USA) vol.32 (1985) p.1101]
[9] B.R.Nag ["in Electron Transport in Compound Semiconductors" (Ger, Springer-Verlag, Berlin, 1975)]
[10] D.E.Hill [J. Appl. Phys. (USA) vol.41 (1970) p.1815]
[11] K.H.Zschauser [Proc. of the Fourth Int. Symp. on GaAs and Related Compounds, Boulder, CO, 1972, Ed. C.Hilsum, Conf. Ser. no.71]
[12] A.L.Mears, R.A.Stradling [J. Phys. C (GB) vol.4 (1972) p.L22]
[13] H.H.Wieder ["Laboratory Notes on Electrical and Galvanomagnetic Measurements" (Elsevier, Amsterdam, 1971) ch.5]
[14] D.C.Look, P.W.Yu, J.Ehret, Y.K.Yeo, R.Kwor [in "Semi-Insulating III-V Materials", (Evian 1982) (Shiva, Nantwich, 1982) p.372]
[15] C.M.Wolfe, G.E.Stillman [Appl. Phys. Lett. (USA) vol.18 (1971) p.205]
[16] D.C.Look [EMIS Datareview RN=15451 (Dec 1985) "Hole mobility in doped and implanted GaAs"]

6.2 HOLE MOBILITY IN DOPED AND ION-IMPLANTED GaAs

by D.C.Look

December 1985
EMIS Datareview RN=15451

Besides the scattering of holes due to lattice vibrations [1] there is also scattering due to imperfections, i.e. impurities or defects. In most cases, ionized imperfections are much more important than those which are neutral, and ionized-impurity scattering is reasonably well described by the Brooks-Herring treatment. However, as discussed in [1] a detailed calculation of hole mobility is much more complicated than for electron mobility [2], so that, in the former case, only semi - quantitative agreement between theory and experiment is possible.

Wiley [3] has reviewed the situation up through 1975 and has plotted hole mobility vs. hole concentration from many different workers [4-10]. Of course, the ionized-impurity concentration N(I) can be related to the hole concentration (p) only if the compensation ratio (K) is known, and in most of these cases it is not. Nevertheless, a rough fit can be given to the data by assuming p approx. N(I), (i.e. K approx. zero), and then applying Matthiessen´s rule: 1/mu = approx. 1/mu(L) + 1/mu(I). The line that Wiley [3] has drawn through the 300 deg K data can be approximately fitted by the following formula:

$$\mu(Hp) = \frac{1}{2.5 \times 10^{-3} + 1.1 \times 10^{-21}\,[\ln(1+b) - b/(1+b)]\,p(H)} \qquad (1)$$

where $b = 7.5 \times 10^{19}/p(H)$ and where "H" designates a Hall-effect measurement without correction for Hall factor. Eqn (1) should be reasonably valid for p(H) below about 10^{19} cm^{-3}. The temperature dependence, over the range 77-400 deg K, can be roughly represented by

$$\mu(Hp) = \frac{1}{5.0 \times 10^{-9}\,T^{2.3} + 5.5 \times 10^{-18}\,[\ln(1+b) - b/(1+b)]\,P(H)/T^{1.5}} \qquad (2)$$

where $b \approx 8.3 \times 10^{14}\,T^{2}/p(H)$

Eqn (2) is derived from the lattice-limited formula in ref [1] and from the expected temperature dependence of mu(I). It must be emphasized that these formulae are useful only for purposes of estimation and comparison. For example, the degree to which a measured mu(Hp) falls below that suggested by Eqn 1 or Eqn 2 may well be a measure of the compensation in that sample.

More recent data may be compared with the predictions of these formulae. For example, in a 300 deg K study of Mg implanted GaAs, Yeo et al [11] found that mu(Hp) is approx. 90 $cm^2/V.s$ at the concentration peak of p equal to approx. 5×10^{18} cm^{-3}, whereas the formulas above would have predicted mu(Hp) as being approx. 80 $cm^2/V.s$. For Zn-doped material, Emel´yanenko et al [12] found that mu(p) is approx. 1.0×10^{3} $cm^2/V.s$ at p approx. 1.2×10^{16} cm^{-3} and T = 100 deg K, while the prediction would have been 1.8×10^{2} $cm^2/V.s$. This difference could easily be explained by compensation but other factors may also enter. For extremely heavy doping, p approx. 8 X 10*19 cm*-3 in Zn-implanted material [13], it was found that mu(Hp) is about 45 cm*2/V.s at 300 deg K whereas the prediction would have been approximately 55 cm*2/V.s. Of course, the holes are degenerate at this concentration, and the Born approximation, used in the Brooks-Herring formulation, is no longer valid, so that our formulas cannot be taken too seriously. However, as is seen, they still can explain the observed data to a degree which makes them useful for estimation purposes.

REFERENCES

[1] D.C.Look [EMIS Datareview RN=15450 (Dec 1985) "Hole mobility in undoped GaAs"]

[2] B.R.Nag [in "Electron Transport in Compound Semiconductors" (Sringer-Verlag, Berlin, 1975)]

[3] J.D.Wiley [in "Semicond. & Semimet." (USA) Ed. R.K.Willardson & A.C.Beer (Academic Press. New York) vol.10 (1975) p.91]

[4] O.V.Emel´yanenko, T.S.Lagunova, D.N.Nasledov [Sov. Phys. Solid State (USA) vol.2 (1960) p.176]

[5] Sh.M.Gasanli, O.V.Emel´yanenko, V.K.Ergakov, F.P.Kesamanly, T.S.Lagunova, D.N.Nasledov [Sov. Phys. Semicond. (USA) vol.5 (1972) p.1641]

[6] D.E.Hill [Phys. Rev. (USA) vol.133 (1964) p.A866]

[7] D.E.Hill [J. Appl. Phys. (USA) vol.41 (1970) p.1815]

[8] F.D.Rosi, D.Meyerhofer, R.V.Jensen [J. Appl. Phys. (USA) vol.31 (1960) p.1105]

[9] F.E.Rosztoczy, F.Ermanis, I.Hayashi, B.Schwartz [J. Appl. Phys. (USA) vol.41 (1970) p.264]

[10] J.Vilms, J.p.Garrett [Solid State Electron. (GB) vol.15 (1972) p.443]

[11] Y.K.Yeo, Y.S.Park, P.W.Yu [J. Appl. Phys. (USA) vol.50 (1979) p.3274]

[12] O.V .Emel´yanenko, D.N.Nasledov, N.A.Urmanov [Sov. Phys. Semicond. (USA) vol.10 (1976) p.72]

[13] K.Taatabaie-Alavi, A.N.M.Masum Choudhury, C.G.Fonstad [Appl. Phys. Lett. (USA) vol.43 (1983) p.505]

6.3 HOLE MOBILITY IN GaAs, TEMPERATURE DEPENDENCE

by A.R.Adams

December 1985
EMIS Datareview RN=15457

The low field Hall mobility of holes in high purity GaAs has been measured as a function of temperature by Hill [1], by Mears and Stradling [2], and by Zschauer [3]. They observed a room temperature mobility of 400 cm*2/V.s at 300 deg K rising to about 9000 cm*2/V.s at 77 deg K. In his excellent review of the mobility of holes in III-V compounds Wiley [4] concludes that a temperature dependence of the form mu = 400 X (300/T)*2.3 gives best agreement with experiment in agreement with Rosi et al [5].

Due to the complex nature of the valence band a full theoretical treatment including band warping, non-parabolicity and the presence of light holes is very difficult. Wiley concludes, however, that a reasonable fit to experiment can be achieved assuming non-polar optical and acoustic deformation potential scattering, and scattering by polar optical phonons.

Adding the influence of impurities and the temperature dependence of ionized impurity scattering given by the Brooks-Herring theory assuming a Matthiessen´s rule Blakemore [6] concludes that close to room temperature the Hall mobility may be written in the form

$$\mu(H) = \frac{1}{(2.5 \times 10^{-3}) \times (T/300)^{2.3} + (4 \times 10^{-21}) \times N_i \times (300/T)^{1.5}} \ \mathrm{cm^2/V.s}$$

The "drift" or "conductivity" mobility, mu(D) which is normally required in device analysis, is related to the Hall mobility, mu(H) by mu(D) = mu(H)/r. Mears and Stradling [2] measured the apparent change in mu(H) caused by increasing the magnetic field to 4 T. Assuming they reached the high field limit where r approaches 1.00 their results give values of r approx. 1.25 at 300 deg K and r approx. 1.5 at 77 deg K for Hall mobility measurements at low magnetic fields. We therefore conclude that the drift mobility of holes in GaAs rises from mu(D) = 320 cm*2/V.s at 300 deg K to about 6000 cm*2/V.s at 77 deg K. This is less fast than the Hall mobility and writing mu(D) = 320 X (300/T)*2.2 apppears appropriate for pure material.

REFERENCES

[1] D.E.Hill [J. Appl. Phys. (USA) vol.41 (1970) p.1815]
[2] A.L.Mears, R.A.Stradling [J. Phys. C (GB) vol.4 (1971) p.L22]
[3] K.H.Zschauer [Proc 4th Int. Symp. GaAs and Related Compounds, Boulder (1972)]
[4] J.D.Wiley [Semiconductors and Semimetals Ed by R.K. Willardson,

A.C.Beer (Academic, New York, 1975) vol.10 ch.2 p.91]
[5] F.D.Rosi, D.Meyerhofer, R.V.Jensen [J. Appl. Phys. (USA) vol.31 (1960) p.1105]
[6] J.S.Blakemore [J. Appl. Phys. (USA) vol.53 (1982) p.R123]

6.4 HOLE MOBILITY IN GaAs, PRESSURE DEPENDENCE

by A.R.Adams

December 1985
EMIS Datareview RN=15452

Adams et al [1] observed that the Hall mobility of holes in p-type GaAs grown by LPE on semi-insulating substrates increases with pressure at a rate of 0.31%/kbar in the range 0-8 kbar. This increase contrasts with the observed decrease in electron mobility with pressure [2]. Analysis of the hole mobility is difficult due to the complexities of the valence band. The light hole effective mass will increase with pressure in a similar way to the electron effective mass. However, because of its relatively large density of states, most of the holes are in the heavy hole band and therefore changes in the light hole band have little effect on the average hole mobility. The heavy hole effective mass itself is relatively insensitive to pressure. The dominant effects therefore appear to be:

(a) The energy of the polar phonons increases with pressure. This decreases their density and hence decreases polar phonon scattering.
(b) An increase in the density of GaAs with pressure which decreases the acoustic phonon scattering.

Combining these effects using Matthiessen's rule, Adams et al [1] were able to obtain reasonable agreement with experiment.

REFERENCES
----

[1] A.R.Adams, L.G.Shantharama [accepted for publication in Physica B]
[2] D.Lancefield, A.R.Adams, B.J.Gunney [Appl. Phys. Lett. (USA) vol.45 (1984) p.1121]

6.5 HOLE EFFECTIVE MASS IN GaAs, PRESSURE DEPENDENCE

by A.R.Adams

December 1985
EMIS Datareview RN=15453

No direct measurement has yet been made of the pressure dependence of the effective mass of holes in GaAs. However, from a study of the pressure dependence of tunnel diode characteristics Alekseeva et al [1] deduce a value

$$\frac{d\,[\ln m^*(h)]}{dP} = -(9 \pm 1) \times 10^{-3}/\text{kbar}$$

Alekseeva et al have also calculated $dm^*(h)/dP$ from the valence band parameters deduced from the pseudopotential calculations of Neumann et al [2] from which they conclude that

$$\frac{d\,[\ln m^*(h)]}{dP} = -5.0 \times 10^{-4}/\text{kbar}$$

More recently Adams et al [3] have measured the pressure dependence of the hole mobility from which they conclude that

$$\frac{d\,[\ln m^*(h)]}{dP} = -1.5 \times 10^{-4}/\text{kbar}$$

Clearly there is considerable scatter in the few values available. However, it is clear that the pressure dependence of heavy holes is considerably less than that for electrons as expected from k.p theory. Based on this theory one would expect the light hole effective mass to vary with pressure by the same percentage as the electron effective mass.

REFERENCES

[1] Z.M.Alekseeva, A.P.Vyatkin, G.F.Karavaev, N.P.Krivorotov [Phys. Status Solidi b (Germany) vol.88 (1978) p.321]

[2] H.Neumann, I.Topol, K.R.Schulze, E.Hess [Phys. Status Solidi b (Germany) vol.56 (1973) p.K55]

[3] A.R.Adams, L.G.Shantharama [accepted for publication in Physica B (Netherlands)]

6.6 HOLE DIFFUSION LENGTH AND COEFFICIENT IN n-TYPE GaAs

by D.R.Wight

January 1986
EMIS Datareview RN=15454

The definitions of diffusion length (L) and diffusion coefficients (D) as well as the criteria for their measurement as bulk properties are identical to those described for minority electrons in p-GaAs [4]. The interferences from surfaces, dislocations and photo-recycling processes are also similar in the two material types as well as the use of absorption coefficient data to predict L´ and the mobility data to predict D´ (see ref [4]).

Experimentally there is a less complete picture for n-type material but a number of careful and systematic studies have been made. For example, Hwang [1,2] used the analysis of luminescence spectra to obtain L for both Bridgman and Czochralski ingot materials and Casey et al [3] have used the analysis of short circuit photocurrent in p-n junction structures to determine L in liquid phase epitaxial material. Their data are summarised in the table along with the calculated values for D (using tau data in ref [2]) and the predicted values L´ and D´.

TABLE: hole diffusion length and diffusion coefficient in n-GaAs

at 300 deg K

n cm*-3	L 10*-4 cm	L´ 10*-4 cm	D cm*2/s	D´ cm*2/s
2 X 10*16	1.8	17	1.6	7.5
1 X 10*17	1.8	7	1.6	6.3
1 X 10*17 (LPE)	4.8	7		6.3
3 X 10*17 (LPE)	3.8	5.7		5.5
5 X 10*17	1.8	5.0	1.6	5.0
2 X 10*18	1.2	3.7	1.4	3.5
2 X 10*18	1.2	3.7		3.5
6.5 X 10*18	0.3	3.3		2.5
6.5 X 10*18	0.02	3.3	0.006	2.5

For L, the agreement is tolerable for the LPE materials but is progressively worse as the doping exceeds 10*18 cm*-3. This probably reflects the disagreement seen in the hole lifetime data at these carrier concentrations [5]. The ingot data are similar to the LPE data only at 2 X 10*18 cm*-3 and L values are much lower than the LPE or L´ values at lower and higher carrier concentrations due to the presence of non-radiative defect centres which increase catastrophically as the donor solubility limit is approached.

Considering the values for D, it should be noted that the lifetime was not measured in the LPE samples so no D values can be calculated. D and D´ values in ingot materials, like the tau data in [5], are fairly close at 2 X 10*18 cm*-3 and are progressively lower than predicted as n decreases. This almost certainly reflects the fact that the mobility is being held to a low value by defect scattering centres. At concentrations near the donor limit (approx. 7 X 10*18 cm*-3) the effect is catastrophic.

If we review this situation it is clear that with the hindsight of the information from p-GaAs no complete or accurate analysis of L and D has emerged in high quality n-GaAs. But it would appear that because the MOCVD material in [5] exhibits long lifetimes close to the radiative limit, all of the features of agreement and the photon-recycling process could now be demonstrated together with a test of the theoretical description of lifetime in degenerate n-GaAs.

REFERENCES

[1] C.J.Hwang [J. Appl. Phys. (USA) vol.40 (1969) p.3731]

[2] C.J.Hwang [Phys. Rev. B (USA) (1972) p.1355]

[3] H.C.Casey, B.I.Miller, E.Pinkas [J. Appl. Phys. (USA) vol.44 (1979) p.1281]

[4] D.R.Wight [EMIS Datareview RN=15612 (Jan 1986) "Electron diffusion length and coefficient in p-type GaAs"]

[5] D.R.Wight [EMIS Datareview RN=15456 (Jan 1986) "Hole lifetime in n-type GaAs"]

6.7 HOLE LIFETIME IN n-TYPE GaAs

by D.R.Wight

January 1986
EMIS Datareview RN=15456

Minority carrier lifetime in n-GaAs at 300 deg K has been less extensively studied than lifetime in p-GaAs. The reasons for this are twofold: (1) the luminescence efficiency of p-type materials was initially found to be higher than that for n-type materials, so light-emitting devices prompted the study of p-type recombination mechanisms; (2) the properties of electrons in GaAs in general attract more attention because of their much superior transport properties when compared to electrons in silicon. Systematic measurements on ingot materials have been reported by Hwang [1] and MOCVD materials by t´Hooft et al [2]. The comparison with theory is complicated by other non-radiative recombination mechanisms and photon recycling processes in exactly the same fashion as in p-GaAs [4] and indeed at majority carrier densities below 3 X 10*17 cm*-3 the radiative limit predicted from absorption edge data is identical for minority electrons and holes. At higher densities the absorption edge changes in shape in different fashions for p- and n-type materials and the predicted electron lifetimes are longer than the hole lifetimes [3].

The table lists these experimental and predicted lifetime values and at lower doping (with exceptions) the agreement is reasonable, becoming worse at high doping levels where it is suspected that the onset of donor insolubility creates defects that quench the bulk lifetime, tau. The validility of the theoretical prediction in such degenerate material is also perhaps questionable.

TABLE: hole lifetimes in n-GaAs

n cm*-3	tau ns	predicted tau ns
2 X 10*16	20	450
2.5 X 10*16 (MOCVD)	260	380
3 X 10*16 (MOCVD)	227	300
5 X 10*16 (MOCVD)	117	180
1.4 X 10*18	15	40
3 X 10*18	7	50
7 X 10*18	0.8	80

The exception is the low doped ingot datum where it is suggested that recombination at spurious impurity or native defects dominates the radiative processes and quenches the lifetime to about 20 ns, independent of doping in the range 10*16 - 10*18 cm*-3 in the materials used.

REFERENCES

[1] C.J.Hwang [J. Appl. Phys. (USA) vol.42 (1971) p.4408]
[2] G.W.t´Hooft, C. Van Opdorp, H.Veenvliet, A.T.Vink [J. Cryst. Growth (Netherlands) vol.55 (1981) p.173]

[3] D.R.Wight ["Gallium Arsenide: Materials, Devices and Circuits" Ed Howes & Morgan (Wiley, 1985) ch.1]
[4] D.R.Wight [EMIS Datareview RN=15612 (Jan 1986) "Electron diffusion length and coefficient in p-type GaAs"]

CHAPTER 7

ENERGY GAP OF GALLIUM ARSENIDE

7.1 Direct band gap of GaAs, temperature dependence

7.2 Direct band gap of GaAs, pressure dependence

7.3 Optical deformation potential for holes at the GAMMA point in GaAs

7.4 Optical gap of amorphous GaAs

7.1 DIRECT BAND GAP OF GaAs, TEMPERATURE DEPENDENCE

by A.R.Adams

December 1985
EMIS Datareview RN=15491

Since GaAs is a direct gap material it is possible to do very precise optical measurements of the band gap Eo. These have included absorption [1-5], reflectance [6-8] and photoluminescence [9-11]. The observed variation of Eo is non-linear with temperature but closely follows the empirical equation proposed by Varshni [12]. Thurmond [13] showed that in the temperature range 0-1000 deg K the experimental results fitted

$$E_o(T) = 1.519 - \frac{(5.405 \times 10^{-4})\, T^2}{(T + 204)} \quad \text{eV}$$

The estimated standard deviation over this temperature range is 3 meV, but Blakemore [14] observes that the fit is much better than this at room temperature and below.

The variation of Eo with temperature can be associated theoretically with the dilation of the lattice and with electron-phonon interactions. The former effect can be calculated from the pressure dependence of Eo, the bulk modulus and the change in the lattice constant with temperature. This effect accounts for a 27 meV decrease in Eo from 0 to 300 deg K [15] the rest apparently being due to the electron-phonon interactions.

REFERENCES

[1] M.D.Sturge [Phys. Rev. (USA) vol.127 (1962) p.768]
[2] D.D.Sell [Phys. Rev. B (USA) vol.6 (1972) p.3750]
[3] H.C.Casey, D.D.Sell, K.W.Wecht [J. Appl. Phys. (USA) vol.46 (1975) p.250]
[4] M.B.Panish, H.C.Casey [J. Appl. Phys. (USA) vol.40 (1969) p.163]
[5] M.A.Afromowitz, D.Redfield [Proc. 9th Int. Conf. Phys. Semicon., Moscow (1968) p.98]
[6] D.D.Sell, H.C.Casey, K.W.Wecht [J. Appl. Phys. (USA) vol.45 (1974) p.2650]
[7] J.L.Shay [Phys. Rev. B (USA) vol.2 (1970) p.803]
[8] D.D.Sell, S.E.Stokowski, R.Dingle, J.V.DiLorenzo [Phys. Rev. B. (USA) vol.7 (1973) p.4568]
[9] M.A.Gilleo, P.T.Bailey, D.E.Hill [J. Lumin. (Netherlands) vol.1/2 (1970) p.562]
[10] F.Urbach [Phys. Rev. (USA) vol.92 (1953) p.1324]
[11] D.Bimberg, W.Schairer [Phys. Rev. Lett. (USA) vol.28 (1972) p.442]
[12] Y.P.Varshni [Physica (Netherlands) vol.34 (1967) p.149]
[13] C.D.Thurmond [J. Electrochem. Soc. (USA) vol.122 (1975) p.1133]
[14] J.S.Blakemore [J. Appl. Phys. (USA) vol.53 (1982) p.R123]
[15] L.G.Shantharama, A.R.Adams, C.N.Ahmad, R.J.Nicholas [J. Phys. C (GB) vol.17 (1984) p.4429]

7.2 DIRECT BAND GAP OF GaAs, PRESSURE DEPENDENCE

by EMIS Group

July 1985
EMIS Datareview RN=15500

The direct energy gap of GaAs increases with applied pressure and for pressures above 35 kbar GaAs becomes an indirect gap material. This is because the X6 conduction band minimum falls in relation to the L6 and GAMMA (6) minima as the pressure is increased [1].

Zurawsky et al [2] measured the energy gap of [100] oriented n-GaAs by monitoring photocurrent vs optical excitation energy and obtained a pressure dependence of 12.2 meV/kbar in the 0-10 kbar range. Wolford and Bradley [14] carried out photoluminescence and photoluminescence excitation spectroscopy on high purity MOCVD n-type GaAs at 4.2 deg K in a diamond anvil and obtained a value of 10.73 +- 0.05 meV/kbar in the 0-40 kbar range.

Welber et al [3] subjected GaAs to hydrostatic pressures up to 180 kbar and obtained the following expression for the absorption edge Eo in eV at room temperature:

Eo = 1.45 + 0.0126 P - 3.77 X 10*-5 X P*2
where P is the pressure in kbars.

Pollak and Cardona [4] report the following piezo-electroreflectance measurements:

Crystal direction	[001]	[110]	[111]
dEg/dP (meV/kbar)	11.5	12.0	11.5

Other data on energy gap vs uniaxial stress are in refs [5-7]. Energy gap vs hydrostatic pressure data are in refs [8-13].

REFERENCES

[1] J.S.Blakemore [J. Appl. Phys. (USA) vol.53 no.10 (Oct 1982) p.R123-81]

[2] W.P.Zurawsky, J.E.Littman, H.G.Drickamer [J. Appl. Phys. (US) vol.54 no.6 (June 1983) p.3216-9]

[3] B.Welber, M.Cardona, C.K.Kim, S.Rodriguez [Phys. Rev. B (USA) vol.12 no.12 (15 Dec 1975) p.5729-38]

[4] F.H.Pollak, M.Cardona [Phys. Rev. (USA) vol.172 no.3 (15 Aug 1968) p.816-37]

[5] I.Balslev [Solid State Commun. (USA) vol.5 (1967) p.315]

[6] R.N.Bhargava, M.I.Nathan [Phys. Rev. (USA) vol.161 (1967) p.695]

[7] J.Wajda, M.Grynberg [Phys. Status. Solidi (Germany) vol.37 no.K55 (1970)]

[8] R.Bendorius, A.Shileika [Solid State Commun. (USA) vol.8 (1970) p.1111]

[9] J.Feinleib, S.Groves, W.Paul, R.Zallen [Phys. Rev. vol.131 (1963) p.2070]

[10] G.E.Fenner [J. Appl. Phys. (USA) vol.34 (1963) p.2955]
(USA) vol.11 (1959) p.140]

[11] A.L.Edwards, T.E.Slyhouse, H.G.Drickamer [J. Phys. Chem. Solids (USA) vol.11 (1959) p.140]

[12] A.R.Hutson, A.Jayaraman, A.S.Coriell [Phys. Rev. (USA) vol.155 (1967) p.786]

[13] A.Jayaraman, M.E.Sikorski, J.C.Irvin, G.H.Yates [J. Appl. Phys. (USA) vol.38 (1967) p.4454]

[14] D.J.Wolford, J.A.Bradley [Solid State Commun.(GB) vol.58 no.12 (Mar 1985) p.1069-75]

7.3 OPTICAL DEFORMATION POTENTIAL FOR HOLES AT THE GAMMA POINT IN GaAs

by W.Potz

August 1985
EMIS Datareview RN=14413

The optical deformation potential d(0) for holes at the GAMMA point is usually determined by fitting of experimental data on hole transport to theory. By such a procedure, P.Lawaetz [1] and J.D.Wiley [2] obtained d(0) = 41.0 eV. Cardona et al [3] performed Raman experiments and reported d(0) = 48.0 eV.

Potz and Vogl [4] obtain d(0) as 29.1 eV and 37.0 eV by a pseudopotential and a tight-binding calculation, respectively. Blacha et al [5] calculate d(0) = 36.4 eV. Different definitions for optical-phonon deformation potentials and their relation to d(0) are given in [4] and [5].

REFERENCES

[1] P.Lawaetz [Thesis, The Technical University of Denmark, Lyngby, (unpublished)]
[2] J.D.Wiley [in Semiconductors and Semimetals, ed by R.K.Willardson and A.C.Beer (Academic, New York, 1975) vol.10 p.91]
[3] M.Cardona, H.Grimsditsch, D.Olego [in "Light Scattering in Solids" by J.L.Birman, H.Z.Cummins, and K.K.Rebane (Plenum, New York, 1979)]
[4] W.Potz, P.Vogl [Phys. Rev. B (USA) vol.24 (1981) p.2025]
[5] A.Blacha, H.Presting, M.Cardona [Phys. Status Solidi b (Germany) vol. 126 (1984) p.11]

7.4 OPTICAL GAP OF AMORPHOUS GaAs

by EMIS Group
August 1985

EMIS Datareview RN=15502

The optical gap of amorphous GaAs is generally less than the energy gap of the crystalline material, by an amount which is very sensitive to the method and circumstances of preparation. In the following survey all band gap values quoted are for samples at room temperature.

Segui et al [1] grew hydrogenated amorphous GaAs films by glow discharge of an AsH3 + Ga (CH3)3 mixture diluted in He. Optical absorption measurements indicate a band gap of 1.34 eV compared with 1.43 eV for crystals oriented in the [100] direction.

Cheorghiu et al [2] flash evaporated the GaAs crystalline powder onto glass or silica plates under ultra high vacuum conditions. Reflectance and transmission measurements indicated an optical gap of 1.0 eV for films deposited at 100-300 deg K. For annealed stabilised films the band gap was 1.15 eV.

Alimoussa et al [3] prepared amorphous GaAs films by RF sputtering onto pyrex substrates at below 303 deg K C in an argon atmosphere with added hydrogen. Optical absorption measurements showed the variation of optical gap with hydrogen partial pressure P(H). Total pressure P(T) was approx 10*-2 torr. Optical gaps of 1.06 eV, 1.41 eV and 1.45 eV (saturation value) were found for P(H)/P(T) = 0,20 and 40% respectively.

Matsumoto et al [4] prepared films by molecular beam deposition onto substrates of SiO2, Si or Al/SiO2. The film structure and composition depended on the substrate temp. 313-543 deg K and the molecular beam intensity ratio. From optical absorption measurements the optical energy gap was estimated to be 1.0 eV for sample composition ratio (As/Ga) = 1.03, decreasing to 0.75 eV for (As/Ga) = 1.71.

REFERENCES

[1] Y.Segui, F.Carrere, A.Biu [Thin Solid Films (Netherlands) vol.92 no.4 (June 1982) p.303-7]

[2] A.Cheorghiu, M.L.Theye [Philos. Mag. B (GB) vol.44 no.2 (Aug 1981) p.285-306]

[3] L.Alimoussa, H.Carchano, J.P.Thomas [J. Phys. Colloq. (France) vol.42 noC-4 (Oct 1981) p.683-6]

[4] N.Matsumoto, K.Kumabe [Jpn. J. Appl. Phys. (Japan) vol.19 no.9 (Sept 1980) p.1583-90]

CHAPTER 8

OPTICAL FUNCTIONS OF GALLIUM ARSENIDE
(Complex Refractive Index, Dielectric Function and Absorption Coefficient)

8.1 Optical functions of intrinsic GaAs: General remarks

8.2 Optical functions of intrinsic GaAs: Static and IR dielectric constants

8.3 Optical functions of intrinsic GaAs: Reststrahlen region (0-60 meV)

8.4 Optical functions of intrinsic GaAs: Transparent region (0.1-1.4 eV)

8.5 Optical functions of intrinsic GaAs: Direct bandgap region (1.3-1.5 eV)

8.6 Optical functions of intrinsic GaAs: Visible to near-UV

8.7 Optical functions of intrinsic GaAs: Vacuum UV (up to 155 eV)

8.8 Table of optical functions of intrinsic GaAs: Refractive index and absorption coefficient vs energy (0-155 eV)

8.9 Extrinsic effects on the optical functions of GaAs

8.1 OPTICAL FUNCTIONS OF INTRINSIC GaAs: GENERAL REMARKS

by D.E.Aspnes

December 1985
EMIS Datareview RN=15430

Optical functions of GaAs have already been discussed in two excellent recent reviews [1,2]. The article by Palik [1] is devoted entirely to optical functions and contains tables of the complex refractive index at selected energies from 0.00124 to 155 eV. Results are included up to 1983. A wide-ranging survey of the physical properties of GaAs, covering nearly every aspect of the material and including not only assessments of the best parameters available but also discussions of relevant basic physical phenomena, has been performed by Blakemore [2]. This review, published in 1982, contains 405 references. The reader interested in more detail is encouraged to consult these works.

Optical functions include the complex dielectric function e = e1 + ie2 where e denotes permittivity; the complex refractive index n´ = n + ik = SQRT e, where n and k are the ordinary refractive index and the extinction coefficient, respectively; and the absorption coefficient = 4 pi k/Y, where Y is the wavelength of the light. The extinction coefficient is occasionally defined as n´ = n(1 + k). The signs of the imaginary parts in the above expressions are defined according to the physics convention. The optics and engineering conventions are obtained by complex conjugation; e2 and k remain positive definite in both representations.

These quantities are often termed optical constants, but this is misleading because they are all strongly dependent on the wavelength of light. The use of the term "constant" should be restricted to the static dielectric constant, the zero-frequency limit of e1 and the infrared dielectric constant, the same limit with the ionic (lattice) contributions removed.

In principle only one set of optical functions is needed since they are all interrelated. However, the different sets are useful for different purposes. Wave propagation is described by the propagation vector k = fh/c, where f and c are the frequency and vacuum speed of light, respectively. Thus n and k are the logical optical functions to describe wave propagation. On the other hand, the fundamental response of a solid to an electric field is described by the standard equations D = eE = E + 4 pi P, where D and E are the macroscopic (observable) displacement and electric fields, respectively, and P is the dipole moment per unit volume. To a good approximation macroscopic dipole moments per unit volume are additive, whence e1 and e2 provide a convenient means of dealing with modifications (usually extra sources of polarization, such as impurities and free carriers) to the intrinsic dielectric response. Also e1 and e2 can be given simple analytic representations over certain frequency ranges, thereby allowing efficient calculation of optical functions to supplement tabulated values[3].

REFERENCES
----

[1] E.D.Palik [in "Handbook of Optical Constants of Solids" ed. by E.D.Palik (Academic, New York, 1985) p.429-44]
[2] J.S.Blakemore [J. Appl. Phys. (USA) vol.53 (1982) p.R123]
[3] D.E.Aspnes [EMIS Datareview RN=15437 (Dec 1985) "Table of optical functions of intrinsic GaAs: refractive index and absorption coefficient vs energy"]

8.2 OPTICAL FUNCTIONS OF INTRINSIC GaAs: STATIC AND IR DIELECTRIC CONSTANTS

by D.E.Aspnes

December 1985
EMIS Datareview RN=15431

Values of the static and infrared dielectric constants (e) were reviewed by Stillman et al [1] in 1977 and have been reassessed by Blakemore [2]. Briefly, Stillman's recommendation of e(o) = 12.9 was supported by Rode [3], who used dielectric and optical considerations to deduce a value of 12.91, and by Neidert [4] who found a value 12.90 +- 0.07 from room-temperature microwave measurements. The value of e(o) recommended by Blakemore is slightly smaller, but still within the range of experimental uncertainty of the above values:

$$e(o) = 12.40 (1 + 1.2 \times 10^{-4} T K^{-1}) \quad (1a)$$
$$= 12.85 \text{ at } 300 \text{ deg K} \quad (1b)$$

where the temperature dependence was taken from the microwave data of Champlin et al [5,6]. The linear temperature variation applies from 100 to 600 deg K, and Eqns (1) are valid from DC to 10^{11} Hz (energy range of 0 to 0.0004 eV; wavenumber range of 0 to 3 cm^{-1}; wavelength range from 330 microns to infinity). The zero-frequency refractive index corresponding to e(o) = 12.85 is given in the table [7].

The infrared dielectric constant cannot be measured directly but must be inferred from the static dielectric constant and phonon energies via the Lyddane-Sachs-Teller relation: $e(o)/e(\infty) = [w(LO)/w(TO)]^2$ where w(LO) = 292.1 cm^{-1} (36.21 meV) and w(TO) = 268.7 cm^{-1} (33.31 meV) are the energies of the q = 0 longitudinal and transverse optic phonons, respectively [8]. The values of w(LO) and w(TO) cited here were determined by Palik [9] as an average of parameters obtained in the analysis of infrared reflectance data in the reststrahlen region [10]. The Palik values are within experimental uncertainty (+- 0.5 cm^{-1}) of those recommended by Blakemore [2] a difference too small to influence the accuracy to which e(infinity) can be calculated. Including the temperature dependence of w(LO) and w(TO) leads to [2]

$$e(\infty) = 10.60 (1 + 9 \times 10^{-5} T K^{-1}) \quad (1b)$$
$$= 10.88 \text{ at } 300 \text{ deg K} \quad (2b)$$

REFERENCES

[1] G.E.Stillman, G.M.Wolfe, J.O.Dimmock [in "Semicond. & Semimet." (USA) vol.12 Eds R.K.Willardson, A.K.Beer, (Academic Press, New York 1977) p.169]
[2] J.S.Blakemore [J. Appl. Phys. (USA) vol.53 (1982) p.R133]
[3] D.L.Rode [in "Semicond. & Semimet." (USA) vol.10 Eds R.K.Willardson, A.K.Beer (Academic Press, New York 1975) p.1]
[4] R.E.Neidert [Electron. Lett. (USA) vol.16 (1980) p.244]
[5] K.S.Champlin, G.H.Glover [Appl. Phys. Lett. (USA) vol.12 (1968) p.231]
[6] T.Lu, G.H.Glover, K.S.Champlin [Appl. Phys. Lett. (USA) vol.13 (1968) p.404]

[7] D.E.Aspnes [EMIS Datareview RN=15437 (Dec 1985) "Table of optical functions of intrinsic GaAs: refractive index and absorption coefficient vs energy]

[8] Energy (E) to wavelength (W) conversion is done according to E = hc/nW = 1.2395086 eV microns, where h = 6.626176(36) X 10*-34, c = 2.99792438 X 10*-10 cm s*-1, 1 eV = 1.6021892(46) X 10*-19 j and n (ordinary refractive index of dry air) = 1.000277 at 15 deg C, 760 mm Hg. The fundamental constants are from [CRC Handbook of Chemistry and Physics, 64th edition, Ed. C.Weast (CRC Press, Boca Raton, 1983) p.F198]. Data for n is from [ibid. p.E364]

[9] E.D.Palik (in "Handbook of Optical Constants of Solids", Ed. E.D.Palik (Academic Press, New York, 1985) p.429-44]

[10] D.E.Aspnes [EMIS Datareview RN=15432 (Dec 1985) "Optical functions of intrinsic GaAs: reststrahlen region (0-60 meV)"]

8.3 OPTICAL FUNCTIONS OF INTRINSIC GaAs: RESTSTRAHLEN REGION (0-60 meV)

by D.E.Aspnes

December 1985
EMIS Datareview RN=15432

The reststrahlen region of the infrared contains the strong first-order phonon structure associated with the LO and TO lattice vibrations. Palik [1] has fit the dispersion relation

$$e = (n + ik)^2 = e(\infty) \left[1 + \frac{w(LO)^2 - w(TO)^2}{w(TO)^2 - w^2 + i\,\Gamma\,w}\right]$$

to the reflectance data of Kachare et al [2] and of Holm et al [3]. These data were obtained from samples chemically polished to remove all traces of surface damage.

In this expression e denotes permittivity; n the ordinary refractive index; k the extinction coefficient; w(LO) = 292.1 cm^{-1}, w(TO) = 268.7 cm^{-1}, the wavenumbers of the longitudinal and transverse optical phonons; e(infinity) = 11.0; and GAMMA = 2.4 cm^{-1}. Values of n calculated from e in this equation agree with those determined from other data to within 0.03 over the frequency range from 0 to 480 cm^{-1} (energy range 0 to 60 meV). Values at selected energies are given in the table [4]. The zero-frequency limit of the equation is only 0.6% higher than the ordinary refractive index calculated from e(infinity), fairly remarkable agreement in view of the different approaches used to obtain these quantities.

Values of k calculated from the above agree with those from other sources to within 0.0002 up to approximately 320 cm^{-1} (40 meV), where multiphonon absorption begins to dominate the spectrum. Absorption data on high purity samples have been reported up to 800 cm^{-1} (about 100 meV) by Cochran et al [5]. These values have been included in the table [4] and have been used to calculate corresponding values of k. Because multiphonon absorption is weak, not exceeding a = 55 cm^{-1} from 45 to 65 meV or a = 2 cm^{-1} from 70 to 100 meV, values of n are relatively unaffected. Sample-dependent local-mode structure from low-mass substitutional impurities such as Li, Be, B, C, and Si also appear in this spectral range.

REFERENCES

[1] E.D.Palik [in "Handbook of Optical Constants of Solids", Ed. E.D.Palik (Academic Press, New York 1985) p.429-44]

[2] A.H.Kachare, W.G.Spitzer, F.K.Euler, A.Kahan [J. Appl. Phys (USA) vol.45 1974) p.2938]

[3] R.T.Holm, J.W.Gibson, E.D.Palik [J. Appl. Phys. (USA) vol.48 (1977) p.212]

[4] D.E.Aspnes [EMIS Datareview RN=15437 "Table of optical functions of intrinsic GaAs: refractive index and absorption coefficient vs energy"]

[5] W.Cochran, S.J.Fray, F.A.Johnson, J.E.Quarrington, W.Williams [J. Appl. Phys. Suppl. (USA) vol.34 (1961) p.2102]

8.4 OPTICAL FUNCTIONS OF INTRINSIC GaAs: TRANSPARENT REGION (0.1-1.4 eV)

by D.E.Aspnes

December 1985
EMIS Datareview RN=15433

Except for sample-dependent deep-level absorption in impure material, GaAs is transparent from 0.1 to 1.4 eV (wavenumbers from 807 to 11290 cm*-1, wavelengths from 12.4 to 0.885 microns). Taking advantage of this transparency. Pikhtin and Yas´kov [1] have obtained highly accurate values of n by the minimum beam deviation method. They expressed their results analytically as

$$n^2 = 1 + \frac{A}{\pi} \ln \frac{E(1)^2 - E^2}{E(0)^2 - E^2} + \frac{G(1)}{E(1)^2 - E^2} + \frac{G(2)}{E(2)^2 - E^2} + \frac{G(3)}{E(3)^2 - E^2} \qquad (1)$$

where E is the photon energy. The E(i) are characteristic oscillator energies with values E(0) = 1.428 eV, E(1) = 2.0 eV, E(2) = 5.1 eV, and E(3) = 0.0333 eV, and A and the G(i) are oscillator strengths with values A = 0.7 (E(0)/eV)*-1/2 = 0.5858, G(1) = 39.194 eV*2, (G2) = 136.08 eV*2, and G(3) = 0.00218 eV*2. The oscillators associated with E(0), E(1), E(2) and E(3) roughly summarize absorption processes arising from the first 3 direct gaps in the electronic interband density of states and the lattice absorption, respectively. The temperature dependence of the refractive index n was measured by Marple [2] and was found to obey the relation

$$n(w,T) = n(w)\,(1 + 4.5 \times 10^{-5}\ T\ K^{-1}) \qquad (2)$$

where w is the angular frequency, which is identical to that given by Eqn (2a) in [3] recalling that n*2 = e, the complex permittivity. The values of n calculated from the equation in ref [4] and from Eqn (1) above agree to within 0.02 at 0.60 eV.

REFERENCES

[1] A.N.Pikhtin, A.D.Yas´kov [Sov. Phys. Semicond. (USA) vol.12 (1978) p.622]
[2] D.T.F.Marple [J. Appl. Phys. (USA) vol.35 (1964) p.1241]
[3] D.E.Aspnes [EMIS Datareview RN=15431 "Optical functions of intrinsic GaAs: static and IR dielectric constants"]
[4] D.E.Aspnes [EMIS Datareview RN=15432 "Optical functions of intrinsic GaAs: reststrahlen region (0-60 meV)"]

8.5 OPTICAL FUNCTIONS OF INTRINSIC GaAs: DIRECT BANDGAP REGION (1.3-1.5 eV)

by D.E.Aspnes

December 1985
EMIS Datareview RN=15434

Within approximately 0.1 eV of the fundamental direct gap (1.424 eV at room temperature), the optical functions are dominated by excitonic effects. Owing to the extreme sharpness of the n = 1 exciton absorption structure, they are strong functions of both temperature and crystal quality. An extensive investigation of both effects has been performed by Sell and co-workers [1-3]

The room temperature data of Sell et al [1] for n and of Casey et al [2] for alpha are listed from 1.3 to 1.5 eV in the table [4]. To lowest order, temperature effects can be approximated by shifting these data in energy according to the temperature dependence of the fundamental direct gap Eo itself. This follows the Varshni equation:

$$E_o(T) = 1.519 - \frac{5.4 \times 10^{-4}\, T^2}{T + 204} \text{ eV}$$

T is the absolute temperature and the coefficients have been determined by Thurmond [3]. However, this does not describe the dramatic sharpening of the lineshape at temperatures below 30 deg K [3]. The n = 1 exciton feature is also sensitive to the presence of free carriers, becoming noticeably affected near 10^{16} cm^{-3} and washing out completely by 5.9×10^{17} cm^{-3} [2].

The n data of Sell [1] et al join rather smoothly to the refractive index values given by Eqn (1) in [5]. However, the k data are about 10% lower than those of Sturge [6] for reasons that are not clear but may be related to improvements in crystal quality in the more recent work.

REFERENCES

[1] D.D.Sell, H.C.Casey, K.W.Wecht [J. Appl. Phys. (USA) vol.45 (1974) p.2650]
[2] H.C.Casey, D.D.Sell, K.W.Wecht [J. Appl. Phys. (USA) vol.46 (1975) p.250]
[3] C.D.Thurmond [J. Electrochem. Soc. (USA) vol.122 (1975 p.1133]
[4] D.E.Aspnes [EMIS Datareview RN=15437 (Dec 1985) "Table of optical functions of intrinsic GaAs: refractive index and absorption coefficient vs energy"]
[5] D.E.Aspnes [EMIS Datareview RN=15433 (Dec 1985) "Optical function of intrinsic GaAs: transparent region (0.1-1.4 eV)"]
[6] M.D.Sturge [Phys. Rev. (Netherlands) vol.127 (1962) p.768]

8.6 OPTICAL FUNCTIONS OF INTRINSIC GaAs: VISIBLE TO NEAR-UV

by D.E.Aspnes

December 1985
EMIS Datareview RN=15435

There is no shortage of optical functions data in the spectral region above the fundamental absorption edge, but relative discrepancies of the order of 50% or more are not uncommon. The reason for these discrepancies is easily understood. The penetration depth of light over much of the visible to near-UV spectral range is only of the order of 100 A, whence surface preparation, already important in the infrared, becomes crucial. Since optical functions are never measured directly but must be inferred from the attenuation of light intensity (reflectometry) or the change in polarization state (ellipsometry) upon reflection, the derived functions are only accurate to the extent that the sample conforms to an idealized representation, usually the two-phase model consisting of a clean, uniform, isotropic substrate with a mathematically sharp interface to the ambient. Thus surface residues left by improper cleaning after etching, or surface oxides and contamination overlayers resulting from air exposure, can cause significant systematic errors. The importance of proper sample preparation has become more widely appreciated, and progress has occurred mainly along these lines. Recent results for Ge, for which the data base is larger, now show optical function agreement to within 1 or 2% regardless of the optical technique used to obtain the data [1].

The most comprehensive set of optical function data for GaAs in the visible to near-UV was reported by Aspnes and Studna [1] who obtained them using spectroscopic ellipsometry. The advantage of this technique relative to reflectrometry is that the optical functions can be calculated on a wavelength-by-wavelength basis from ellipsometric data without requiring a Kramers-Kronig transformation, which thereby eliminates a source of systematic errors involving extrapolations of reflectance data into experimentally inaccessible wavelength regions. In addition, the ellipsometer was used to monitor sample preparation in real time to ensure that the surfaces were of as high a quality as could be achieved with simple wet-chemical etching and stripping techniques (Br2 in methanol followed by a water rinse).

These data are listed in the table in [2]. The n values match fairly well to those of Sell et al but the k values show a 10% discontinuity with respect to those of Casey et al [3]. The reason is due to the difficulty of rotating-analyzer spectroellipsometers to accurately determine small values of k; consequently, the k values of Sell et al should be considered more accurate.

Temperature effects in the visible to near-UV result primarily from the shift and thermal broadening of the various critical point structures. Direct measurements have not yet been performed for GaAs, but they should not be significantly different from recently published results for Ge [4].

REFERENCES
----

[1] D.E.Aspnes, A.A.Studna [Phys. Rev. B (USA) vol.27 (1983) p.985]
[2] D.E.Aspnes [EMIS Datareview RN=15437 (Dec 1985) "Table of optical functions of intrinsic GaAs: refractive index and absorption coefficient vs energy"]
[3] H.C.Casey, D.D.Sell, K.W.Wecht [J. Appl. Phys. (USA) vol.46 (1975) p.250]
[4] L.Vina, S.Logothetidis, M.Cardona [Phys. Rev. B (USA) vol.30 (1984) p.1979]

8.7 OPTICAL FUNCTIONS OF INTRINSIC GaAs: VACUUM UV (UP TO 155 eV)

by D.E.Aspnes

December 1985
EMIS Datareview RN=15436

The only complete set of optical function data that presently exist above 6 eV are those calculated by Philipp and Ehrenreich by a Kramers-Kronig analysis of normal incidence reflectance data taken from 0 to 25 eV [1]. These results are also given in the table in ref [2]. However, these results show characteristic distortions of up to 30% due to the presence of inadvertent surface overlayers upon comparison to the spectroellipsometric results in the visible to near-UV, a consequence of the inability to assess surface preparation directly with the vacuum ultraviolet reflectometer. The corrections necessary to eliminate these artifacts have been applied to Si data determined in the same experimental program [3] but they have not yet been applied to GaAs.

Above 25 eV, the reflectance of GaAs becomes too low for accurate measurement, but thin-film geometries allow the absorption coefficient to be determined in transmission. Cardona and co-workers have reported absorption coefficient data from 22 to 155 eV for polycrystalline GaAs deposited on carbon films and microscope slides previously coated with KCl [4], the results of which are also given in [2]. Differences between the Philipp-Ehrenreich [1] and the Cardona et al [4] results for k in the region of overlap indicate that the packing fraction of GaAs in the polycrystalline films was less than unity. However, no other data are currently available in this or higher spectral ranges.

REFERENCES

[1] H.R.Philipp, H.Ehrenreich [Phys. Rev. (USA) vol.129 (1963) p.1550] See listing in B.O.Seraphin, H.E.Bennett [in "Semicond & Semimet. (USA) Eds. R.K.Willardson, A.C.Beer (Academic Press, New York, 1967) p.499]

[2] D.E.Aspnes [EMIS Datareview RN=15437 (Dec 1985) "Table of optical functions of intrinsic GaAs: refractive index and absorption coefficient vs energy"]

[3] H.R.Philipp [J. Appl. Phys. (USA) vol.43 (1972) p.2835]

[4] M.Cardona, W.Gudat, B.Sonntag, P.Y.Yu [in "Proc. 10th Internat. Conf. Phys. Semicond.", Cambridge, Eds S.P.Keller, J.C.Hensel & F.Stern (US Atomic Energy Commission, Oak Ridge, Tennessee, 1970) p.209]

8.8 TABLE OF OPTICAL FUNCTIONS OF INTRINSIC GaAs: REFRACTIVE INDEX AND ABSORPTION COEFFICIENT VS ENERGY (0-155 eV)

by D.E.Aspnes

December 1985
EMIS Datareview RN=15437

TABLE: Intrinsic optical functions for single crystal GaAs at room temperature for selected energy values. n is the refractive index, k is the extinction coefficient and alpha is the absorption coefficient.

Wavelength in A = approx. 12395/(energy in eV).

Energy (eV)	n	k	1000 alpha (cm*-1)	Ref
0.000	3.584	-	-	[1]
0.000	3.6053	0.0000	0.000	[2]
0.005	3.6117	0.0003	0.000	"
0.010	3.633	0.001	0.003	"
0.015	3.675	0.002	0.003	"
0.020	3.758	0.003	0.007	"
0.025	3.946	0.009	0.023	"
0.030	4.643	0.048	0.15	"
0.031	5.085	0.09	0.28	"
0.032	6.050	0.23	0.76	"
0.0325	7.169	0.51	1.67	"
0.033	10.123	2.04	6.8	"
0.03322	12.413	6.49	21.9	"
0.03339	7.531	11.70	39.6	"
0.0335	4.054	10.67	36.2	"
0.034	0.841	5.96	20.5	"
0.035	0.302	2.86	10.1	"
0.0355	0.266	1.93	7.0	"
0.036	0.344	0.99	3.63	"
0.038	2.092	0.054	0.21	"
0.040	2.545	0.022	0.09	"
0.045	2.929	0.004	0.020	[3]
0.050	3.067	0.002	0.012	"
0.055	3.137	0.006	0.030	"
0.060	3.180	0.003	0.020	"
0.065	3.208	0.009	0.055	"
0.070	3.227	0.000	0.002	"
0.080	3.253	0.	0.	"
0.080	3.234	0.	0.	[4]
0.090	3.249			"
0.100	3.260			"
0.150	3.283			"
0.200	3.292			"
0.250	3.298			"
0.300	3.303			"
0.400	3.313			"

0.500	3.324			"
0.600	3.338			"
0.700	3.354			"
0.800	3.374			"
0.900	3.397			"
1.000	3.423			"
1.100	3.455			"
1.200	3.492			"
1.300	3.539			"
1.300	3.535	-	-	[5,6]
1.310	3.541	-	-	"
1.320	3.546	-	-	"
1.330	3.553	-	-	"
1.340	3.558	-	-	"
1.350	3.566	-	-	"
1.360	3.573	0.	0.01	"
1.370	3.581	0.	0.02	"
1.380	3.590	0.	0.03	"
1.390	3.600	0.	0.09	"
1.400	3.611	0.001	0.21	"
1.410	3.628	0.003	0.46	"
1.420	3.646	0.026	4.00	"
1.430	3.640	0.056	8.1	"
1.440	3.635	0.053	7.7	"
1.450	3.634	0.059	8.7	"
1.460	3.634	0.062	9.1	"
1.470	3.635	0.066	9.8	"
1.480	3.636	0.068	10.2	"
1.490	3.638	0.071	10.7	"
1.500	3.640	0.074	11.2	"
1.500	3.666	0.080	12.21	[7]
1.600	3.700	0.091	14.83	"
1.700	3.742	0.112	19.28	"
1.800	3.785	0.151	27.49	"
1.900	3.826	0.179	34.45	"
2.000	3.878	0.211	42.79	"
2.100	3.904	0.240	51.15	"
2.200	4.013	0.276	61.46	"
2.300	4.100	0.320	74.56	"
2.400	4.205	0.371	90.34	"
2.500	4.333	0.441	111.74	"
2.600	4.492	0.539	142.02	"
2.700	4.694	0.696	190.53	"
2.800	4.959	0.991	281.33	"
2.900	5.052	1.721	505.75	"
3.000	4.509	1.948	592.48	"
3.100	4.373	2.146	674.17	"
3.200	3.938	2.288	742.21	"
3.300	3.709	2.162	723.09	"
3.400	3.596	2.076	715.28	"
3.500	3.531	2.013	714.20	"
3.600	3.495	1.965	717.14	"
3.700	3.485	1.931	724.14	"
3.800	3.501	1.909	735.28	"
3.900	3.538	1.904	752.62	"
4.000	3.601	1.920	778.65	"
4.100	3.692	1.969	818.23	"
4.200	3.810	2.069	880.86	"
4.300	3.939	2.260	984.86	"
4.400	4.015	2.563	1143.26	"
4.500	3.913	2.919	1331.28	"

4.600	3.769	3.169	1477.66	"
4.700	3.598	3.452	1644.29	"
4.800	3.342	3.770	1834.18	"
4.900	2.890	4.047	2009.92	"
5.000	2.273	4.084	2069.81	"
5.100	1.802	3.795	1981.86	"
5.200	1.599	3.484	1836.14	"
5.300	1.499	3.255	1748.74	"
5.400	1.430	3.079	1685.29	"
5.500	1.383	2.936	1636.68	"
5.600	1.349	2.815	1597.99	"
5.700	1.325	2.710	1565.73	"
5.800	1.311	2.625	1543.07	"
5.900	1.288	2.557	1528.86	"
6.000	1.264	2.472	1503.20	"
6.000	1.395	2.048	1245.	[8]
6.2	1.424	1.976	1241.	"
6.6	1.247	2.047	1369.	"
7.0	1.063	1.838	1304.	"
8.0	0.899	1.435	1163.	"
9.0	0.901	1.136	1036.	"
10.0	0.913	0.974	987.	"
12.0	0.895	0.791	962.	"
14.0	0.840	0.602	854.	"
16.0	0.850	0.411	666.	"
18.0	0.936	0.324	591.	"
20.0	1.025	0.212	430.	"
22.0	1.043	0.240	535.	"
22.0	-	0.148	330.	[9]
26.0	-	0.0872	230.	"
28.0	-	0.0683	194.	"
30.0	-	0.0648	197.	"
40.0	-	0.0426	173.	"
50.0	-	0.0430	218.	"
60.0	-	0.0389	237.	"
70.0	-	0.0376	267.	"
80.0	-	0.0353	286.	"
90.0	-	0.0323	295.	"
100.0	-	0.0294	298.	"
110.0	-	0.0278	310.	"
120.0	-	0.0245	298.	"
130.0	-	0.0224	295.	"
140.0	-	0.0206	292.	"
150.0	-	0.0193	293.	"

REFERENCES

[1] J.S.Blakemore [J. Appl. Phys. (USA) vol.53 (1982) p.R123]

[2] E.D.Palik [in "Handbook of Optical Constants of Solids", Ed. E.D.Palik (Academic Press, NewYork,1985) p.429]

[3] W.Cochran, S.J.Fray, F.A.Quarrington, W.Williams [J. Appl. Phys. Suppl. (USA) vol.34 (1961) p.2102]

[4] A.N.Pikhtin, A.D.Yas´kov [Sov. Phys. Semicond. (USA) vol.12 (1978) p.622]

[5] D.D.Sell, H.C.Casey, K.W.Wecht [J. Appl. Phys. (USA) vol.45 (1974) p.2650]

[6] H.C.Casey, D.D.Sell, K.W.Wecht [J. Appl. Phys. (USA) vol.46 (1975) p.250]

[7] D.E.Aspnes, A.A.Studna [Phys. Rev. B (USA) vol.27 (1983) p.985]

[8] H.R.Philipp, H.Ehrenreich [Phys. Rev. (USA) vol.129 (1963)

p.1550]. See listing in B.O.Seraphin, H.E.Bennett [in "Semicond. & Semimet" (USA), Eds R.K.Willardson, A.C.Beer (Academic Press, New York, 1967) p.499]

[9] M.Cardona, W.Gudat, B.Sonntag, P.Y.Yu [in "Proc 10th Internat. Conf. Phys. Semicond", Eds S.P.Keller, J.C.Hensel & F.Stern (US Atomic Energy Commission, Oak Ridge, Tennesse, 1970) p.209]

8.9 EXTRINSIC EFFECTS ON THE OPTICAL FUNCTIONS OF GaAs

by D.E.Aspnes

December 1985
EMIS Datareview RN=15438

The most important extrinsic effects, and the only ones that will be discussed here, are those due to free carriers. The primary results of free carriers are the appearance of plasma structure and additional absorption in the infrared, and the shift of the optical gap with the onset of conduction band degeneracy at a free electron concentration of about 4 X 10*17 cm*-3 (Burstein shift). However, if present in concentrations greater than 10*19 cm*-3, free carriers shift and broaden critical point structures across the entire visible through near-UV spectra.

For n-type material, the results of Spitzer and Whelan [1] indicate that for energies below 2500 cm*-1 (0.3 eV), free-carrier absorption can be expressed approximately as

a = approx. 7.5 X 10*-20 n(c) Y cm*-1;   Y at or above 4 microns   (1)

where n(c) is the free electron concentration (cm*-3) and Y is the wavelength (microns). For higher energies the absorption is dominated by intraconduction band transitions between GAMMA and L and is given approximately as

a = approx. 6 X 10*-18 n(c) cm*2;   Y at or below 4 microns   (2)

Agreement is found to doping levels up to 5 X 10*18 cm*-3. For p-type material, the results of Braunstein and Kane [2] showed that the infrared spectrum is dominated by intravalence band transitions with a threshold near 0.35 eV. However, considerable absorption was also observed at lower energies.

The contribution DELTA e(f) of free electrons to the general dielectric response can be represented by the simple Drude model (w = angular frequency):

DELTA e(f) = - 4 pi n(c) q*2/[m(c)´w(w + i GAMMA)]   (3)

where m(c)´ = 0.0632 m(e) is the conduction band effective mass of GaAs at room temperature (here, because of character set limitations, q denotes the charge on an electron). For sufficiently large free electron densities this term cancels the infrared dielectric constant arising from bound electrons, thereby impedance-matching the material to the ambient and dropping the reflectance to very low values. The reflectance minimum occurs near the plasma frequency (energy), which is given approximately by

$$Ep*2 = \frac{n(c)\ q*2\ h*2}{pi\ m(c)'\ e(infinity)} \qquad (4a)$$

$$= 2.0 \text{ X } 10*-21\ n(c)\ cm*3\ eV*2 \qquad (4b)$$

The plasma energy increases as the square root of carrier concentration and passes above the first-order phonon structure of the reststrahlen band for n(c) of the order of 10*18 cm-3. However, the conduction band effective mass increases with the increasing degeneracy of the electron gas due to the curvature of the conduction bands, so Eqn (4) is only an approximation.

The Burstein shift in GaAs has been studied in detail by Hwang [3]. The fundamental direct gap begins to shift to higher energies at the onset of degeneracy and increases by 87 meV at n(c) = 3.8 X 10*18 cm*-3.

REFERENCES

[1] W.G.Spitzer, J.M.Whelan [Phys. Rev. (USA) vol.114 (1959) p.59]
[2] R.Braunstein, E.O.Kane [J. Phys. & Chem Solids (GB) vol.23 (1962) p.1423]
[3] C.J.Hwang [J. Appl. Phys. (USA) vol.40 (1969) p.3731]

CHAPTER 9

ELECTRO-OPTIC PROPERTIES OF GALLIUM ARSENIDE

9.1 Electro-optical coefficients of GaAs

9.2 Franz-Keldysh effect parameters of GaAs

9.1 ELECTRO-OPTICAL COEFFICIENTS OF GaAs

by D.J.Robbins

December 1985
EMIS Datareview RN=15460

There has been increased interest recently in the electro-optic properties of semiconducting materials, both in theroetical (parametrised) models [1-3] and in experimental measurement [4-7] as a result of the development of phase modulators utilising this effect. In zinc-blende structures the quadratic electro-optic (Kerr) effect is usually negligible. This is the case for GaAs [1] and leaves the linear electro-optic (Pockels) effect, which is characterised by a single coefficient r(23,1) = r(41). Generally, results from 1.3 micron phase modulators in GaAs are consistent with a value of r(41) approx. - 1.5 X 10*-12 m/V but do not provide a precise measurement.

Usually experimental measurements of r(41) are performed at low frequencies compared to acoustic resonator frequencies so that it is the unclamped or constant stress value $r^T(41)$ which is measured.

$$r^T(41) = r^S(41) + r^P(41)$$

where $r^S(41)$ is the clamped or constant strain value. The difference between $r^T(41)$ and $r^S(41)$ is $r^P(41)$ the piezoelectrically induced electro-optic coefficient. This distinction is important because it appears that the large discrepancies between many early measurements of the electro-optic coefficient may have been caused by residual strain in the samples. Additional problems are caused by multiple reflections and the effects of photoconductivity [4].

Older work, much of which is reviewed in [5], yields a scatter of values between -1.0 X 10*-12 m/V and -1.6 X 10*-12 m/V for a wavelength range 1.0 micron to 10.6 micron. Theoretical predictions suggest that there should be little wavelength dependence within this range [2]. Recent papers by Suzuki and Tada [3] and Sugie and Tada [4] contain the most comprehensive and reliable studies to date. These were carried out on high resistivity GaAs:Cr samples which had a sufficiently high absorption coefficient to render mutliple reflections negligible. Their results, which are in good agreement with measurements due to Walsh [8], are tabulated below.

Values of $r^S(41)$ in the table were obtained from $r^T(41)$ and measurements of the elastooptic coefficient p(44).

Ref	Wavelength micron	n(o)	T r (41) 10*-12 m/V	S r (41) 10*-12 m/V
6	0.9	3.57	-1.6	
4	1.064	3.48	-1.17	-1.33
5	1.15		-1.43+-0.07	
4	1.21	3.43	-1.25	-1.41
4	1.31	3.41	-1.28	-1.46
4	1.50	3.38	-1.36	-1.53
5	3.39		-1.24+-0.04	
5	10.6		-1.51+-0.05	

Indirect support for a value approx. equal to 1.5 X 10*-12 m/V at 3.39 micron arises from the first-order Raman coefficient which is related to r(41) and to the second harmonic generation coefficient d(41) [9].

Measurements of the temperature dependence of r(41) [6,7] indicate a linear variation in the range 100-300 degK, with:

r(41)(T) = r(41) (100degK) [1 + (3.2+-0.5) X 10*-3 (T-100)]

where T is the temperature [6]. In [6] values of n*3 r(41) for the energy range 1.2 eV to 1.45 eV are given. At 1 micron r(41) approx. equal to 1.16 +- .09 X 10*-12 m/V in good agreement with [4]

$$r(41)SI = \frac{r(41)\ esu}{3 \times 10^{4}}$$

At higher energies r(41) rises steeply towards the band edge in general agreement with theoretical prediction [2].

REFERENCES

[1] C.C.Shih, A.Yariv [J. Phys. C (GB) vol.15 (1982) p.825-46]

[2] S.Adachi, K.Oe [J. Appl. Phys. (USA) vol.56 no.1 (July 1984) p.74-80]

[3] S.Adachi, K.Oe [J. Appl. Phys. (USA) vol.56 no.5 (Sept 1984) p.1499-1504]

[4] S.Sugie, K.Tada [J. Appl. Phys. (USA) vol.15 no.3 (March 1976) p.421-31]

[5] N .Suzuki, K.Tada [J. Appl. Phys. (USA) vol.23 no.8 (Aug 1984) p.1011-6]

[6] Yu.V.Shaldin, D.A.Belogurov [Sov. J. Quantum Electron (USA) vol.6 no.8 (Aug 1976) p.897-901]

[7] T.G.Okroashvili [Opt. & Spektrosk. (USSR) vol.47 no.4 (Oct 1979) p.442-3]

[8] T.E.Walsh [RCA Rev (USA) vol.27 (1966) p.323]

[9] S.C.Varshney, A.A.Gundjian [J. Appl. Phys. (USA) vol.52 no.10 (Oct 1981) p.6301-5]

9.2 FRANZ-KELDYSH EFFECT PARAMETERS OF GaAs

by D.J.Robbins

January 1986
EMIS Datareview RN=15461

Although the Franz-Keldysh effect has been studied widely in various guises, particularly in electroreflection (see, for example, a number of articles in [1]), only a few direct measurements of the absorption edge shift with applied field have been made. Since the applied field is generally obtained by making measurements within a depletion region are one experimental difficulty is that of ensuring a uniform field. As a result the absorption coefficient, alpha, is often given as a function of voltage so that the results remain device specific. An exception to this is work by Stillman et al [2,3] and more recently by Kingston et al [4,5]. In [3] high purity epitaxial layers were used with a net electron concentration of n = approx. 10*13 cm-3, in a Schottky barrier device which was fully depleted at zero bias. It appears that good qualitative agreement with experiment as regards shape of the absorption edge is obtained over a range of near band gap photon energies hf, with Eg-hf in the range -40 meV to 120 meV, using expressions due to Thermalingham [6] and Callaway [7] for direct radiative transition rates in the presence of an electric field. It seems therefore appropriate to include that expression here. As a function of field E (V/cm) and photon energy hf (eV) :

$$\alpha(hf, E) = \frac{E^{1/3}}{n(hf)} \sum_{j=1}^{2} \alpha(j)\, Z$$

$$\text{where } Z = [\mathrm{mod}\,(Ai'\,\beta(j))]^{2} - \beta(j)[\mathrm{mod}\,Ai\,(\beta(j))]^{2}$$

The summation SIGMA is over the valence bands, and for GaAs

alpha (1) = 0.23 X 10*4
alpha (2) = 0.105 X 10*4
beta (1) = 4.645 X 10*4 (Eg-hf)E*(-2/3)
beta (2) = 5.36 X 10*4 (Eg-hf)E*(-2/3)

Here n(hf) is the refractive index, Ai an Airy function and Ai' its derivative. As shown in [3] this expression compares favourably with experimental results for fields of approx. 4-6 X 10*4/cm. The coefficients beta(j) depend upon the effective masses. The coefficients alpha(j) involve in addition the optical matrix element and are therefore scaled to obtain a good fit with experiment. In fact alpha(j) and beta(j) depend only weakly on material parameters which are similar for a range of III-V materials so that the expression above is useful for other materials. This is supported by work on 1.3 micron InGaAsP [4]. Despite the fact that a real absorption edge will involve transition to shallow levels and the above equation derives from a band-band theory it still appears to work even for a donor concentration of 1 X 10*16/cm3 [4]. However, it is generally true that the theoretical curve under-estimates the required field for a given band edge shift. For example, it appears that the theoretical curve for E = approx. 4 X 10*4 V/cm actually corresponds to an applied field of E = approx. 5 X 10*4 V/cm. More simply, at 50 meV below the band edge (Eg approx 1.42 eV) with wavelength = approx. 0.905 micron, alpha = approx. 4.4 X 10*3 (E - 3 X 10*4) cm-1 for E in the range 4 X 10*4 - 2 X 10*5 V/cm [5]. The reader is referred to ref [6] for precise definitions of alpha and beta.

At energies above the absorption edge the change in absorption with field exhibits the well known Franz-Keldysh oscillations. The variation in period of the oscillation as a function of field is well characterised and is discussed by a number of authors. For GaAs see, for example, Silbrstein and Pollack [8] and Bobyler et al [9]. More general data for alpha (hf,E) appears not to be available.

Electroabsorption also exhibits a polarisation dependence because of the valence band anisotropy induced by the field which causes light polarised along the field to be absorbed more strongly than light polarised normal to the field. A study in AlGaAs/GaAs heterostructures is given by Reinhart [10]. However, the electroabsorption constant is related to the applied voltage in [10] rather than field.

Finally, electroabsorption has been used to characterise extrinsic effects in high resistivity GaAs:O at 300 deg K [11] and GaAs:Cr [12], by measuring defect and impurity level electroabsorption spectra. The spectral structure may be compared with defect and impurity behaviour observed in other photoelectronic experiments but it remains difficult to identify particular band-defect transitions, and this has not been persued in the more recent literature.

REFERENCES

[1] D.F.Blossey, P.Handler, D.E.Aspnes, N.Bottka [in "Semiconductors and Semimetals" vol.9, Eds P.K.Willardson & A.G.Beer]

[2] G.E.Stillman, C.M.Wolfe, J.A.Rossi, J.P.Donnelly [Appl. Phys. Lett. (USA) vol.25 no.11 (1974) p.671-3]

[3] G.E.Stillman, C.M.Wolfe, C.O.Bozler, J.A.Rossi [Appl. Phys. Lett. (USA) vol.28 no.9 (1976) p.544-6]

[4] R.H.Kingston [Appl. Phys. Lett. (USA) vol.34 no.11 (1979) p.744-6]

[5] R.H.Kingston, F.J.Leonberger [IEEE J.Quantum Electron. (USA) vol.19 no.3 (1983) p.1443-51]

[6] K.Tharmalingham [Phys. Rev. (USA) vol.130 (1963) p.2204-6]

[7] J.Callaway [Phys. Rev. (USA) vol.130 (1963) p.594]

[8] R.P.Silberstein, F.H.Pollack [J. Vac. Sci. & Technol. (USA) vol.17 no.5 (1980) p.1052-6]

[9] B.A.Bobylev, A.F.Kravchenko, A.S.Terekhov [Sov. Phys. Semicond. (USA) vol.7 no.11 (1974) p.1381-3]

[10] F.K.Reinhart [Appl. Phys. Lett. (USA) vol.22 no.8 (1973) p.372-4]

[11] A.D.Jonath, R.H.Bube [Surface Sci. (Netherlands) vol.37 (1973) p.167-74]

[12] T.Nishino, T.Yanagida, Y.Hamakawa [Jpn. J. Appl. Phys. (Japan) vol.11 (1972) p.1221]

CHAPTER 10

INFRA-RED (IR) ABSORPTION BY GALLIUM ARSENIDE

10.1 IR absorption bands of bulk GaAs

10.2 Two-photon IR absorption coefficient of GaAs

10.1 IR ABSORPTION BANDS OF BULK GaAs

by M.R.Brozel

December 1985
EMIS Datareview RN=15419

Infrared absorption below the band edge in GaAs takes many forms. In undoped material the absorption spectrum consists of one and two phonon "lattice bands". Single phonon absorption bands extend from 150 to 340 cm*-1 while multi-phonon structures are visible up to about 550 cm*-1 with prominent maxima around 380 cm*-1 and 520 cm*-1 [1,2,29,30].

When impurities are introduced, three dominant absorption processes can be stimulated:

1. Free carrier absorption due to the excitation of damped oscillations of conduction electrons or holes.
2. Absorption due to localised vibrational modes (LVM); high energy vibrational modes of relatively light impurity atoms which do not couple directly to phonon modes.
3. Absorption due to electronic excitation both within ions and also between defect electrons and the valence or conduction bands.

1 FREE CARRIER ABSORPTION

In this process light excites conduction electrons into higher band states from which they thermalize losing energy to the lattice.

There are three contributions to the absorption coefficient due to the interaction of electrons with (a) optical phonons, (b) acoustic phonons and (c) ionized impurities. The measured absorption coefficient as a function of wavelength, f(Y), is therefore a complicated function of wavelength of the form

$$f(Y) = aY^{3/2} + bY^{5/2} + cY^{7/2}$$

where a, b, and c correspond to the mechanisms detailed above. Values of f(Y) have been tabulated [3] and allow information about respective interactions to be deduced and the roles of particular defects, such as ionized impurities, to be assessed [4]. In p-type material, inter-valence band transitions cause similar absorption. These effects have been discussed by Braunstein et al [32].

2 LVM ABSORPTION

Certain defects have vibrational modes of energy higher than the maximum of the lattice optical phonons. They can, therefore, only couple to the lattice by second order interactions. For this reason, their lifetimes are long, and as a result, their vibrational frequencies are well defined, and they give rise to sharp absorption bands which can act as fingerprints for the impurities. In general

LVM-active defects are considerably lighter than the host atoms, but the criteria for the existence and properties of LVM absorption bands are complicated [2].

The table below represents LVM modes in GaAs that may be observed directly in non-conducting material. In conducting GaAs, free carrier absorption (previous section) is dominant and must be removed by either high energy electron irradiation [2] or by lithium or copper diffusion [3]. These treatments produce irradiation damage centres and the creation of Li- or Cu-related complexes respectively. Neither of these are present in the original material and are discussed in detail elsewhere [2,6,7,8] although several Li-complexes are included for completeness. Because LVM modes are vibrational in character and frequencies of particular defects are controlled by their atomic masses, the relevant isotopes are indicated.

(Note on notation in table: Be-9(Ga) denotes Be isotope of mass 9 on a Ga site):

Label	Specific Defect Identification	LVM energy (cm*-1)	Calibration Factor [11,18] (cm*-1)	Notes	Ref
Li	Li-6	389	-	uncomplexed	[6]
	Li-7	364	-	"	[6,2]
Li complexes (see end of table)					
Be	Be-9(Ga)	482	-	shallow acceptor	[10]
B	B-10(Ga)	540	-	isoelectronic; produced by decomposition of boron oxide in LEC process	[11]
	B-11(Ga)	517	6 X 10*16		
"boron-induced resonant mode"	B(Ga)	123	-		[24,28]
B(2)	B-10(As)	628	Similar to B(Ga) ?	boron acceptor centre	[9]
	B-11(As)	601			
B(2)-Si	B-11(As) - Si-28(Ga)	570.9 B mode		doubly doped; B + Si	[2]
		661.0 " "			
		349 Si mode	-	" "	[24]
B(2)-Si	B-10(As) - Si-28(Ga)	596.0 B mode	-		[12]
		684.8 " "	-		
		352 Si mode	-		[24]
B(2)-Se	B-11(As)-Se(As)	576.4 B mode	-	doubly doped; B + Se	[25]
		609.4 " "	-		
		621.7 " "	-	" "	

B(2)-Se	B-10(As)-Se(As)	601.0 " " 636.5 " " 649.0 " "	- - -		[25]
B(2)-Te	B-11(As)-Te	580.7 " " 606.5 " " 622.6 " "	- - -	doubly doped; B + Te " "	[25]
B(2)-Te	B-10(As)-Te	605.0 " " 633.1 " " 649.6 " "	- - -		[25]
C	C-12(As)	582	2.4 X 10*16 3-4.1 X 10*16 1.1 X 10*16 0.95 X 10*16	major residual contaminant in LEC GaAs	[11] [12] [13] [14]
N	N-14(As)	480	-	N ion-implanted; anneals out above 900 K	[15]
O	O-16	800	-	only known report of band	[16]
Mg	Mg-24(Ga) Mg-25(Ga) Mg-26(Ga)	331 326 321.5	- - -	shallow acceptor	[17]
Al	Al-27(Ga)	362	3.1 X 10*16	contaminant in Ga metal	[11]
Si donors	Si-28(Ga) Si-29(Ga) Si-30(Ga)	384 378 373	6 X 10*16	dominant defect in undoped Bridgman-grown GaAs	[18]
Si acceptors	Si-28(Ga)	398	6.8 X 10*16 about 10*17	[Si(As)] is usually approx 0.1 [Si(Ga)]	[11,1] [9]
Si pair- bands	Si-28(Ga) - Si-28(As)	393 464	-	present when [Si] above 10*17 cm*-3	[19,2]
Si defect complexes	Si-28 - X Si-28 - Y	369 367	- -	present when [Si] above 10*18 cm*-3. X and Y are native defects	[19]
Si - Zn	Si-28(Ga) - Zn(Ga)	378 382 395	- - -	Si + Zn doped	[21]

Si - Ge complex	Si-28(Ga) - Ge(As)	373	-	Si + Ge doped	[22]
		403	-		
Si - Cu complex	Si-28(Ga) - Cu(Ga)	374	-	major complex	[23]
		376	-	in Cu-diffused	
		399	-	Si doped GaAs	
P	P-31(As)	355.4	-	iso-electronic	[1]
Mg-Li	many absorption bands		-	Li-diffused	[17]
Mg-Te	many absorption bands		-	doubly doped	
Mg-Se	many absorption bands		-	doubly doped	
Si-28 - Li-6		470	-	Li-diffused	[6]
		481	-		
		488	-		
Li-Li	many absorption bands		-	Li-diffused	[26]
Li-6 - Mn		419	-		[27]
		413	-		
		391	-		
Li-7 - Mn		391	-		[27]
		386	-		
		365	-		
Li-6 - Cd		423	-		[27]
		401	-		
		377	-		
Li-7 - Cd		395	-		[27]
		375	-		
		354	-		
Li-6 - Zn		433	-		[27]
		404	-		
		385	-		
		361	-		
Li-7 - Zn		405	-		[27]
		378	-		
		361	-		
		340	-		

3 ELECTRONIC ABSORPTION BANDS

These are due to excitation of electrons within defect centres or are due to transitions between defect centres and either the valence or conduction band. Much of this data has been obtained from ref [31]. Because these bands are, in general, very temperature-dependent the table below gives the temperature at which the bands were originally measured.

3.1 Excitonic absorption

Centre	Energy (meV)	Temperature deg K	State	Ref
	Eg - 14.5	10	2P(3/2)	[33]
Mn	Eg - 10.5	10	2P(5/2) (GAMMA8)	
	Eg - 8.8	10	2P(5/2) (GAMMA7)	
C,Mg,Be	Eg - 7	1.5	1S	[34]
Mg	Eg - 2 (doublet)	1.5	2 hole replicas of	
Be	Eg - 2.2 (doublet)	1.5	exciton recombination	[34]
C	Eg - 3.5 (doublet)	1.5	at neutral acceptor centres	
Cu(Ga) V(As)	156 meV	6	Cu "C" centre	[35,36]

3.2 Ionization of neutral acceptors

Centre		Wavelength	Temp deg K	Ref
Cr3+(d3)	broadband	2.5 micron to band edge	300	[37,38]
Mn3+(d4)	broadband	14 microns to band edge (peak at 8 microns)	10	[33]
Co3+(d6)	broadband	8 microns to band edge (peak at 3 microns)	18	[39]
		" "	77	[40]
Cu3+(d8)		7.5 microns	-	[35]

3.3 Intra-centre transitions involving shallow acceptor centres

Shallow acceptors give rise to photoconductivity peaks. These are associated with intracentre excitation and subsequent thermal ionization from the excited states to the valence band. Whilst these effects were originally observed by low-temperature photoconductivity in thin epitaxial samples, these transitions will produce sharp absorption bands in thicker samples.

Impurity	Ground State Energy (meV)	Transition Energies (cm*-1) 1S(3/2)-2P(3/2)	1S(3/2)-2P(5/2) (GAMMA8)	1S(3/2)-2P(5/2) (GAMMA7)	Ref
C	26.9	122.3	156.0	170.9	[51]
Mg	28.7	137.9	169.9	186.0	[51]
Zn	30.6	156.3	187.2	202.0	[51]
Si	34.8	-	220.1	235.1	[51]

There are several weaker higher order bands in addition (see reference 51).

3.4 Intra-centre transitions in transition metal ions

Zero phonon lines, unless otherwise stated, due to intra 3d-shell transitions of isolated transition metal ions and measured at 4 deg K. Vibronic sidebands are not included in the table if zero phonon lines (ZPL) are visible and reference should be made to the original works in order to assess the significance of these.

Impurity	Wavenumbers cm*-1	Assignment	Ref
Ni+	4618	$^2T_2$ to ^2E (isolated Ni)	[54]
Ni - S	4427	nearest neighbour pair	[41]
Ni - Se	4410	" " "	[41]
Ni - Te	4369	" " "	[41]
Ni - Si	4699	" " "	[41]
Ni - Sn	4627	" " "	[41]
Co2+(d7)	4039	$^4A_2(F)$ to $^4T_1(F)$	[41]
	7000-8500	$^4A_2(F)$ to $^4T_1(F)$; no ZPL	[42]
		$^2E(G)$ to $2T_1(G)$	[43]
	11317	$^4A_2(F)$ to $^4T_1(P)$	[43]
Fe2+(d6)	3002	^5E to $^5T_2$	[42]
	2988	-	[44,55]
Cr2+(d4)	6619 6612.5 6610	$^5B_2(^5T_2)$ to $^5A_1(^5E)$	[44,55]
V2+(d3)	8800	$^4T_1(F)$ to $^4T_1(P)$; no ZPL	[46]

3.5 Important intra- and inter-centre transitions not involving transition metal ions

"78 meV" LEC GaAs grown from Ga-rich melt. Origin unknown, possibly Ga(As)	575 590	2P (GAMMA8) 2P (GAMMA7)		[47] [47]
EL2 origin unknown, associated with As(Ga)	6000 to band edge (broadband)	?	ZPL not visible above 10 deg K	[48-50,52,53,56]

REFERENCES

[1] R.C.Newman [in "Infrared Studies of Crystal Defects", Taylor and Francis (1973) p.154]

[2] G.A.Gledhill, J.F.Angress, M.R.Brozel, R.C.Newman [Proc. Int. Conf. on Lattice Dynamics (1977) p.388]

[3] W.Walukiewicz, L.Lagowski, L.Jastrzebski, M.Lichtensteiger, H.C.Gatos [J. Appl. Phys. (USA) vol.50 (1979) p.899]

[4] L.Jastrzebski, L.Lagowski, W.Walukiewicz, H.C.Gatos [J. Appl. Phys. (USA) vol.51 (1980) p.2301]

[5] R.T.Chen, V.Rana, W.G.Spitzer [J. Appl. Phys. (USA) vol.51 (1980) p.1532]

[6] P.C.Leung, J.Fredrickson, W.G.Spitzer, A.Kahan, L.Bouthillette [J. Appl. Phys. (USA) vol.45 (1974) p.1009]

[7] M.R.Brozel, R.C.Newman [J. Phys. C: (GB) vol.11 (1978) p.3135]

[8] J.Maguire, R.C.Newman, I.Grant, D.Rumsby, R.M. Ware [J. Phys.D (GB) vol.18 (1985) p.2029]

[9] J.Woodhead, R.C.Newman, I.Grant, D.Rumsby, R.M.Ware [J. Phys. (GB) vol.16 (1985) p.5523]

[10] K.Laithwaite, R.C.Newman, P.D.Greene, [J. Phys.C (GB) vol.8 (1975) p.L77]

[11] M.R.Brozel, J.B.Clegg, R.C.Newman [J. Phys.D (GB) vol.11 (1978) p.1331]

[12] W.M.Theis, K.K.Bajing, C.W.Litton, W.G.Spitzer [Appl. Phys. Lett. (USA) vol.4 (1982) p.70]

[13] A.T.Hunter, H.Kimura, J.P.Bukus, H.V.Winston, O.J.Marsh [Appl. Phys. Lett. (USA) vol.44 (1982) p.74]

[14] Y.Homma, Y.Ishu, T.Kobayashi, J.Osaka [J. Appl. Phys. (USA) vol.57 (1985) p.2931]

[15] A.H.Kachare, W.G.Spitzer, A.Kahan, F.K.Euler, T.A.Whatley [J. Appl. Phys. (USA) vol.44 (1973) p.4393]

[16] Z.L.Akkerman, L.A.Borisova, A.F.Kravchenko, [Sov. Phys. Semicond. (USA) vol.10 (1976) p.590]

[17] P.C.Leung, L.H.Skolnik, W.P.Allred, W.G.Spitzer [J. Appl. Phys. (USA) vol.43 (1972) p.4096]

[18] J.Woodhead, R.C.Newman, A.K.Tipping, J.B.Clegg, J.A.Roberts, I.Gale, [J. Phys. D (GB) vol.18 (1985) p.1575]

[19] M.R.Brozel, R.C.Newman, B.Ozbay [J. Phys. C. (GB) vol.12 (1979)

p.L785]
[20] W.M.Theis, W.G.Spitzer [J. Appl. Phys. (USA) vol.56 (1984) p.890]
[21] W.G.Spitzer [in "Advances in Solid State Physics" vol.1 Pergamon/Vieweg, New York (1971) p.44]
[22] M.R.Brozel, K.Laithwaite, R.C.Newman, B.Ozbay [J. Cryst. Growth (Netherlands) vol.50 (1980) p.619]
[23] O.G.Lorimor, W.G.Spitzer [J. Appl. Phys. (USA) vol.37 (1966) p.2509]
[24] K.Laithwaite, R.C.Newman, J.F.Angress, G.A.Gledhill [Inst. Phys. Conf. Ser. No.33a, (1977) p.133]
[25] S.R.Morrison, R.C.Newman, F.Thompson [J. Phys. C (GB) vol.5 (1972) p.L46]
[26] M.E.Levy, W.G.Spitzer [J. Phys. C (GB) vol.6 (1973) p.3223]
[27] O.G.Lorimor, W.G.Spitzer [J. Appl. Phys. (USA) vol.38 (1967) p.3008]
[28] J.F.Angress, G.A.Gledhill, R.C.Newman [J. Phys. & Chem. Solids (GB) vol.41 (1980) p.341]
[29] E.S.Koteles, W.R.Datars [Can. J. Phys. (Canada) vol.54 (1976) p.1676]
[30] W.Cochran, S.J.Fray, F.A.Johnson, J.E.Quarrington, N.J.Williams, [J. Appl. Phys. (USA) vol.32 (1961) p.2102]
[31] Landolt-Bornstein vol.17 Group III, "Crystal and Solid State Physics" pub. Springer-Verlag (1982)
[32] R.Braunstein, E.O.Kane [J. Phys. & Chem. Solids (GB) vol.23 (1962) p.1423]
[33] R.A.Chapman, W.G.Hutchinson [Phys. Rev. Lett. (USA) vol.18 (1976) p.443]
[34] D.J.Ashen, P.J.Dean, D.T.J.Hurle, J.B.Mullins, A.M.White [J. Phys. & Chem. Solids (GB) vol.36 (1975) p.1041]
[35] F.Willmann, M.Blatte, H.J.Queisser, J.Treusch [Solid State Commun. (USA) vol.9 (1971) p.2281]
[36] H.J.Queisser, C.S.Fuller [J. Appl. Phys. (USA) vol.37 (1966) p.4895]
[37] G.M.Martin, M.J.Verheijke, J.A.T.Jansen, G.Poiblaud [J. Appl. Phys. (USA) vol.50 (1979) p.467]
[38] G.M.Martin, G.Jacob, G.Poiblaud, A.Goltzene, C.Schwab [Inst. Phys. Conf. Ser.No.59 (1980) p.281]
[39] J.M.Baranowski, M.Grynberg, E.M.Magerramov [Phys. Status Solidi b (Germany) vol.50 (1972) p.433]
[40] D.G.Andrianov, A.S.Savelev, N.I.Suchkova, E.P.Rashevskaya, M.A.Filippov [Sov. Phys. Semicond. (USA) vol.11 (1977) p.426]
[41] H.Ennen, U.Kaufmann, J.Schneider [Solid State Commun. (USA) vol.34 (1980) p.603]
[42] J.M.Baranowski, J.W.Allen, G.L.Pearson [Phys. Rev. (USA) vol.160 (1967) p.627
[43] A.M.Hennel, S.M.Uba [J. Phys. C (GB) vol.11 (1978) p.4565]
[44] G.K.Ippolitova, E.M.Omelyanovskii [Sov. Phys. Semicond. (USA) vol.33 (1975) p.156]
[45] B.Clerjaud, A.M.Hennel, G.Martinez [Solid State Commun. (USA) vol.33 (1980) p.983]
[46] A.V.Vasil´ev, G.K.Ippolitova, E.M.Omelyanovskii, E.M.Ryskin [Sov. Phys. Semicond. vol.10 (1976) p.713]
[47] K.E.Elliott, D.E.Holmes, R.T.Chen, C.G.Kirkpatrick [Appl. Phys. Lett. (USA) vol.40 (1982) p.898]
[48] G.M.Martin, A.Mitonneau, A.Mircea [Electron. Lett. (GB) vol.13 (1977) p.191]
[49] G.M.Martin [Appl. Phys. Lett. (USA) vol.39 (1981) p.747]

[50] G.M.Martin [Phys. Scr. (Sweden) vol.T1 1982) p.38]
[51] R.F.Kirkman, R.A.Stradling, P.J.Lin-Chung [J. Phys. C (GB) vol.11 (1978) p.419]
[52] M.R.Brozel [EMIS Datareview RN=15401 (Nov 1985)] (discusses EL2 mapping using 2 dimensional IR absorption measurements)
[53] K.Ishida, A.Yahata, T.Kikuta [Jpn. J. Appl. Phys. (Japan) vol.24 (1985) p.L250]
[54] W.Drozdzewicz, A.M.Hennel, Z.Wasilewski, B.Clerjaud, F.Gendron, C.Porte, R.Germer [Phys. Rev. B (USA) vol.29 (1984) p.2438]
[55] B.Clerjaud [J. Phys. C. (GB) vol.18 (1985) p.3615]
[56] M.Kaminska, M.Skowronski, J.Lagowski, J.M.Parsey, H.C.Gatos [Appl. Phys. Lett. (USA) vol.43 (1983) p.302]

10.2 TWO-PHOTON IR ABSORPTION COEFFICIENT OF GaAs

by I.W.Boyd

July 1985
EMIS Datareview RN=15211

A significant amount of work has gone into the absorption processes in GaAs and in particular the two-photon absorption mechanism has received considerable attention. There are two key review papers on the field to which the readers attention is drawn [1,2]. The paper by Seiler et al [1] tabulates in chronological order the main details of 36 previous publications in the field, both calculations and theory. The modified Pidgeon-Brown model [3,4] is shown to quantitatively agree with the spectra obtained.

The paper of Bechtel and Smith [2] summarises the results of experimental absorption measurements that have resulted in the estimation of the two-photon absorption coefficient in GaAs at a wavelength of 1.06 micron. More recent measurements have been published by Lotem et al [5], Bosacchi and Bessey [6], Stewart and Bass [7], and by Boggess et al [8]. The two photon coefficient at this wavelength has been measured to be more than 6000 cm/GW and to be as low as 22 cm/GW.

It now appears that the true value of this coefficient is quite close to the lower end of this range. There is in fact a chronological trend of the measured value towards about 30 cm/GW, and this is directly a consequence of improved knowledge of the behaviour of the laser radiation within the GaAs itself. The two main problems in isolating the two-photon coefficient, beta, were:

(1) The strong influence of free carrier absorption in the material. The routine availability of picosecond duration radiation over the last decade has resulted in more favourable experimental circumstances in which to perform these intensity dependent studies.

(2) There is now a rapidly growing understanding of the strong nonlinear refraction of the laser beam that accompanies the nonlinear absorption during these optical absorption experiments. This phenomenon which can be more accurately described as self-defocussing, actually causes the transmitted radiation to rapidly diverge after leaving the sample and to miss the collecting detector. In fact, optical limiting devices have been constructed using this effect [8].

In detail, it has been found that the actual number of free carriers generated by two-photon absorption can quite readily modify the refractive index of the material itself, and therefore the propagation of the remainder of the laser radiation. This now well known self-action phenomenon leads to a negative contribution to the refractive index across the laser-excited region. In the centre, where the carrier plasma is most dense, the refractive index is reduced by the largest amount. One may consider this effect to be similar to the introduction of a negative lens into the material that rapidly defocusses the laser beam. This `fan-out´ of the radiation over an area much larger than conventional linear propagation formulae would predict has in the past resulted in a lot of radiation actually transmitted by the GaAs missing the detector. Therefore, a misleadingly large absorption coefficient was inferred. In more recent times, where researchers have been aware of such problems and taken experimental precautions against these effects, the much lower values of the two-photon coefficient have been measured.

Another problem in extracting from transmission measurements has involved the isolation of contribution to the material absorption due to free carrier absorption. Since two-photon absorption is an intensity dependent process, then in order to obtain an increased contribution to the total absorption by two-photon absorption it is advantageous to use laser pulses as short as possible. For equivalent amounts of energy, a shorter laser pulse will induce more two-photon absorption than a longer pulse, while the total amount of free-carrier absorption will remain approximately constant. Therefore, picosecond pulses are prefered to nanosecond pulses for these measurements.

In conclusion, the value of beta for GaAs at one micron is quite reasonably averaged out amongst several of the most recent experiments to be 30 -+ 8 cm/GW.

REFERENCES

[1] D.G.Seiler, C.L.Littler, D.Heiman [J. Appl. Phys. (USA) vol.57 (1985) p.2191]

[2] J.H.Bechtel, W.L.Smith [Phys. Rev. B (USA) vol.13 (1976) p.3515]

[3] C.R.Pidgeon, R.N.Brown [Phys. Rev. (USA) vol. 146 (1966) p.575]

[4] M.Weiler, R.L.Aggarwal, B.Lax [Phys. Rev. B (USA) vol.17 (1978) p.3269]

[5] H.Lotem, J.H.Bechtel, W.L.Smith [Appl. Phys. Lett. (USA) vol.28 (1976) p.389]

[6] B.Bosacchi, J.S.Bessey [J. Appl. Phys. (USA) vol.49 (1978) p.4609]

[7] A.F.Stewart, M.Bass [Appl. Phys. Lett. (USA) vol.37 (1980) p.1040]

[8] T.F.Boggess, A.L.Smirl, S.C.Moss, I.W.Boyd, E.W.van Stryland [IEEE J. Quantum Electron. (USA) vol.21 (1985) p.482]

CHAPTER 11

PHOTOCONDUCTIVITY SPECTRA OF GALLIUM ARSENIDE

11.1 IR Photoconductivity spectra of bulk SI GaAs

11.2 Far-IR photoconductivity spectra of shallow donors in GaAs epilayers: General remarks

11.3 Far-IR photoconductivity spectra of shallow donors in LPE GaAs

11.4 Far-IR photoconductivity spectra of shallow donors in VPE GaAs

11.5 Far-IR photoconductivity spectra of shallow donors in MOCVD GaAs

11.6 Far-IR photoconductivity spectra of shallow donors in MBE GaAs

11.7 Far-IR photoconductivity spectra of shallow donors in GaAs doped by neutron transmutation

11.1 IR PHOTOCONDUCTIVITY SPECTRA OF SI BULK GaAs

by D.C.Look

January 1985
EMIS Datareview RN=15480

We will define semi-insulating (SI) GaAs as material with greater than 10*7 ohm cm resistivity. The only KNOWN impurities or defects which produce such resistivities are Cr (Cr++/Cr+++ transition) and EL2 (possibly related to As+++(Ga)/As++++(Ga) transition). Thus, we will discuss material dominated by one or both of these imperfections.

Photoconductivity (PHC) can occur by a direct transition of an electron or hole from a bound imperfection state to the relevant band, or by an indirect transition to a higher-lying state and then to a band. The former process is characterized by a "threshold" in the PHC spectrum, and the latter process, by a "peak". Threshold and peak energies from several studies are given in the table below. It is assumed samples doped with oxygen are really controlled by the defect EL2.

To summarize the data for Cr-doped SI GaAs, thresholds are observed at 0.5 and 0.7 eV, varying with temperature, and peaks at 0.82 and 0.88 eV, independent of temperature. The threshold at 0.7 eV is attributed to the transition Cr++ to Cr+++ + e(CB) while the peaks at 0.82 and 0.88 eV are associated with the intracentre process

$$Cr^{++}(^5T_2) \text{ to } Cr^{++}(^5E) \text{ to } Cr^{+++} + e(CB)$$

The situation is less certain for EL2, since this defect has not been unambiguously identified, but the peaks at 1.04 and 1.17 eV are probably associated with an intracenter transition which somehow provides an electron to the conduction band possibly through an Auger process [15].

Deep Centre	Temp. deg K	Threshold (eV)	Peaks (eV)	Ref
Cr	400	0.63	-	[1]
Cr	77	0.76	0.87	[2]
	300	below 0.68		
Cr	300		0.87	[3]
Cr	90-200		0.9	[4]
Cr	78		0.88	[5]
	300		0.90	
Cr	82	0.73	0.88	[6]
	295	0.68	0.88	
Cr	78	0.7	0.88	[7]
	295	0.5,0.7	0.87	[8]

Cr	2	0.8	0.9	[9,10]
Cr	77	0.5,0.7	0.9	[11]
Cr	2	0.8	0.82	[12]
Cr	80	0.75	0.87	[13]
EL2	82	0.8	1.0	[14]
	295	0.7,1.0		
EL2	4	0.82,1.03,1.27	1.04,1.17	[15]

The magnitude of the PHC for Cr-doped SI GaAs at 295 deg K ranges from about 10*-8 - 10*-7 mho cm*-1 at the 0.88 eV peak, and from 10*-7 - 10*-6 mho cm*-1 in the bandgap region for a photon flux of 5 X 10*14 cm*-2 sec cm*-1 [8]. The 77 deg K response in the peak region can be about two orders of magnitude higher [6]. Carrier lifetimes are typically 10*-9 - 10*-8 sec at 295 deg K [1] but can be much longer at low temperatures due to carrier trapping. Spatial fluctuations of potential can influence this process [16]. The PHC can also be quenched by applying a second light source. At 82 deg K the quenching can be nearly 100% with a 1.1 eV source [6].

For SI GaAs dominated by EL2 the overall magnitudes of the PHC spectra are not greatly different from those of the Cr-doped material, although the shapes are more gradual, especially at 295 deg K [14]. Perhaps the most distinctive feature of low-temperature EL2-related spectra is a quenching band at 1.1-1.2 eV [13,14] which presumably is related to a metastable state of EL2. The quenching band disappears above 125 deg K [14].

REFERENCES

[1] B.R.Holeman, C.Hilsum [J. Phys. Chem. Solids (GB) vol.22 (1961) p.19]

[2] G.A.Allen [J. Phys. D (GB) vol.1 (1968) p.593]

[3] D.R.Heath, P.R.Selway, C.C.Tooke [J. Phys. D (GB) vol.1 (1968) p.29]

[4] E.M.Omel´yanovskii, L.Ya-Pervova, E.P.Rashevskaya, V.I.Fistul´ [Sov. Phys. Semicond. (USA) vol.5 (1971) p.484]

[5] A.A.Gutkin, A.A.Lebedev, G.N.Talalakin, T.A.Shaposhnikova [Sov. Phys. Semicond. (USA) vol.6 (1972) p.928]

[6] A.L.Lin, R.H.Bube [J. Appl. Phys. (USA) vol.47 (1976) p.1859]

[7] E.M.Omel´yanovskii, A.N.Pantyukhov, L.Ya.Pervova, V.I.Fistul´, Yu.A.Visil´ev [Sov. Phys. Semicond. (USA) vol.9 (1976) p.1267]

[8] D.C.Look [Solid State Commun. (GB) vol.24 (1977) p.825]

[9] L.Eaves, T.Instone, M.J.Cardwell, P.R.Jay, D.J.Stirland, H.C.Wright [Proc. Sec. "Lund" Int. Conf. (St. Maxime, France, 1979]

[10] L.Eaves, T.Englert, T.Instone, C.Uihlein, P.J.Williams, H.C.Wright ["Semi-Insulating III-V Materials", Nottingham [ed G.J.Rees (Shiva Publishing (GB) 1980) p.145]

[11] J.C.Manifacier, J.Bonnafe, J.Jimenez-Lopez, J.P.Fillard [Proc. "Lund" Int. Conf., St. Maxime, France, 1979]

[12] L.Eaves, P.J.Williams, C.Uihlein [J. Phys. C (GB) vol.14 (1981) p.L693]
[13] J.Jimenez, M.A.Gonzalez, J.A.DeSaja, J.Bonnafe [J. Mat. Sci. (GB) vol.19 (1984) p.1207]
[14] A.L.Lin, E.Omel'yanovskii, R.H.Bube [J. Appl. Phys. (USA) vol.47 (1976) p.1852]
[15] N.Tsukada, T.Kikuta, K.Iishida [Jpn. J. Appl. Phys. (Japan) vol.24 (1985) p.L302]
[16] B.Pistoulet, P.Girard, G.Hamamdjian [J. Appl. Phys. (USA) vol.56 (1984) p.2275]

11.2 FAR-IR PHOTOCONDUCTIVITY SPECTRA OF SHALLOW DONORS IN GaAs EPILAYERS: GENERAL REMARKS

by R.A.Stradling

January 1985
EMIS Datareview RN=15481

In GaAs the shallow donor energies are small and rather well described by hydrogenic theory so that the chemical shifts of the ground-state energies of the individual donor species amount to only a few per cent or about 0.1 meV (approx. 1 cm*-1). Far-infrared spectroscopy of the transitions between bound states provides a very high-resolution technique for studying chemical shifts and identifying residual contaminants. A number of mechanisms, notably the `photothermal´ effect [6,10] enable bound-to-bound transitions to be detected by photoconductivity, with the result that very high sensitivity can be achieved. Consequently very thin high-purity samples can be easily studied. (Stillman et al [10] also provide an excellent review of early work on photoconductivity from donors in GaAs). The residual impurity concentrations are usually in the 10*14 - 10*15/cm*3 range, even in the best material presently available. The resultant line widths are such that the central cell structure is barely resolved, even in the highest-purity samples. A variety of techniques can be employed to narrow the line including the following:

(i) The application of a magnetic field narrows the lines through the shrinkage of the electronic wavefunction. This has the additional beneficial effect of increasing the central-cell splittings.

(ii) Illumination with bandgap radiation generates electrons and holes which are captured and neutralise a proportion of the ionised donor and acceptor sites, thereby reducing the random electric field which broadens the lines [7]. However, because donar - acceptor pair recombination is very rapid in direct-gap semiconductors, a significant line narrowing by intrinsic radiation is only found in the highest-purity samples.

(iii) A reduction in temperature between 4 and 2 deg K can frequently narrow the lines, even though the total number of ionised sites material shun donor sites located near to ionised acceptors at the lowest temperatures [4,8] and the lines are frequently found to narrow on reducing the temperature below 4 deg K because of the increasingly dipolar character of the electric fields [5]. When this line narrowing is found, a further substantial improvement is often observed on introducing intrinsic radiation.

(iv) The application of hydrostatic pressure increases the bandgap and hence also the effective mass, donor binding energy and chemical shifts, thereby also improving the resolution. The application of pressure of 8.5 kbar also produces an anticrossing of the deepest lying of the shallow donors with states associated with higher-order conduction band minima [12]. At higher pressures the donors are lost from the experimental spectrum because of the decrease in oscillator strength, thereby simplifying the identification of the remaining donors. In addition, a remarkable line narrowing is sometimes found in the region of crossover.

Care must be taken in interpreting the far infrared spectrum because all the fine structure does not arise necessarily from different donor species. In addition, the overlap with adjacent, incompletely resolved lines can act to shift the line positions by an appreciable fraction of the observed line width. There also exists the possibility of distortion of individual line shapes by such effects as `channel spectra´ (i.e. standing-wave resonances within the sample) and polarisation phenomena produced inadvertently by the various optical components and the sample itself. There can be unresolved substructure due to non-parabolicity effects arising from the spin-up and spin-down impurity states having slightly different transition energies.

While permitting very high sensitivity, the use of photoconductivity for detection introduces other possibilities for the distortion of the lines. It has been noted that the line positions observed in photoconductivity and in absorption may differ by up to half a linewidth when tunnelling between adjacent impurity sites rather than the normal photothermal mechanism is responsible for the generation of the photoconductivity signal [3]. It should also be noted that any photoconductive line represents a convolution of absorption energy relaxation and recombination processes. In addition, the use of intrinsic light, while being beneficial in reducing the number of ionised sites and hence the line width, introduces a further complication. The current paths will certainly be altered by the presence of photoexcited carriers, thereby increasing the contributions from regions close to the surface. In an inhomogeneous sample this can change the relative heights of individual components.

Very recently a `notch effect´ has been reported for strong lines where the centre of some lines appears to invert. The mechanism involves the change of dielectric function close to a strong absorption peak. Great care must be taken to avoid misidentifying a `notched´ line as two central-cell components [11].

Important pointers to the validity of interpreting the component peaks of any 1s - 2p transitions in terms of an equivalent number of donor species lie within the tracking of the components with measuring frequency or magnetic field and the reproducibility of individual donor lines from sample to sample and of different 1s - 2p lines in the same sample.

The so-called `signature curve´ of energy against magnetic field has been studied in detail by Asfar et al [1] for a number of different donors.

The shallow donors can exist also in a negatively charged state (D- ion) and recently extensive structure due to D- ions has been reported in VPE and MOCVD samples by Armistead et al [2]. Small differences are found which are attributable to differing chemical shifts from the different donor species introduced by the two techniques. The effects of temperature, optical excitation compensation and DC electric field are studied [9,13]. A particular novelty of D- spectra in GaAs is that spectra can be followed in the high magnetic field regime where the dimensionless parameter characterising the field (i.e. hf(c)/2R*) is greater than unity.

REFERENCES

[1] M.N.Afsar, K.J.Button, G.L.McCoy [Proc. Int. Symp. on GaAs (Inst. Phys. Conf. Series vol.56) (1980) p.547]

[2] C.J.Armistead, S.P.Najda, R.A.Stradling, J.C.Maan [Solid State Commun (USA) vol.53 (1985) p.1109]

[3] A.C.Carter, G.P.Carver, R.J.Nicholas, J.C.Portal, R.A.Stradling [Solid State Commun. (USA) vol.24 (1977) p.55]

[4] J.Golka [Phys. Rev. (USA) vol.8 (1973) p.3895]

[5] J.Golka, J.Trylski, M.S.Skolnick, R.A.Stradling, Y.Couder [Solid State Commun. (USA) vol.22 (1977) p.623]

[6] S.M.Kogan, B.I.Sedunov [Sov. Phys.-Solid State (USA) vol.8 (1966) p.1898]

[7] D.M.Larsen [Phys. Rev. (USA) vol.8 (1973) p.535]

[8] D.M.Larsen [Phys. Rev. B (USA) vol.11 (1975) p.3904]

[9] S.J.Najda [Ph.D Thesis University of St. Andrews (1985)]

[10] G.E.Stillman, C.M.Wolfe, J.O.Dimmock [Semicond. & Semimet. (USA) vol.12 (1977) p.169]

[11] G.E.Stillman, T.S.Low, B.Lee [Solid State Commun. (USA) vol.53 (1985)]

[12] R.A.Stradling [Festkorperprobleme (Germany) vol.XXV (1985) p.591]

[13] C.Trager, D.A.Cowan, R.A.Stradling [Physica B (Netherlands) vol.134 (1985) p.250]

11.3 FAR-IR PHOTOCONDUCTIVITY SPECTRA OF SHALLOW DONORS IN LPE GaAs

by R.A.Stradling

January 1986
EMIS Datareview RN=15483

Liquid phase epitaxy (LPE) can produced material with liquid nitrogen mobilities as good as that grown by VPE. Sulphur (X2) is generally the dominant residual donor. In contrast to VPE the donor line attributed to silicon (X1), if detected at all, appears at concentrations an order of magnitude lower and donor X3 has not been observed as a contaminant in undoped material. Two other donor lines are observed regularly in high purity material. The first of these has been identified as Sn by Stradling et al [1,2], and the second is thought to be lead by [3,2]. In nominally undoped material grown at Cornell University the Sn line has been reported to be as strong as that due to S(X2) ref [4], whereas in samples prepared at Standard Telecommunications Laboratories or at the Max Planck Institute at Stuttgart, the Sn line was always considerably weaker than X2, while Pb varied from being markedly stronger than Sn and comparable with X2 to being an order of magnitude weaker than X2 and even weaker than X1 (Si), although comparable with Sn.

REFERENCES

[1] R.A.Stradling, L.Eaves, R.A.Hoult, N.Miura, P.E.Simmonds [Proc. Int. Conf. on GaAs (Inst. Phys. Conf. Series) (1972) p.65]

[2] C.M.Wolfe, G.E.Stillman, D.M.Korn [Proc. Int. Sym. on GaAs (Inst. Phys. Conf. Series b vol.33) p.120]

[3] R.A.Coles, R.A.Hoult, R.F.Kirkman, R.A.Stradling [J. Phys. D (GB) vol.11 (1978) p.945]

[4] T.S.Low, G.E.Stillman, C.M.Wolfe [Proc. Int. Symp. on GaAs (Inst. Phys. Conf. Series) (1981) p.143]

11.4 FAR-IR PHOTOCONDUCTIVITY SPECTRA OF SHALLOW DONORS IN VPE GaAs

by R.A.Stradling

January 1986
EMIS Datareview RN=15484

The high purity and control achieved by chloride VPE ($AsCl_3$-Ga-H_2) vapour phase system enabled much of the early donor identification to be achieved with this material. However, hydride VPE (AsH_3-Ga-HCl-H_2) had advantages in the growth of alloys and liquid nitrogen mobilities as good as those achieved with the chloride technique can now be obtained.

Samples prepared by both the hydride and the chloride VPE processes generally show the same set of donors X1, X2 and X3, if the material has been prepared without intentional doping. As shown by Skromme et al [1], reducing the Ga/As ratio enhances the intensities of X1 and X3 with respect to X2 which is consistent with the interpretation of X1 as substitutional Si reached through studies with MBE material (see ref [2] on MBE samples) and with X3 as a Group IV donor (carbon or germanium) but see the discussion [3] on the alternative possibility that complexes involving Ga vacancies or As defects could be involved. X2 was initially identified as Si. However, when material with liquid nitrogen mobilities in excess of 10^5 $cm^2/V.s$ was produced by the MOCVD growth technique, it was found that significant amounts of the X2 donor were not introduced with this technique. Consequently, back-doping experiments could be performed with sulphur which produced a strong X2 peak without substantially broadening or shifting the other components in the central cell spectrum. Skromme et al [1] found that sulphur was always the dominant donor with hydride material. With the conventional chloride process, Ozekhi et al [4] found that either silicon or sulphur could become the most abundant donor. These workers also studied samples in which N_2 was employed as the carrier gas instead of H_2. These samples showed no sign of X1. Ozekhi et al [4] also suggested that the initial assignment of X1 as associated with a gallium vacancy and X2 as silicon was erroneous, as was subsequently confirmed by the back-doping experiments with MBE and MOCVD material. Skromme et al [1] find the same residual donors (Si and S) and acceptors (Zn and C) in InP samples produced in the same hydride growth system and suggest that common sources for the contaminants are involved.

Much of the early donor identification was performed with material grown with the $AsCl_3$:Ga:H_2 growth technique and back-doped by this method. Wolfe et al (1976) doped with Ge, Sn, Pb, Se and Te also attempted to dope with oxygen by adding Ga_2O_3 to the melt. The positions of the Sn and Se donors agreed with the results of Stradling et al [5] for LPE material. The chemical shifts derived for the Ge and Pb donors were similar to the results of Cooke et al [6], but differences of the order of 0.02 meV in the reported values probably arise from broadening effects. The order of the chemical shifts and their magnitude at zero magentic magnetic field are believed to be as follows Low et al [7], Armistead et al [8]: Pb (0.03 meV); Si (0.05 meV); Se (0.06 meV); Sn (0.07 meV); S (0.1 meV); Te (0.14 meV); Ge (0.18 meV).

REFERENCES
----

[1] B.J.Skromme, T.S.Low, T.J.Roth, G.E.Stillman, J.K.Kennedy, [J. Electron. Mater. (USA) vol.12 (1983) p.433]

[2] R.A.Stradling [EMIS Datareview RN=15482 "Far-IR photoconductivity of shallow donors in MBE GaAs"]

[3] R.A.Stradling [EMIS Datareview RN=15485 "Far-IR photoconductivity of shallow donors in MOCVD GaAs"]

[4] M.Ozekhi, K.Kitahara, K.Nakai, A.Shibatomi, K.Dazai, S.Okawa, O.Rynzan [Jpn. J. Appl. Phys. (Japan) vol.16 (1977) p.1617]

[5] R.A.Stradling, L.Eaves, R.A.Hoult, N.Miura, P.E.Simmonds, [Proc. Int. Symp. on GaAs (Inst. Phys. Conf. Series vol.17) (1972) p.65]

[6] R.A.Cooke, R.A.Hoult, R.F.Kirkman, R.A.Stradling [J. Phys (GB) vol.11 (1978) p.945]

[7] T.S.Low, G.E.Stillman, C.M.Wolfe [Proc. Int. Symp. on GaAs (Inst. Phys. Series vol.63)]

[8] C.J.Armistead, P.Knowles, S.P.Najda, R.A.Stradling [J. Phys. (GB) vol.17 (1984) p.6415 }

11.5 FAR-IR PHOTOCONDUCTIVITY SPECTRA OF SHALLOW DONORS IN MOCVD GaAs

by R.A.Stradling

January 1986
EMIS Datareview RN=15485

The first reported photoconductive spectrum for MOCVD (alkyl) material was with a relatively impure sample (liquid nitrogen mobility of 50,000 cm*2/V.s) which nevertheless showed central cell structure consisting of two peaks whose amplitudes were of the ratio of 10 to 7 which were well resolved because the two donor species involved happened to be the deepest and shallowest of the shallow donor contaminants frequently found in epitaxial material [1]. The deepest and slightly more abundant X3 (in the nomenclature of Stillman [2]) was originally identified as carbon but may be germanium in view of its appearance when neutron transmutation is used to introduce donors [3]. Hess et al [4], however, note that X3 is frequently found as the dominant donor in samples where photoluminescence shows only carbon and occasionally Zn acceptors and no significant concentrations of germanium acceptors. They also comment that the absence of germanium acceptors could arise from the high As to Ga ratios employed in growth which could force the amphoteric group IV impurities to take up the Ga site and act as acceptors. By the same argument, however, concentrations of gallium vacancies and arsenic antisite defects are also accepted and the possibility that X3 is a complex involving these defects cannot be excluded.

The shallower donor reported by Cooke et al [1] is thought to be Pb. With a series of samples of different origin Low et al [5] found X3 to be the dominant donor in all cases, with only traces of Pb together with Si, S and Sn. In extremely high mobility layers obtained by fractional distillation of both the AsH3 and trimethylgallium, Hess et al [4] show that donor X3 accounts for more than 90% of the total shallow donors present.

This control of the donor species present in high purity MOCVD enabled Low et al [6] to show that donor X2 was associated with sulphur which is generally only present as a minority donor in MOCVD material. Sulphur doping was accomplished by introducing H2S into the gas steam. It should be noted, however, that the back-doped samples produced very broad photothermal spectra with the central cell structure of the 1s-2p minus line shifted somewhat to lower energy compared with the normal positions of the X2 and X3 donors and the peaks on the 1s-2p plus transition shifted correspondingly to higher energy by the interaction with charged impurities as discussed by Larsen. Low et al [6] also point out that in an earlier report with a VPE sample the S donor lay between the X2 and X3 donors and this result almost certainly arose because of distortion of the X2 line shape by the `notch effect´, which arises with strong lines and results from a variation of the optical constants in the region of such lines as discussed by Stillman et al [2].

REFERENCES
----

[1] R.A.Cooke, R.A.Hoult, R.F.Kirkman, R.A.Stradling [J. Phys. D (GB)

vol.11 (1978) p.945]
[2] G.E.Stillman, T.S.Low, B.Lee [Solid State Commun. (USA) vol.53 (1985)]
[3] R.A.Stradling [EMIS Datareview RN=15486 (January 1986) "Far-IR photoconductivity spectra of shallow donors in GaAs doped by neutron transmutation "]
[4] K.L.Hess, P.D.Dapkus, H.M.Manasevit, T.S.Low, B.J.Skromme, G.E.Stillman [J. Electron. Mater. (USA) vol.11 (1982) p.1115]
[5] T.S.Low, G.E.Stillman, C.M.Wolfe [Proc. Int. Symp. on GaAs (Inst. Phys. Conf. Series) (1981) vol.63 p.143]
[6] T.S.Low, G.E.Stillman, T.Nakanisi, T.Udagawa, C.M.Wolfe [Appl. Phys. Lett. (USA) vol.41 (1982) p.183]

11.6 FAR-IR PHOTOCONDUCTIVITY SPECTRA OF SHALLOW DONORS IN MBE GaAs

by R.A.Stradling

January 1986
EMIS Datareview RN=15482

The very low donor backgrounds achievable in the present generation of molecular beam epitaxial reactors has provided the clue to the identity of the "X1" donor species [8] almost invariably found in VPE grown material. Previously this donor was thought to be sulphur, but by intentional doping of p-type MBE material X1 has been positively identified as silicon by Low et al [6]. This group also reports the presence of a background of sulphur at a level of 2 X 10*14/cm*3 in layers grown in a Varian Gen II with no other intentional dopant than silicon and an elemental As source. With material grown using an elemental As source and using Sn as the n-type dopant, Si is present, but at a lower concentration than S; whereas if AsH3 is cracked in a SiO2 furnace to provide As, the amplitude of the Si peak increases with increasing furnace temperature.

Traces of Pb were found in several samples by Low et al [4,6]. Kuchar et al [3] find only the Si intentionally introduced in a high purity sample grown by a Gen II but report sulphur to be dominant in a sample produced by an earlier model of MBE reactor (Varian 360). The latter result was confirmed by Armistead et al [1] who also found sulphur to be the most abundant donor in material produced by a similar reactor without intentional doping.

The quality of material which can be produced by MBE techniques is such that shallow donor spectra can now be obtained with fine structure due to different donor states in thin quantum well material. The donor energies are modified in quantum wells because of:

(i) the quantum confinement effect which deepens the donor energies; and
(ii) the change in symmetry of the wavefunctions imposed by the potential barrier at the interface.

For a donor located at the interface in a single heterojunction the ground state wavefunction is predominantly 2p rather than 2s because of symmetry considerations. The binding energy is therefore reduced by a factor of approximately four in the absence of a magnetic field Initial results as reported first by Jarosik et al [2] and reviewed by Shanabrook [7] show three distinct donor lines in a magnetic field. Two of these are thought to be due to donors located at the centre and edges of the well. The origin of the third peak is unclear, but could arise from a chemically-shifted donor.

REFERENCES
----

[1] C.J.Armistead, A.M.Davidson, P.Knowles, S.P.Najda, R.A.Stradling, R.J.Nicholas, S.J.Sessions [Physics (USA) vol.177 (1983) p.289]
[2] N.C.Jarosik, B.D.McCombe, K.R.Y.Shanabroo, J.Comas, J.Ralston, G.Wicks [Phys. Rev. Lett. (USA) vol.54 (1985) p.1283]
[3] F.Kuchar, R.Meisels, G.Weimann, H.Burkhard [Appl. Phys. A

(Germany) vol.33 (1984) p.83]
[4] T.S.Low, G.E.Stillman, C.M.Wolfe [Proc. Int. Conf. on GaAs (Inst. Phys. Conf. Series vol.63) (1981) p.143]
[5] T.S.Low, G.E.Stillman, D.M.Collins, C.M.Wolfe, S.Tiwari, L.F.Eastman [Appl. Phys. Lett. (USA) vol.40 (1982) p.1034]
[6] T.S.Low, G.E.Stillman, A.Y.Cho, A.R.Calawa [Appl. Phys. Lett. (USA) vol.40 (1982) p.611]
[7] R.V.Shanabrook [Invited talk at Int. Conf. on Electronic Properties of Two Dimensional Systems 1986]
[8] G.E.Stillman, C.W.Wolfe, J.O.Dimmock [Semicond. & Semimet. (USA) vol. 12 (1977) p.169]

11.7 FAR-IR PHOTOCONDUCTIVITY SPECTRA OF SHALLOW DONORS IN GaAs DOPED BY NEUTRON TRANSMUTATION

by R.A.Stradling

January 1986
EMIS Datareview RN=15486

Neutron transmutation doping (NTD) of semiconductors provides a convenient method for introducing donor impurities in a controlled manner. The technique involves bombarding the crystal with a flux of thermal neutrons. A small percentage of the lattice atoms capture a neutron and decay into shallow donors. This doping method is well understood in the elemental semiconductors, Si and Ge, and is now applied on a large scale in the semiconductor industry.

NTD doping of GaAs by thermal neutrons was first reported by Mirianoshvilli and Nanobashvili [1]. Thermal neutrons are captured by a small proportion of the GaAs lattice nuclei, depending on the relative abundance of the various isotopes of Ga and As and their different capture cross-sections. The irradiated atoms decay from an excited state to form a stable donor atom: i.e. a lattice atom is transmuted to a donor impurity. The radioactive decay rate probabilties and relative donor yields are well documented (Vesaghi [2]), 40% of the donors formed are Ge-70, Ge-72 and 60% are Se-76, due to the different thermal neutron capture cross-sections of the respective lattice atoms. After irradiation, a considerable amount of crystalline damage occurs and the samples are found to have very high resistivities. Thermal annealing of the radiation damaged crystal is required in order to activate a high proportion of the transmuted atoms as electrically active shallow donors.

FIR photoconductivity from shallow donors in NTD GaAs has been reported by Stoelinga et al [3] and by Afsar et al and also has been studied by Najda et al [4]. All groups agree that the intensity in the region of the X3 donor is considerably enhanced and attribute this to the introduction of germanium. Se is either not observed or appears less strongly than would be expected from the respective cross-sections for neutron capture of Ge and Se.

REFERENCES

[1] S.M.Mirianoshvilli, D.I.Nanobashvili [Sov. Phys. Semicond. (USA) vol.4 (1971) p.1612]
[2] M.A.Vesaghi [Phys. Rev. B (USA) vol.25 (1982) p.5436]
[3] J.M.M.Stoelinga, D.M.Larsen, W.Walnkiewicz, R.L.Aggarwal, C.O.Bozler [J. Phys. & Chem. Solids (GB) vol.39 (1978) p.873]
[4] S.P.Najda, R.A.Stradling, F.Kuchar [to be published]

CHAPTER 12

PHOTOLUMINESCENCE SPECTRA OF GALLIUM ARSENIDE

12.1 Photoluminescence spectra of undoped SI GaAs

12.2 Photoluminescence spectra of Cr doped SI GaAs

12.3 Photoluminescence spectra of undoped LEC p-type GaAs

12.4 Photoluminescence spectra of LPE GaAs

12.5 Photoluminescence spectra of VPE GaAs

12.6 Photoluminescence spectra of MOCVD GaAs

12.7 Photoluminescence spectra of MBE GaAs

12.8 Photoluminescence spectra of Group II doped GaAs

12.9 Photoluminescence spectra of Group IV doped GaAs

12.10 Photoluminescence spectra of Group VI doped GaAs

12.1 PHOTOLUMINESCENCE SPECTRA OF UNDOPED SI GaAs

by EMIS Group

August 1985
EMIS Datareview RN=15133

Low temperature (usually 4.2 deg K) luminescence spectra of semi-insulating (SI) liquid encapsulated Czochralski (LEC) grown GaAs are used in characterising this material, which is employed as an integrated circuit substrate. In particular, they are important for determining the nature of the deep instrinsic defect EL2, situated 0.75 eV below the conduction band [1]. The compensation mechanism between EL2 and a shallow acceptor, probably residual C, is thought to be responsible for the SI properties of undoped GaAs. EL2 related photoluminescence (PL) intensity has been found to be greatest in As-rich samples [2-4].

0.65 eV PL BAND

There is a well studied PL band at 0.65 eV, also referred to as the 0.635 or 0.63 eV band. Refs [5-7] attribute the band to oxygen on As sites or an oxygen-related complex. Later work [8-15] associates this directly with the EL2 defect. However, Samuelson et al [16] throw into question the direct relationship between this band and EL2. PL spectra of samples (EL2 concn of 2 X 10*16/cm*3) before and after quenching by persistent light irradiation suggest that EL2 acts only as an intermediate step in the excitation process.

PL, infra-red absorption and resistivity investigations [14] of a large number of LEC SI GaAs wafers having either 0.65 eV or 0.80 eV as the dominant band revealed that the 0.65 eV dominant crystals have a Fermi level close to the conduction band (CB), while for the 0.80 eV dominant crystals it is further from the CB.

0.68 eV PL BAND

More recently a PL band at 0.68 eV and a 1 micron absorption band have been studied [17-21]. Measurements of PL temperature dependence [19] show the 0.68 eV band to be due to a radiative transition between the EL2 centre and the valence band. The presence of both normal and metastable states of EL2 was inferred from the occurrence of the persistent PL quenching (fatigue) effect. Tajima [21] also correlated this peak with EL2.

A scanning electron microscope cathodoluminescence study of undoped SI LEC GaAs ingots by Warwick and Brown [22] has revealed the spatial distribution of the 0.68 eV emission to a resolution of 3 microns. The authors conclude from the emission image and other experimental evidence that polygonised dislocation arrays (cell walls) attract or getter, rather than generate, the EL2 defect centres responsible for the emission.

0.80 eV PL BAND

The 0.68 eV emission has been accompanied by a broad band at 0.77 eV with peak energy 0.775 eV and a half width of 0.25 eV [20]. Windcheif et al [23] designated this peak at 0.8 eV and observed shoulders between 0.6 and 0.7 eV. It was associated with the As(Ga) antisite defect by comparison with the EPR signal of the defect. The onset of a band at 1.0 eV corresponds to the second ionisation energy of the Ga(As) double donor. Yu et al [20] concluded that possible origins of the emission were deep centres at about 0.45 or 1.07 eV from the CB edge, assigned to a doubly ionised anti-site donor As++(Ga) at E(c)-1.0 eV and a donor at E(c)-0.42 eV respectively.

The 0.80 eV PL band has been variously associated with the As(Ga) antisite defect [23], the V(Ga) defect or its complex [24] and the EL2+ charge state [14].

PL PEAK DUE TO SURFACE DAMAGE

Surface damage by saw cutting, mechanical polishing and scribing has been found to cause a new PL band at 1.4 eV in the low temperature (10 deg K) spectrum, regardless of growth method, doping or conductivity type [25]. The band results from mechanically induced radiative centres at or near the surface, although deep radiative and nonradiative centres may also be created. The damage extends 10-20 microns into the crystal and anneals out at or above 673 deg K. The damaged layer has to be removed by chemical polishing before use in devices.

NEW PL TECHNIQUE

Tajima [26] has developed a new PL method called twofold excitation modulated PL spectroscopy. A modulation technique is used to analyse variations induced by below-gap as well as above-gap excitation. This is useful for characterising deep levels. Using the technique the author determined that high concns of nonradiative deep centres (1 X 10*16/cm*3) are present in LEC but not in boat grown crystals.

REFERENCES

[1] E.M.Martin et al [J. Appl. Phys. (USA) vol.51 (1980) p.2841]

[2] T.Kikuta et al [Jpn. J. Appl. Phys. Part 2 (Japan) vol.22 no.8 (Aug 1983) p.L541-3]

[3] D.E.Holmes et al [IEEE Trans. Electron Devices (USA) vol. ED-29 (1982) p.1045]

[4] J.Lagowski et al [Appl. Phys. Lett. (USA) vol.40 (1982) p.342]

[5] W.I.Turner, G.D.Pettit, N.G.Ainslie [J. Appl. Phys. (USA) vol.34 (1963) p.3274]

[6] B.Deveaud, P.N.Favennec [Inst. Phys. Conf. Ser. (GB) no.45 (1978) p.492]

[7] P.W.Yu, D.C.Walters [Appl. Phys. Lett. (USA) vol.41 (1982) p.863]

[8] A.Mircea-Roussel, S.Makram-Ebeid [Appl. Phys. Lett. (USA) vol.38 (1981) p.1007]

[9] P.Leyral, G.Guillot [in Proc. of the Second Conf. on Semi-Insulating III-V Materials, ed. by S.Makram-Ebeid, B.Tuck (Shiva, England, 1982) p.166-71]

[10] P.Leyral, G.Vincent, A.Nouailhat, G.Guillot [Solid State Commun. (USA) vol.42 (1982) p.67]

[11] B.V.Shanabrook, P.B.Klein, E.M.Swiggard, S.G.Bishop [J. Appl.

Phys. (USA) vol.54 (1983) p.336]
[12] M.Levinson [Phys. Rev. B vol.28 (1983) p.3660]
[13] M.Tajima [Jpn. J. Appl. Phys. Part 2 (Japan) vol.23 no.9 (Sept 1984) p.L690-3]
[14] T.Kikuta, H.Emori, T.Fukuda, K.Ishida [Int. Conf. on Solid State Devices and Materials, Kobe (1984). Extended Abstracts p.173-6]
[15] B.V.Shanabrook, P.B.Klein, S.G.Bishop [Physica B (Netherlands) vol.117 (1983) p.173]
[16] L.Samuelson, P.Omling, H.G.Grimmeiss [Appl. Phys. Lett. (USA) vol.45 no.5 (1 Sept 1984) p.521-3]
[17] M.S.Skolnick, D.A.O.Hope, B.Cockayne [in Semi-Insulating III-V Materials, Kah-nee-ta ed. by D.C.Look, J.S.Blakemore (Shiva, England 1984) p.446-52]
[18] T.Kikuta, T.Katsumata, T.Obokata, K.Ishida [Inst. Phys. Conf. Ser. No.74 (1985) p.47]
[19] P.W.Yu [Appl. Phys. Lett (USA) vol.44 no.3 (Feb 1984) p.330-2]
[20] P.W.Yu [Phys. Rev. B (USA) vol.29 no.4 (Feb 1984) p.2283-5]
[21] M.Tajima [Appl. Phys. Lett. (USA) vol.46 no.5 (1 March 1985) p.484-6]
[22] C.A.Warwick, G.T.Brown [Appl. Phys. Lett. (USA) vol.46 no.6 (15 Mar 1985) p.574-6]
[23] J.Windcheif et al [Appl. Phys. (Germany) vol.A30 no.1 (Jan 1983) p.47-9]
[24] J.Lagowski, H.C.Gatos, J.M.Parsey, K.Wada, M.Kaminska, W.Walukiewicz [Appl. Phys. Lett. (USA) vol.40 (1982) p.342]
[25] V.Swaminathan, M.S.Young, R.Caruso [J. Appl. Phys. (USA) vol.57 no.4 (15 Feb 1985) p.1387-90]
[26] M.Tajima [Proc. 13th Int. Conf. on Defects in Semiconductors (Aug 1984) (Metallurgical Society of AIME, 1985) p.997-1003]

12.2 PHOTOLUMINESCENCE SPECTRA OF Cr-DOPED SI GaAs

by EMIS Group

July 1985
EMIS Datareview RN=15140

The main feature of the low temperature Cr-related photoluminescence (PL) spectra of semi-insulating (SI) GaAs:Cr is a sharp zero-phonon line at 0.839 eV accompanied by phonon bands on the low energy side. This has been well studied because of the importance of SI single crystals as substrates for integrated circuits and integrated optics, and because Cr gives rise to an acceptor level (Cr++/Cr+++) near midgap responsible for the SI properties of GaAs:Cr when the Cr concn exceeds the background donor level.

However, some controversy surrounds the precise assignment of the 0.839 eV line, although it is generally attributed to a trigonal centre. Following the first experimental evidence for this explanation presented by Killoran et al [1] and, independently, by Eaves et al [2], Barrau and co-workers [3-5] obtained high resolution optical and Zeeman spectra which suggest that the line results from the transition

^{5}E to $^{5}T_{2}$ (symmetry group Td)

whereas Picoli et al [6,7] and Eaves & Uihlein [8,9] proposed the transition

$^{5*}E$ to ^{5}E (symmetry group C3v)

Fujiwara et al [10] assume that the luminescence centre has [111] C3v axial symmetry, which appears now to be generally accepted, and conclude from systematic studies of high temperature thermal annealing of GaAs:Cr under various excess pressures of As that the centre is a complex of Cr on a Ga site with an As vacancy as the nearest neighbour.

Deveaud et al [11] report a zero-phonon line at 0.844 eV for GaAs:Si,Te:Cr samples and report the first PL excitation measurements. The line results from an internal transition of the Cr++ - Te complex. There is also a line at 0.820 eV due to the isolated Cr++. The Cr++ - Si complex was not observed because its excited state is predicted to be above the conduction band minimum.

Yu [12] gives data on bands in the 0.57 eV region after low temperature heat treatment of GaAs:Cr. Deveaud et al [11] suggest that these may be due to internal transitions at Cr+++ - donor complexes.

Residual stress in wafers subject to mechanical or thermal treatment can greatly affect the performance and reliability of LSI devices. Fujiwara et al [13] show that the residual stress can be characterised from the way it splits and shifts the 0.839 eV line. PL can also be used to determine the uniformity of wafers. Yokogawa et al [14] investigated the effect of whole ingot annealing on the uniformity of wafers cut from LEC low-Cr doped ingots. The 1.49 eV emission line was found to be intensified by a factor of up to 100 as a result of the annealing. Moreover, the difference in PL intensity between the seed and tail was greatly reduced as were the radial and axial variations of the PL intensity distribution.

Finally, PL measurements in the 1.41 eV zero phonon line have been used to study surface layers formed on SI GaAs:Cr substrates after heat treatment [15].

REFERENCES

[1] N.Killoran, B.C.Cavenett, W.E.Hagston [Proc. Semi-Insulating III-V Conf., ed. by G.J.Rees (Nottingham,1980)]

[2] L.Eaves, T.Englert, T.Instone, C.Ulhlein, P.J.Williams, H.C.Wright [Proc. Semi-Insulating III-V Conf., (Nottingham, 1980)]

[3] J.Barrau, D.X.Thanh, M.Brousseau, J.C.Brabant, F.Voillot [Solid State Commun. (USA) vol.44 no.3 (Oct 1982) p.395-9]

[4] D.Thebault, J.Barrau, G.Armelles, M.Brousseau [J. Phys. C (GB) vol.17 no.11 (20 Apr 1984) p.2011-8]

[5] J.Barrau, M.Brousseau, J.P.Austen, C.A.Bates [J. Phys. C (GB) vol.16 (1983) p.4581-98]

[6] G.Picoli, B.Deveaud, D.Galland [Proc. Semi-Insulating III-V Conf., ed. by G.J.Rees Nottingham (1980) p.254-8]

[7] G.Picoli, B.Deveaud, B.Lambert, A.Chomette [J. Phys. Lett. (France) vol.44 p.L85]

[8] L.Eaves, C.Uihlien [J. Phys. C (GB) vol.15 (1982) p.6257-69]

[9] L.Eaves, C.Uihlein [Phys. Rev. B (USA) vol.26 p.4473-84]

[10] Y.Fujiwara, A.Kojima, T.Nishino, Y.Hamakawa [J. Lumin (Netherlands) Pt I vol.31/32 (Dec 1984) p.451-3]

[11] B.Deveaud, B.Lambert, G.Picoli, G.Martinez [J. Appl. Phys. (USA) vol.55 no.12 (15 June 1984) p.4356-60]

[12] P.W.Yu [in Proc. of the 2nd Semi-Insulating III-V Materials Conf. (Evian, France 1982) ed. by S.M.Ebeid, B.Tuck p.308]

[13] Y.Fujiwara, A.Kojima, T.Nishino, Y.Hamakawa, K.Yasutaka, M.Umeno, H.Kawabe [Int. Conf. on Solid State Devices and Materials (Kobe, 1984) p.177-80]

[14] M.Yokogawa, S.Nishine, M.Sasaki, K.Matsumoto, K.Fuijta, S.Akai [Jpn. J. Appl. Phys. (Japan) vol.23 no.5 (May 1984) p.L339-41]

[15] V.Swaminathan, N.E.Schumaker, J.L.Zilko [J. Lumin. (Netherlands) vol.22 no.2 (Jan/Feb 1981) p.153-70]

12.3 PHOTOLUMINESCENCE SPECTRA OF UNDOPED LEC p-TYPE GaAs

by EMIS Group

July 1985
EMIS Datareview RN=15137

P-type material is formed if the melt stoichiometry becomes Ga rich during liquid encapsulated Czochralski (LEC) growth of GaAs [1].

The photoluminescence spectrum at 4.2 deg K contains two dominant emission bands at about 1.49 eV and 1.441 eV [2]. The former consists of two structures at 1.493 and 1.490 eV which are due to free-electron to neutral-acceptor and neutral-donor to neutral-acceptor pair transitions respectively. The neutral acceptor is C on an As site.

The 1.441 eV band has two phonon replicas due to emission of one and two LO phonons separated by 36 and 72 meV from the ZPL. The band is interpreted as the unresolved neutral-donor neutral-acceptor pair and free-electron neutral-acceptor transition involving a deep acceptor, the former dominant at temperatures up to about 30 deg K and the latter dominant above this temperature.

A study of the temperature dependence of this emission band between 2 and and 300 deg K gave an activation energy of 77+-2 meV for the acceptor. The acceptor involved has been suggested to arise from the cation antisite double acceptor Ga(As) [3]. The first and second ionisation energies for this centre were found to be 77 and 230 meV from the valence band edge. The centre is favoured by a high concentration of As vacancies as a result of Ga-rich growth.

Hetzler et al [5] carried out selective excitation photoluminescence and electronic Raman scattering on LEC samples grown from As deficient melts, by which they identified two excited states of the acceptor, which they refer to as having an activation energy of 78 meV. The excitation energies measured are 62.9 and 66.9 meV above the ground states. It was thought that they correspond to transitions from the 1s2 ground state to the split 1s12s1 excited states of a double acceptor. Bishop et al [7] observed the 78 meV acceptor level in samples grown from stoichiometric melts which suggests that a Ga-rich melt is not a necessary condition for its occurrence.

Near the intrinsic region of the PL spectrum weak emission lines are observed at 1.5117, 1.5124, 1.5102 and 1.5071 eV [2], and 1.283 eV [7]. The 1.5124 eV line is assigned as the unresolved exciton emission bound to C(As). The 1.5071 eV band is always present with the 1.441 eV line, possibly being due to an exciton bound to the Ga(As) acceptor. The relative intensities of the 1.443 and 1.284 eV lines of samples from two different suppliers (Rockwell and Westinghouse) have been found to be identical for a wide range of excitation intensity [3,7]. Elliot [8] observed a strong PL band at about 1.32 eV in a Si-compensated sample from a Ga-rich melt (Rockwell R41-T sample). Bishop et al [7] studied the same band in the same sample and found it to be clearly distinguishable from the 1.283 eV band of Yu et al [3] by virtue of its low temperature peak energy, its strongly diminishing intensity with increasing temperature and its decay characteristics.

Reynolds et al [6] had previously reported six sharp lines between 1.508 and 1.515 eV which were assigned to donor-acceptor complexes associated with an antisite defect. Two of these lines correspond to the 1.5117 and 15102 eV emissions.

In conclusion, the main emission bands are at 1.49 and 1.44 eV, and p-type conductivity in undoped LEC GaAs is controlled by the cation antisite double acceptor Ga(As) along with C(As).

REFERENCES

[1] D.E.Holmes et al [Appl. Phys. Lett. (USA) vol.40 (1982) p.46]
[2] P.W.Yu, D.C.Reynolds [J. Appl. Phys. (USA) vol.53 no2. (Feb 1982) p.1263-5]
[3] P.W.Yu et al [Appl. Phys. Lett. (USA) vol.41 (1982) p.532]
[4] P.W.Yu [Phys. Rev. B (USA) vol.27 no.12 (June 1983) p.7779-81
[5] S.R.Hetzler, et al [Appl. Phys. Lett. (USA) vol.44 no.8 (Apr 1984) p.793-5]
[6] D.C.Reynolds et al [J. Appl. Phys. (USA) vol.51 (1980) p.4842]
[7] S.G.Bishop, B.V.Shanabrook, W.J.Moore [J. Appl. Phys. (USA) vol.56 no.6 (15 Sept 1984) p.1785-90]
[8] K.R.Elliott [Appl. Phys. Lett. (USA) vol.42 (1983) p.274]

12.4 PHOTOLUMINESCENCE SPECTRA OF LPE GaAs

by J.Medland and EMIS Group

December 1985
EMIS Datareview RN=15415

The photoluminescence (PL) spectra of liquid phase epitaxy (LPE) layers is an important aid to the detection of impurities and defect levels as well as to the assessment of crystalline quality.

EXCITON PEAKS

The dominant excitonic emissions from high purity LPE layers (Nd + Na up to 10*15/cm*3) [3,4] ar given below:

PEAK (eV)		ASSIGNMENT
1.5153	$(F,X)_{n=1}$	free exciton, n=1 state of upper polariton branch
c.1.515	(F,X)	free exciton, lower polariton branch
1.515-1.5146	(D(o)X)*	excited states of (D(o),X) or excitons bound to complexes involving neutral donors
1.5141	(D(o)h) and/or $(D(o),X)_{n=1}$	exciton bound to neutral donor or valence band to neutral donor
1.5133	(D(+),X)	exciton bound to ionized donor
1.5128	J=1/2	exciton bound to neutral acceptor
1.5124	J=3/2	
1.5122	J=5/2	
1.5097(weaker)	$(D(o),X)_{n=2}$	two-electron transition of (D(o),X) with donor left in the n=2 state

Ref [3] shows how the relative intensity of the (D(o),X) and (A(o),X) lines varies with doping type and compensation.

IMPURITY RELATED PEAKS

Silicon and carbon are the main residual impurities in undoped LPE layers [1,5,6]. The following lines measured at low excitation density are commonly present in high purity layers:

1.4931	conduction band to C(As)	[1,4,7-9]
1.4891	neutral donor to C(As)	[4]
1.4851	CB to Si(As)	[4,5,10,11]
1.4814	neutral donor to Si(As)	[4]

Other impurities (as well as Si and C) were identified by Skromme et al [4] using variable temperature PL and photothermal ionisation spectroscopy. Most samples were grown in a graphite slider-boat. The following PL lines were identified using low excitation intensities of 2.4 mW/cm*2 (CB = conduction band):

Mg 1.4914 eV CB to neutral acceptor (7 deg K)
Mg 1.4875 eV neutral donor to neutral acceptor (2 deg K)
Ge 1.4781 eV CB to neutral acceptor (7 deg K)
Ge 1.4746 eV neutral donor to neutral acceptor (2 deg K)

S is the dominant donor in LPE material, with smaller quantities of Sn, Pb and Si donors [4]. Mn and Cu acceptors, at approx 1.406 and 1.355 eV respectively [4] are also present with their associated phonon replicas and an exciton bound to a neutral tin acceptor centre is seen in samples containing large amounts of tin.

Since the conduction band to acceptor (or free-to-bound (FB)) and donor acceptor pair transitions (DAP) are only accurate to approx +-0.5 meV and approx. +-1 meV respectively, depending on factors such as impurity concentration, "strain" and, for DAP transitions, excitation power density [2] unambiguous peak identification is only reliable if FB and DAP transitions are clearly distinguished. This is normally accomplished by varying the sample temperature [1,4] and excitation power density [2].

Taniguchi and Ikoma [21] observed an oxygen related peak at 0.64 eV with full width at half maximum of 120 meV as previously reported [22-24]

Morkoc et al [11] report a 25 deg K Zn peak at 1.489 eV for samples grown in a high purity Spectrosil boat, though zinc is not normally seen in LPE layers.

PL DETECTION OF SUBSTRATE/LPE INTERFACE DEGRADATION

A degraded layer (conducting or Si) can occur at the substrate/LPE interface. This is formed at the substrate surface during the 30 minutes or so of heat treatment in a hydrogen gas flow needed prior to moving the substate into the liquid. This can adversely affect device performance by, for example, causing a depletion zone to penetrate into an n-type epilayer.

The degraded layer can be detected by PL emission arising from the As vacancies which are caused by the evaporation of As atoms. Koschel et al [9] reported such emission at 1.412 eV, with a phonon replice at 1.376 eV, for a measurement temperature of 4.2 deg K. Undoped n-GaAs epilayers with a carrier concentration of 7 X 10*16 per cm*3 and a thickness of 0.5 micron were grown on a virgin SI substrate. The line is attributed to a C-acceptor - As-vacancy complex. The layer and the associated PL were virtually removed by etching with an unsaturated (Ga + GaAs) solution before LPE growth. This is known as the "Ga etch-back" technique [12,13], Otsubo et al [14] found a degradation-related peak at 1.40 eV in the 77 deg K PL of variously doped substrates and epilayers. The peak was found to be concentration dependent and strong for n-type material weak for p-type. For times up to 30 minutes and temperatures of approx. 700-900 deg C, degraded layers are formed at the substrate/LPE layer interface (typically 5-10 microns thick) and the thickness is found to be proportional to the square root of the heating time. For their investigations of this phenomenon see refs [15-18,27].

SOLUTION BAKEOUT INVESTIGATED BY PL

Baking of LPE Ga solutions markedly reduces the residual impurities in the undoped epilayers produced [5,17-20]. Garrido et al [5] obtained PL spectra for 800 deg C baking times of up to 48 hours. These, together with Hall effect measurements, strongly suggest a decrease in epilayer contamination by C, and probably Si, as the heating time increases. Source and ambient materials, rather than the graphite crucible, appear to have been responsible for the C contamination.

OTHER ASPECTS

PL has been used as a tool for studying LPE layer electronic surface properties [25]. The possibility of using PL changes under high-intensity excitation to asses the quality of Al2O3 facet coatings on LPE GaAs has been examined and found not be a useful technique in relation to laser reliability [26].

REFERENCES

[1] D.J.Ashen, P.J.Dean, D.T.J.Hurle, J.B.Mullin, A.M.White, P.D.Greene [J. Phys. & Chem. Solids (GB) vol.36 (1975) p.1041]

[2] E.W.Williams, H.B.Bebb ["Semiconductors and Semimetals", Ed. R.K.Willardson & A.C.Beer (Academic Press, New York, 1972) vol.8 p.324]

[3] V.Heim, P.Hiesinger [Phys. Status Solidi b (Germany) vol.66 (1974) p.461]

[4] B.J.Skromme, T.S.Low, G.E.Stillman ["Gallium Arsenide and Related Compounds", 1982, Inst. Phys. Conf. Ser. No.65 (1983) Ch.6, p.485-92]

[5] L.Garrido, J.L.Castano, J.Piqueras [J. Appl. Phys. (USA) vol.56 no.2 (15 July 1984) p.569-71]

[6] J.L.Castano, J.Piqueras [J. Appl. Phys. (USA) vol.54 (1983) p.3422]

[7] G.B.Stringfellow, W.Koschel, F.Briones, J.Gladstone, G.Patterson [Appl. Phys. Lett. (USA) vol.39 (1981) p.581]

[8] T.Chen, B.Sun [Chin. Phys. (USA) vol.2 no.3 (July-Sept 1982) p.643-7]

[9] W.H.Koschel, S.G.Bishop, B.D.McCombe, W.Y.Lum, H.H.Wieder ["Gallium Arsenide and Related Compounds", 1976, Inst. Phys. Conf. No.33a (1977) Ch.2 p.98-104]

[10] D.W.Kisker, H.Tews, W.Rehm [J. Appl. Phys. (USA) vol.54 (1983) p.1332]

[11] H.Morkoc, L.F.Eastman, D.Woodard, [Thin Solid Films (Switzerland) vol.71 no.2 (1 Sept 1980) p.245-8]

[12] L.M.F.Kaufmann, K.Heime, W.G.Burchard [J. Cryst. Growth (Netherlands) vol.34 (1976) p.289]

[13] P.Nordquist, H.Lessoff, E.M.Swiggard [Mater. Res. Bull. (USA) vol.11 (9176) p.939]

[14] M.Otsubo, H.Miki, S.Mitsui [Jpn J. Appl. Phys. (USA) vol.16 no.11

(Nov 1977) p.1957-66]
[15] M.Bugajski, E.Gawronska, J.Petryk, L.Szymanski [Electron Technol. (Poland) vol.10 no.3 (1977) p.61-78]
[16] J.S.Harris, Y.Nannichi, G.L.Pearson [J. Appl. Phys. (USA) vol.40 (1969) p.4575]
[17] E.Munoz, W.L.Snyder, J.L.Moll [Appl. Phys. Lett. (USA) vol.16 (1970) p.262]
[18] T.Ito, M.Takeuchi [Jpn. J. Appl. Phys. (Japan) vol.16 (1977) p.227]
[19] H.Mike, M.Otsubo [Jpn. J. Appl. Phys. (Japan) vol.10 (1971) p.509]
[20] L.Lin, Z.Fang, B.Zhou, S.Zhu, X.Xiang, R.Wu [J. Cryst. Growth (Netherlands) vol.56 (1982) p.533]
[21] M.Taniguchi, T.Ikoma ["Gallium Arsenide and Related Compounds", 1982, Inst. Phys. Conf. Ser. No.65 (1983) Ch.1 p.65-70]
[22] W.J.Turner, G.D.Petit, N.G.Ainsley [J. Appl. Phys. (USA) vol.34 p.3274]
[23] P.W.Yu [Solid State Commun. (USA) vol.32 (1979) p.1111]
[24] A.Mircea-Roussel, S.Makram-Ebeid [Appl. Phys. Lett. (USA) vol.38 (1981) p.1007]
[25] K.Mettler [Appl. Phys. (Germany) vol.12 no.1 (Jan 1977) p.75-82]
[26] V.Swaminathan, W.C.Dautremont-Smith, P.J.Anthony [Mater. Lett. (Netherlands) vol.2 no.3 (Feb 1984) p.179-83]
[27] C.K.Kim, R.M.Malbon, M.Omori, Y.S.Park ["GaAs and Related Compounds", 1978, Inst. Phys. Conf. Ser. No.45 (1979) Ch.4]

12.5 PHOTOLUMINESCENCE SPECTRA OF VPE GaAs

by J.D.Medland

January 1986
EMIS Datareview RN=15601

There are two vapour phase epitaxial (VPE) techniques; hydride VPE (starting materials Ga, HCl, AsH3) and chloride VPE (starting materials AsCl3, H2 and Ga). Both methods involve the same reactants at the growth surface, though the hydride process offers more flexibility of the As/Ga ratio. Low temperature (approx.4 deg K) PL provides a way of identifying (radiative) impurities and defects from their characteristic emission lines, as well as providing an idea of crystalline quality and overall impurity content from peak widths.

EXCITONIC PEAKS

High purity (impurity concentration up to 10*15/cm*3) and strain free GaAs grown by VPE exhibits some (or all) of the following lines as observed by Heim and Hiesinger [1]:

1.5153 eV	(F,X)	n=1 state of free exciton, upper polariton branch
1.515 eV	(F,X)	free exciton - lower polariton branch
1.5146-1.515	(D(o),X)	excited states of exciton bound to neutral donor
1.5141	(D(o),X)	exciton bound to neutral donor
1.5133	(D(t),X)	exciton bound to ionised donor and/or valence band to neutral donor transition (D(o),h)

1.5128	(A(o),X) [weak] J=1/2	excitons bound to neutral acceptor with differing total angular momentum states [2]
1.5124	(A(o),X) [dominant] J=3/2	
1.5122	(A(o),X) J=5/2	

1.5108	´two-electron` transition of a free exciton recombining near a neutral donor
1.5101-1.5014	`two-electron´ transitions of (D(o),X)*
1.5097	`two electron´ transition of (D(o),X), donor left in n=2 state
1.5089	`two electron´ transition of (D(o),C) with donor left in n=3 state

VPE layers showing these transitions are discussed in refs [3,4,5,6,17]. The weak `two-electron´ transitions - only present in the purest materials (with Nd + Nd up to 10*14/cm*3)] - have been used to identify different donors ( which have unresolved n=1 states ); see ref [7]. The so-called `two-hole´ acceptor exciton transitions [3,8,9,10] involving n=2 states of neutral acceptor levels are used for accurate acceptor identifications, often to confirm assignments made with the conduction band (free electrons) to neutral acceptor (bound hole) FB transitions

- see following sections. Excitons bound to deep oxygen donors [3,11] have also been observed at 1.4885 eV, using high excitation powers, and a weak line attributed to excitons bound to a complex involving copper has also been observed at 1.50275 eV [3,12,14,15].

FREE-TO-BOUND (FB) AND DONOR-TO-ACCEPTOR PAIR (DAP) TRANSITIONS

At low excitation power densities and approx. 4 deg K the following transitions can be seen in VPE material (the neutral acceptor involved is given in each case) :

1.4932 eV	FB	C(As)	[3]
1.4915	FB	Be(Ga)	[8]
1.4911	FB	Mg(Ga)	[8]
1.4894	FB	Zn(Ga)	[3]
1.4892	DAP	C(As)	[3]
1.4854	DAP	Zn(Ga)	[3]
1.4850	FB	Si(As)	[8]
1.4848	FB	Cd(Ga)	[8]
1.4816	DAP	Si(As)	[3]
1.4782-1.479	FB	Ge(As)	[3,8]
1.4745	DAP	Ge(As)	[3]
1.406	FB	Mn-related	[3,13]
1.356	FB	Cu-related	[3,14,15]
1.22	complex of a donor associated with a gallium vacancy		[16]

Phonon replicas of the above peaks, especially the deeper acceptors (Mn,Cu) are often seen [3,16], the LO replica 36.5 meV below the original peak, and the TA replica approx. 10 meV below the original peak [5].

SHALLOW ACCEPTORS

The peak positions for the FB emissions depend somewhat on total concentration of neutral and/or ionised impurities, strain and substrate problems (see ref[8]) but for fairly high purity samples (Na + Nd up to 10*15 cm*-3) the positions should be accurate to +-0.5 meV. The DAP peak positions depend implicity on the donor and acceptor concentrations [16,18] but for high purity samples should be accurate to within approx.1 meV. It can be seen that peaks often overlap and variable temperature and excitation power density measurements have to be made to identify transitions [3,8,16].

In undoped material, the residual impurities (which arise from a number of sources; see refs [3,4,17,19]), their degree of incorporation and for the case of amphoteric group IV impurities, whether they form n- or p-type centres, have been studied as a function of growth conditions [3,4,20]. Both hydride and chloride techniques have C and Zn as dominant acceptor impurities with germanium and silicon sometimes present in trace quantities. Germanium and silicon incorporate predominantly as donors (together with the usually predominant S donor).

DEEPER ACCEPTORS

Cu and Mn and their complexes have been seen in layers grown by both VPE techniques, and probably arise from out-diffusion from the substrate or from stainless steel in the growth reactors pipework or gas cylinders [4]. As discussed in detail by Williams and Bebb [16] a broad band at about 1.2 eV is often seen in heavily n-doped GaAs materials and is associated with a complex centre involving a gallium vacancy. This band has distinct properties [16] and it has been observed in undoped VPE material grown on heavily n-doped substrates [3].

OTHER ASPECTS

PL has been used to probe the VPE grown active/buffer layer interface of FET like structures [22] and to observe changes in impurity content of a buffer layer subject to implantation and annealing [23]. Refs [3,4] show how the relative heights of PL peaks due to DAP or FB transitions may be used to estimate relative amounts of different acceptors in the GaAs. For further discussion of the semi-quantitative use of PL see the review by P.J.Dean [24].

REFERENCES

[1] U.Heim, P.Hiesinger [Phys. Status Solidi b (Germany) vol.66 (1974) p.461]

[2] A.M.White, P.J.Dean, B.Day [J. Phys. C (GB) vol.7 (1974) p.1400]

[3] B.J.Skromme, T.S.Low et al [J. Electron. Mater. (USA) vol.12 no.2 (1983) p.433]

[4] J.K.Abrokwah, J.N.Peck et al [J. Electron. Mater. (USA) vol.12 no.4 (1983) p.681]

[5] Landolt-Bornstein [Numerical Data and Functional Relationships in Science and Technology (New Series) vol.7 sub-volume A (Physics of Group IV Elements and III-V Compounds) Springer-Verlag, New York 1982, Ed by O.Madelung p.234]

[6] I.A.Dorrity, D.Hewins, J.D.Medland [Inst. Phys. Conf. Ser. no.74 (1985) p.211]

[7] R.J.Almassy, D.C.Reynolds et al [Solid State Commun. (USA) vol.38 (1981) p.1053-6]

[8] D.J.Ashen, P.J.Dean et al [J. Phys. & Chem. Solids (GB) vol.36 (1975) p.1041-53]

[9] P.W.Boln, R.Bhat, T.D.Harris [Anal. Chem. (USA) vol.56 (1984) p.58-62]

[10] A.M.White, P.J.Dean et al [J. Phys. C (GB) vol.6 (1973) p.L243]

[11] A.M.White, P.J.Dean, D.J.Ashen et al [Proc. 12th Int. Conf. on the Phys. of Semiconductors (1974) p.381]

[12] F.Williman, D.Binberg, M.Blatte [Phys. Rev. B (USA) vol.7 (1973) p.2473]
[13] W.Schairer, M.Schmidt [Phys. Rev. B (USA) vol.10 (1974) p.2501]
[14] Z.G.Wang, H.P.Gislason, B.Monemar [J. Appl. Phys. (USA) vol.58 no.1 (July 1985) p.230]
[15] H.P.Gislason, Z.G.Wang, B.Monemar [J. Appl. Phys. (USA) vol.58 no.1 (July 1985) p.240]
[16] E.W.Williams, H.B.Bebb [Semicond. and Semimet.(USA) vol.8 Eds R.K.Willardson & A.C.Beer (Academic Press, 1972) p.321]
[17] Lin Lan-Ying, Lin Yao-Wang et al [J. Cryst. Growth (Netherlands) vol.56 no.2 (1982) p.344]
[18] R.Dingle [Phys. Rev. (USA) vol.184 no.3 (15 Aug 1969) p.788] vol.17 (1971) p.129]
[19] C.M.Wolfe, G.E.Stillman, E.B.Owens [J. Electrochem. Soc. (USA) vol.17 (1971) p.129]
[20] T.S.Low, B.J.Skromme, G.E.Stillman [Inst. Phys. Conf. Ser. no.65 (Inst. of Phys.,1983) p.515]
[21] T.Nishino et al [Inst. Phys. Conf. Ser. no.65 Albuquerque 1982 (Inst. of Phys., 1983) p.71]
[22] R.J.Almassy, D.C.Reynolds, C.W.Litten, K.K.Baja [J. Electron. Mater. (USA) vol.7 no.2 (1978) p.263-77]
[23] P.K.Bhattacharya, J.K.Rhee et al [J. Appl. Phys. (USA) vol.52 no.12 (Dec 1981) p.7224-31]
[24] P.J.Dean [Prog. Cryst. & Growth Charact. (GB) vol.5 (1982) p.89-174]

12.6 PHOTOLUMINESCENCE SPECTRA OF MOCVD GaAs

by J.Medland and EMIS Group

December 1985
EMIS Datareview RN=15403

Interest in metalorganic chemical vapour deposition (MOCVD) stems from the need to produce thin layers and abrupt interfaces suitable for use in high electron mobility transistors, superlattices and quantum well structures. Low temperature (less than about 10 deg K) photoluminescence (PL) spectra are an effective means of detecting impurities and defects in the layers produced.

EXCITON PEAKS (c. 1.51 eV)

High purity (i.e. Na+Nd up to 10*16/cm*3) MOCVD films, like all epitaxial films, give highly structured narrow emission lines arising mainly from free exciton polaritons and excitons which may have become bound to an impurity site.

Swaminathan et al [1] obtained the following MOCVD exciton peaks close to the band edge (1.5194 eV) at 12 deg K (the film was grown at 589 deg C, 800 r.p.m., with a V/III ratio of 8 and a carrier concentration of 10*15/cm*3 on a semi-insulating substrate):

PL line eV	Probable identification (X = exciton)
1.5156	free X (upper polariton branch)
1.5149-1.5145	excited states of donor X (and including free X, lower polariton branch)
1.5142	X bound to neutral donor
1.5136	X bound to ionised donor, and donor to VB
1.5126	X bound to neutral acceptor
1.5119	defect X
1.5112	defect X

The authors point to the similarity of the positions and widths of the above lines to those of epitaxially produced films [2-6]. It should be noted, however, that these positions differ slightly (probably due to strain in the specimens) from the generally accepted values given in ref [30]. Roth et al [7] report 12 lines in the range 1.5113 to 1.5056 eV for comparable MOCVD samples and suspect that these are closely related to similar lines in MBE films [8,31]

CARBON PEAKS (c. 1.49 eV)

Carbon is the dominant background impurity in MOCVD GaAs prepared using TMG (trimethylgallium, (CH3)3Ga) as the Ga source, but was not revealed in the PL spectra of samples produced using TEG (triethylgallium, (C2H5)3Ga) by Bhat et al [9]. The C probably originates from the adsorption of hydrocarbon radicals onto surface As atoms [10], these radicals being due to fragmentation of the organometallic compounds used. The amount of C incorporated depends on growth temperature, gas phase stoichiometry during growth, operating pressure and substrate orientation (see below).

C could be incorporated as a donor, C(Ga), or as an acceptor, C(As); it is found [32,35] that C(As) predominates and in the light of recent photothermal ionisation experiments [32 and references therein]. Where the "X3 donor" is identified as Ge rather than C, carbon is not expected to incorporate as a donor. The following transitions are generally observed at approx. 4 deg K using low excitation power densities:

1.4935 (+- 0.0005) eV	conduction band (free electron) to (C)As acceptor (bound hole) - FB(C)
1.4895 (+- 0.001) eV	neutral donor to C(As) acceptor pair - DAP(C). Exact position is dependent on concentration and excitation

See refs [1,3,9,12]. Refs [3,9,35] contain a good discussion of observed variations in these peak positions and how to separate out the often overlapping peaks due to zinc.

Kuelch et al [10] conclude from PL spectra at 77 deg K that the contamination level, which is proportional to the ratio of the integrated intensity of the FB(C) peak to the conduction band to valence band peak [3,12,13], is reduced when the growth temperature is lowered e.g. by a factor of 2 when the temperature is lowered from 700 to 600 deg C for a fixed AsH3/TMG ratio of 40. An increase in the AsH3/TMG ratio decreased the C content, as reported previously [1,13-16].Layers on the [111] direction-family As surface had the highest C concentrations and layers on the [111] direction-family Ga surface had the lowest C content.

Takagishi and Mori [11,17] have determined from PL that the C concentration falls as the operating pressure is increased over the range 3 X 10*-3 to 75 Torr. For an AsH3/TMG ratio of 75 there was a conductivity-type conversion at 0.5 Torr. At pressures below this p-type layers resulted from C contamination, while for high pressures n-type layers with low C content were obtained.

This "cross-over" effect from n-type to p-type is characteristic of MOCVD growth and the exact AsH3/TMG ratio at which cross-over occurs depends strongly on source purity, growth temp. and growth rate (small effect if sources pure) as well as overall operating pressure [14].

Blaauw et al [18] found from PL spectra that C incorporation was enhanced when TMAs (trimethylarsenic, (CH3)3As) was used instead of arsine (AsH3). Mori and Takagishi [19] investigated the influence of extra hydrogen carrier gas in TMG/arsine low pressure MOCVD and found that this increased the C incorporation for samples grown at 1.5 torr. Films grown at constant gas velocity had high resistivity and poor PL features.

SILICON PEAKS (c. 1.485 eV)

Si is observed as a residual impurity in MOCVD since it is often present in the organometallic source material AsH3. Si could also come from the SiC coated susceptors or interactions with the growth vessel. Note that Si has been used as a substrate for MOCVD GaAs [20] and as a dopant [21,26]. It can be present as a donor [14] Si(Ga) - which is its predominant incorporation site, and/or an acceptor [18,32].

Blaauw et al [18] report a prominent Si peak at 1.4852 eV when using the adduct TMG - TMAs (containing 9 ppm Si according to the manufacturer's analysis) with arsine. The peak which appears to be associated with Si on As sites and which was previously identified by Ashen et al [3], was not present in samples grown simply from TMG and arsine. Dapkus et al [14] found residual Si (As) by PL and Si(Ga) far-IR photoconductivity analysis of high purity MOCVD GaAs with a total impurity content (including C and Zn) of 5 X 10*14/cm*3 and an electron mobility of 125,000 cm*2/V.s at 77 deg K.

Okamota et al [21] have observed a broad-band peak (1.355 eV) in addition to the band edge peak in the 77 deg K PL of Si-doped GaAs, which they assign to a self-activated complex of V(Ga)-Si(Ga)

1.485 eV FB involving Si(As) - See refs [18,3]
1.482 eV DAP involving Si(As) - see ref [12] and note that exact position is influenced by concentration and excitation power density
See also ref [39] discussion on probable variations in peaks positions and separations of ovelapping peaks.

Zn,Mg,Mn,Ge and Cu PEAKS

Bhat et al [9] point out that volatile Mg compounds (e.g. trialkylmagnesium) may be present in the organometallic starting materials.

ZnAs(2) is used as a starting material in the manufacture of high purity arsine and can also be present in TMG. This explains why zinc is, with C, the dominant acceptor in MOCVD GaAs [32,14,35]. There is evidence that lowering the growth temperature enhances Zn incorporation [22,23].

1.4893 eV FB involving Zn(Ga) - refs [1,3,7,9,12,28,32]
1.4854 eV DAP involving Zn(Ga) - refs [1,7,9,12,28]

See discussion following carbon peak positions for variations in the above peak values.

Germanium, which is probably a contaminant in arsine sources, is incorporated in MOCVD GaAs predominantly as a donor [32], but under the right growth conditions is present as an acceptor Ge(As) revealed by the following 4 deg K PL peaks:-

1.4785 FB involving Ge(As) - refs [3,12,32]

1.4854 eV DAP involving Zn(Ga) - refs [1,7,9,12,28]

See discussion following carbon peak positions for variations in the above peak values.

Germanium, which is probably a contaminant in arsine sources, is incorporated in MOCVD GaAs predominantly as a donor [32], but under the right growth conditions is present as an acceptor Ge(As) revealed by the following 4 deg K PL peaks:-

1.4785 FB involving Ge(As) - refs [3,12,32]
1.4745 DAP involving Ge(As) - refs [12,32]

Again, see discussion following C peak positions for variations in the above peak values. A Mg peak at 1.4911 eV is suspected by Swaminathan et al [1]. Bhat et al [9] observe PL lines which they suspect are due to Mg or Zn. Both TMG and TEG derived samples were used. Bhat et al [9] point out that volatile Mg compounds (e.g. trialkylmagnesium) may be present in the organometallic starting materials [33]. Bearing in mind the previous discussion of peak value variations, Mg-related peaks are:

1.4911 eV FB involving Mg(Ga) - refs [1,3,7,9] and note the differing peak assignments for FB peak from ref [1] suggestion

1.487 eV DAP Mg(Ga) - ref [9]

Blaauw et al [18] and, much earlier, Lee and and Anderson [24] report a line at 1.4089 eV due to Mn on a Ga site. Blaauw et al [18] also detected a PL line at 1.3578 eV due to Cu on Ga sites, previously reported in [12,25]. The authors determined that these impurities were due to the diffusion of Mn and Cu from the SI substrate.

OTHER ASPECTS

Stress (caused by lattice mismatch) induced changes in the PL intensity and energy for MOCVD growth of GaAs on a Si substrate with GaAs/GaAlAs buffer layers is reported in ref [20].

PL spectra of atomic layer epitaxy (ALE) of 100 cycle MOCVD samples are reported by Bedair et al [27]. A sharp PL peak with a full width at half maximum of 11 meV was observed.

Selectively excited PL of the (AX) n=1 lines for the unambiguous identification of acceptors are reported by Bohn et al [28]. See also ref [3,12].
Nonlinearity in the band-to-band recombination PL under high density picosecond photoexcitation was investigated by Lehmen and Ballantyne [29].

A broad luminescent peak at approx. 1.2 eV (at 2 deg K) appearing with increasing sulphur doping and attributed to S(As)-V(Ga) is observed in MOCVD GaAs and thought to be responsible for compensation in n-doped material [34].

REFERENCES

[1] V.Swaminathan, D.L.Van Haren, J.L.Zilko, P.Y.Lu, N.E.Schumaker [J. Appl. Phys. (USA) vol.57 no.12 (15 June 1985) p.5349-53]
[2] D.D.Sell, S.E.Stokowski, R.Dingle, J.V.DiLorenzo [Phys. Rev. B (USA) vol.7 (1973) p.4568]
[3] D.J.Ashen, P.J.Dean, D.T.J.Hurle, J.B.Mullin, A.M.White, P.D.Greene [J. Phys. Chem. Solids (GB) vol.36 (1975) p.1041]
[4] R.Dingle, C.Weisbuch, H.L.Stormer, H.Morkoc, A.Y.Cho [Appl. Phys. Lett. (USA) vol.40 (1982) p.507]
[5] H.Temkin, J.C.M.Hwang [Appl. Phys. Lett. (USA) vol.42 (1983) p.178]
[6] M.Heiblum, E.E.Mendez, L.Osterling [J. Appl. Phys. (USA) vol.54 (1983) p.6982]
[7] A.P.Roth, R.G.Goodchild, S.Charbonneau, D.F.Williams [J. Appl. Phys. (USA) vol.54 (1983) p.3427]
[8] H.Kunzel, K.Plogg [Appl. Phys. Lett. (USA) vol.37 (1980) p.416]
[9] R.Bhat, P.O´Connor, H.Temkin, R.Dingle, V.G.Keramidas ["Gallium Arsenide and Related Compounds", 1981, Inst. Phys. Conf. Ser. No.63 p.101-6]
[10] T.F.Kuech, E.Veuhoff [J. Cryst. Growth (Netherlands) vol.68 no.1 (Sept 1984) p.148-56]
[11] S.Takagishi, H.Mori [Jpn. J. Appl. Phys. (Japan) vol.22 no.12 (Dec 1983) p.L795-7]
[12] B.J.Skromme, T.S.Low, T.J.Roth, G.E.Stillman [J. Electron. Mater. (USA) vol.12 (1983) p.433]
[13] K.Mohammed, J.L.Merz, D.Kasemset [Appl. Phys. Lett. (USA) vol.43 (1983) p.103]
[14] P.D.Dapkus, H.M.Manasevit, H.L.Hess, T.S.Low, G.E.Stillman [J. Cryst. Growth (Netherlands) vol.55 no.1 (Oct 1981) p.10-23]
[15] J.R.Shealy et al ["Gallium Arsenide and Related Compounds",1983, Inst. Phys. Ser.5 p.109]
[16] M.O.Watanabe, K.Morizuka, M.Mashita, Y.Ashizawa, Y.Zohta [Jpn. J. Appl. Phys. (Japan) vol.23 no.2 (Feb 1984) p.103-5]
[17] S.Takagishi, H.Mori [Jpn. J. Appl. Phys. (Japan) vol.23 no.2 (Feb 1984) p.100-2]
[18] C.Blaauw, C.Miner, B.Emmerstorfer, A.J.Springthorpe, M.Gallant [Can. J. Phys. (Canada) vol.63 no.6 (June 1985) p.664-9]
[19] H.Mori, S.Takagishi [Jpn. J. Appl. Phys. (Japan) vol.23 no.12 (Dec 1984) p.877-9]
[20] M.Akiyama, Y.Kawarada, K.Kaminishi [J. Cryst. Growth (Netherlands) vol.68 (1984) p.21]
[21] K.Okamoto, S.Onozawa, T.Imai [J. Appl. Phys. (USA) vol.56 no.10 (Nov 1984) p.2993-5]
[22] S.J.Bass [J. Cryst. Growth (Netherlands) vol.31 (1975) p.172]
[23] H.M.Manasevit, A.C.Thorsen [J. Electrochem. Soc. (USA) vol.119 (1972) p.99]
[24] T.C.Lee, W.W.Anderson, [Solid State Commun. (USA) vol.2 (1964) p.265]
[25] H.J.Queisser, C.S.Fuller [J. Appl. Phys. (USA) vol.37 (1966) p.4895]
[26] E.Veuhoff, T.F.Kuech, B.S.Meyerson [J. Electrochem. Soc. (USA) vol.132 no.8 (Aug 1985) p.1958-61]
[27] S.M.Bedair, M.A.Tischler, T.Katsuyama, N.A.El-Masry [Appl. Phys. Lett. (USA) vol.47 no.1 (1 July 1985) p.51-3]
[28] P.W.Bohn, T.D.Harris, R.Bhat, H.M.Cox [Appl. Spectrosc. (USA) vol.38 no.3 (May/June 1984) p.417-22]
[29] A.Von Lehmen, J.M.Ballantyne [J. Appl. Phys. (USA) vol.58 no.2

(15 July 1985) p.958-62]
[30] U.Heim, P.Hiesigner [Phys. Status Solidi (b) (Germany) vol.66 (1974) p.461]
[31] A.C.Beye, G.Neu [J. Appl. Phys. (USA) vol.58 no.9 (1 Nov 1985) p.3549-55]
[32] T.S.Low, B.J.Skromme, G.E.Stillman [Inst. Phys. Conf. Ser. No.65 (1983)]
[33] B.El.Jani, J.C.Grenet et al [Rev. Phys. Appl. (France) vol.19 (1984) p.7-15]
[34] L.Samuelson et al [J. Cryst. Growth (Netherlands) vol.55 (1981) p.164-72]
[35] K.L.Hess et al [J, Electron. Mater. (USA) vol.11 no.6 (1982) p.1115]

12.7 PHOTOLUMINESCENCE SPECTRA OF MBE GaAs

by J.Medland and EMIS Group

January 1986
EMIS Datareview RN=15139

Photoluminescence (PL) spectra of molecular beam epitaxy (MBE) grown films at low temperature (about 4 deg K) are useful in characterising the films and determining their quality. The positions of the normally excitonic peaks seen in pure material (up to 10*15 cm*-3 of impurities) are those orginally measured by Heim and Hiesigner [1] and confirmed for MBE material in ref [8]:-

PEAK (eV)		ASSIGNMENT
1.5153	(F,X)	n=1 state of free exciton, upper polariton branch
1.515	(F,X)	lower polariton branch
1.515-1.5146	(D(o),X)*	excited states of neutral donor bound exciton
1.5141	(D(o),X)	neutral donor bound exciton
1.5133	(D(+),X)	and D(o)h ionised donor bound exciton and/or neutral donor to valence band
1.5128	J = 1/2	
1.5124	J = 3/2	
1.5122	J = 5/2	

In addition, up to 60 individual features have been observed [15] in the spectral region 1.50 to 1.511 eV. These are known as KP lines after Kunzel and Ploog who first reported 16 sharp lines in this region for samples grown at a substrate temperature of 530 deg C [3].

The dominant impurity incorporated as an acceptor is C(As), from CO contamination of the growth chamber, and from the substrate. This gives rise to peaks measured at low excitations [8,36]:

1.4932 (2 deg K) free electron to bound hole at C(As) acceptor - FB(C)
1.4892 (2 deg K) donor bound electron to acceptor bound hole at C(As)-DAP(C)

NOTE: these peak positions are subject to fluctuations of approx. 0.5 meV [see ref 33] and the DAP peaks positions depend strongly on excitation power density and doping [34]. If doping is such that FB and DAP peaks overlap and are of approx. the same intensity then their positions are shifted toward one another i.e. the energy difference between them decreases.

A broad PL band is often seen in undoped or lightly doped MBE GaAs growing peaks between 1.47 and 1.49 eV [10,11,4] including a peak at 1.479 eV.

The KP lines have been the subject of much study. Recent papers [33,34] indicate that they can be split into two groups: the "g peak" at 1.5111 eV (highest energy emission, shorter lifetime (approx. 1.2 ns), unpolarised) which is present under the widest range of growth conditions [13,34] and dopants (n- and p-type) and the emissions h - v, present mainly in p-type material (longer lifetimes (approx. 2.4 ns and indicating definite polarisation properties). The origin of this emission is still uncertain but the latest suggestions are that the "g peak" is due to an exciton bound to a neutral "acceptor complex defect" - involving the major residual carbon acceptor and an isoelectronic defect [33,34]. The "h - v" emissions are due to exciton recombinations at axially oriented complex defects acting as isoelectronic centres [34]. These isoeletronic impurities may arise from the fact that indium (used to mount substrate to molybdenum block) and boron nitride for the crucible containing the Ga,As and dopant species are present in the growth system. These latest results cast considerable doubt on earlier theories [12,15].

In general KP lines are not observed in non-MBE samples, although lines in the same spectral region emitted by MOCVD layers [20,21] may be related. Neither are they always emitted by MBE samples; they are often absent when As2 molecular beams are used [3], or when substrate temperatures above 600 deg C are used [3].

In attempting to relate the KP lines to the electrical characteristics Rao et al [10] performed low temperature PL measurements on various high quality MBE layers (undoped n- or p-type, Sn-doped n-type, and Be-doped p-type) and found that for most types of layer the 1.5109 and 1.5046 eV lines were prominent, but much more so for p-type than for n-type samples.

1.47 - 1.49 eV REGION

The luminescence in the 1.47 and 1.49 eV region first measured by Briones and Collins has also generated considerable research activity. In their original study [4] the authors show a correlation of the intensity and energy of the luminescence at 1.491 and 1.473 eV with that of the exciton-like KP lines at 1.5109 and 1.5046 eV respectively. These are attributed to complex centres of the type C-V(As) an C-V(Ga) respectively. Koschel et al [23] suggested that a simple vacancy, or carbon-related defect centre differing from a simple carbon acceptor, may be responsible, while Jeong et al [24] proposed interstitial carbon impurities. Akimoto et al [5] found that intensity of KP lines increased as the amount of CO in the background increased and As4/Ga beam flux ratio decreased. They also carried out Fourier transform infrared spectroscopy and inferred the existence of C-O stretching vibrations of polynuclear carbonyls. They suggest that the KP luminescence is due to excitons bound to carbonyls incorporated on As sites. From the correlation between the KP lines and those in the 1.47-1.49 eV region they suggest that luminescence in the latter may be due to a C impurity whose acceptor level is shifted by oxygen. Contour et al [7] also adopted a correlation approach. They performed PL excitation measurements and correlated the 1.5109 eV line, which they attribute to a defect-bound-exciton, with emission at 1.495 eV, thought to result from a free-to-bound acceptor transition. Rao et al [10] also looked for correlations between the KP lines and those in the 1.47-1.49 eV region. They found exceptions to Hayne´s rule [25] upon which, the correlations of Briones and Collins [4] and others are founded. Lines at

1.468, 1.475, 1.48 and 1.496 eV were detected.

Returning to the near bandgap part of the PL spectrum there is a doublet peak at about 1.516 eV. Hopfield [26] and Pekar [27] suggested that exciton-polaritons were responsible (coupling of photons and free excitons). Recently, Koteles et al [28] clearly demonstrated that enhanced scattering of exciton-polaritons at the exciton energy is the cause of the exciton-polarition lineshape observed for pure undoped MBE samples. Bloss et al [29] found that the dip between the two peaks increased as the thickness of the undoped layers, controlled by etching increased, since for a given density of scattering centres the distance an exciton-polariton travels before reaching the sample surface is governed by the layer thickness, and for thicker samples this is greater than the exciton-polariton scattering length.

UP-CONVERSION

Up-conversion or Anti-Stokes emission is observed in MBE and other high purity epitaxial layers [30]. Near band gap emission is generated by below bandgap excitation. The authors explain this as a two-step optical excitation of an electron-hole pair via a deep centre:

Step 1: e excited from VB to deep centre
Step 2: e transfers from deep centre to CB

An electron is thus created in the CB and a hole in the VB while the deep centre remains unchanged. This mechanism depends on native defects, not on intentionally introduced impurities, and could be a useful means of studying deep centres.

SILICON PEAKS

Silicon incorporates primarily as a donor, Si(Ga) [35], but under the right growth conditions can occupy As sites giving rise to peaks at approx 4.2 deg K [11]:-

1.485 eV FB [Si(As)] - from ref [33]
1.4813 eV DAP [Si(As)] - from ref [36]

These peaks are subject to the same variations as described for FB and DAP peaks due to C.

Erickson et al [22] investigate the effect of substrate temperature and As pressure on Si- and C-acceptor peaks at 1.483 eV (Si(o)(Ga), Si(o)(As)) and 1.490 eV (D(o), C(As)) to show the nature of Si incorporation in MBE GaAs grown using As2 beams. The PL spectra together with carrier concn measurements show that the Si dopant exhibits an enhanced acceptor character as the substrate temperature is increased from 570 to 680 deg C at a given As pressure, while for a fixed substrate temperature of 680 deg C the PL peaks suggest that there may be an optimum Asd2 pressure for minimising the incorporation of nonradiative defects and Si acceptors in films grown at this substrate temperature.
Mendez et al [31], using an As4 beam, studied the incorporation of Si as a dopant in GaAs as in function of substrate temperature and As/Ga flux ratio. They too found that the acceptor character of Si increased

with the substrate temperature. Also, the Si self-compensation decreased as the flux ratio was raised from 2 to 6.

Metz et al [32] investigated MBE:Si films grown at low substrate temperature and growth rates. PL spectra and electrical measurements showed a big improvement in quality over films grown at normal temperatures and rates (570 deg C, 1 micron/h). In particular, the following PL peaks were observed:

Substrate Temperature deg C	Growth Rate microns/h	PL Peaks Observed 1.514 eV	1.495 eV
570	1	Yes	Yes
450	1.1	No	Yes
450	0.2	Yes	Yes
380	0.02	Yes	Yes

The 1.514 eV peak was assigned to bound exciton recombination and the 1.495 eV to carbon.

In GaAs grown using cracked AsH3 as the As source, rather than the normal solid As, Mg/Be peaks are dominant:

1.4917 eV FB due to Mg/Be [Ga] measured at 6.2 deg K
1.4880 eV DAP due to Mg/Be [Ga] and low excitation power

Smaller amounts of C and Si (see above for FB/DAP peak positions): be giving FB and DAAP peaks at 1.4778 and 1.4745 eV respectively; Sn, giving an exciton 1.5071 eV and a FB band at 1.3 eV and Mn giving peaks (1.40 eV) [36,11]

REFERENCES
----

[1] U.Heim, P.Hiesinger [Phys. Status Solidi b (Germany) vol.66 (1974) p.461]

[2] M.S.Skolnick, T.D.Harris, C.W.Tu, T.M.Brennan, M.D.Sturge [Appl. Phys. Lett. (USA) vol.46 no.4 (15 Feb 1985) p.427-9]

[3] H.Kunzel, K.Ploog [in "Gallium Arsenide and Related Compounds, 1980", Inst. Phys. Conf. Ser. No.56, p.519-28]

[4] F.Briones, D.M.Collins [J. Electron. Mater.(USA) vol.11 (1982) p.847]

[5] K.Akimoto, M.Dohsen, M.Arai, N.Watanabe [Appl. Phys. Lett. (USA) vol.45 no.9 (1 Nov 1984) p.922-4]

[6] G.B.Scott, G.Duggan, P.Dawson, G.Weimann [J. Appl. Phys. (USA) vol.52 (1981) p.6888]

[7] J.P.Contour, G.Neu, M.Leroux, C.Chaix, B.Levesque, P.Etienne [J. Vac. Sci & Technol.(USA) vol.1 (1983) p.811]

[8] M.Heiblum, E.E.Mendez, L.Osterling [J. Appl. Phys. (USA) vol.54 (1983) p.6982]

[9] H.Temkin, J.C.H.Hwang [Appl. Phys. Lett. (USA) vol.42 (1983) p.178]

[10] E.V.K.Rao, F.Alexandre, J.M.Masson, M.Allovon, L.Goldstein [J. Appl. Phys. (USA) vol.57 no.2 (15 January 1985) p.503]

[11] B.J.Skromme, G.E.Stillman, A.R.Calawa, G.M.Metze [Appl. Phys. Lett. (USA) vol.44 (1984) p.240]

[12] D.P.Halliday, L.Eaves, P.Dawson [Proc 13th Int. Conf. on Defects in Semiconductors, Colorado, USA Aug 1984 (Metallurgical Soc. of AIME, 1985) p.1005-11]
[13] P.J.Dobson, G.B.Scott, W.H.Neave, B.A.Joyce [Solid State Commun. (USA) vol.43 (1982) p.917-9]
[14] P.K.Bhattacharya, H.J.Buhlmann, J.Illegems [J. Appl. Phys. (USA) vol.53 p.6391-8]
[15] D.C.Reynolds et al [Solid State Commun. (USA) vol.52 no.7 (Nov 1984) p.685-8]
[16] W.I.Wang [Appl. Phys. Lett. (USA) vol.44 (1984) p.1149-51]
[17] C.J.Hwang [Phys. Rev. B (USA) vol.8 (1973) p.646-52]
[18] G.C.Osbourne, D.L.Smith [Phys. Rev. B. (USA) vol.20 (1979) p.1556-61]
[19] L.Eaves, D.P.Halliday [J. Phys. C. (GB) vol.17 no.27 (Sept 1984) p.L705-9]
[20] A.P.Roth S.Charbonneau, R.G.Goodchild [J. Appl. Phys. (USA) vol.54 (1983) p.5350-8]
[21] A.P.Roth, S.Charbonneau, R.G.Goodchild, D.F.Williams [J. Appl. Phys. (USA) vol.54 (1983) p.3427-30]
[22] L.P.Erickson et al [J. Appl. Phys. (USA) vol.56 no.8 (15 Oct 1984) p.2231-5]
[23] W.H.Koschel, R.S.Smith, P.Hiesinger [J. Electrochem. Soc. (USA) vol.128 (1981) p.1336]
[24] M.Jeong, J.Shirafuji, Y.Inuishi [Jpn. J. Appl. Phys. (Japan) vol.20 (1981) p.795]
[25] J.R.Haynes [Phys. Rev. Lett. (USA) vol.4 (1961) p.361]
[26] J.J.Hopfield [Phys. Rev. (USA) vol.112 (1958) p.1555]
[27] S.I.Pekar [Sov. Phys. JETP vol.6 (1985) p.685]
[28] E.S.Koteles, J.P.Salerno, W.Bloss, E.M.Brody [17th Int. Conf. on the Physics of Semiconductors, 1985]
[29] W.L.Bloss, E.S.Koteles, E.M.Brody, B.J.Sowell, J.P.Salerno, J.V.Gormley [Solid State Commun. (USA) vol.54 no.1 (Apr 1985) p.103-5]
[30] L.G.Quagliano, H.Nather [Appl. Phys. Lett. (USA) vol.45 no.5 (1 Sept 1984) p.555-7]
[31] E.E.Mendez, M.Heiblum, R.Fischer, J.Klem, R.E.Thomas, H.Morkoc [J. Appl. Phys. (USA) vol.54 no.7 (July 1983) p.4202-4]
[32] G.Metze, A.R.Calawa, J.G.Mavroides [J. Vac. Sci. & Technol.(GB) vol.1 no.2 (Apr-June 1983) p.166-9]
[33] D.J.Ashen, P.J.Dean et al [J. Phys. & Chem. Solids (GB) vol.36 (1975) p.1041-53]
[34] R.Dingle [Phys. Rev. (USA) vol.184 no.3 (15 Aug 1969) p.788]
[35] T.S.Luv, G.E.Stillman, D.M.Collins, C.M.Wolfe, S.Tiwan, L.F.Eastman [Appl. Phys. Lett. (USA) vol.40 (15 June 1982) p.1034]
[36] B.J.Skromme et al [J. Electron. Mater. (USA) vol.12 no.2 (1983) p.433]
(1983)]
[33] B.El.Jani, J.C.Grenet et al [Rev. Phys. Appl. (France) vol.19 (1984) p.7-15]
[34] L.Samuelson et al [J. Cryst. Growth (Netherlands) vol.55 (1981) p.164-72]
[35] K.L.Hess et al [J, Electron. Mater. (USA) vol.11 no.6 (1982) p.1115]

12.8 PHOTOLUMINESCENCE SPECTRA OF GROUP II DOPED GaAs

by B.Hamilton

January 1986
EMIS Datareview RN=15603

SPECTROSCOPIC FEATURES

Shallow impurity photoluminescence (PL) diagnostics is normally carried out at cryogenic temperatures (T below 10 deg K) when particles are thermally stabilised. Three principal transitions are commonly studied: bound excitonic decay [BE], free-to-bound [FB] and donor-acceptor pair [DAP] transitions [1].

Excitonic decay at neutral donors or acceptors provides the sharpest signatures and is favoured for impurity identification, especially in low-doped systems. In particular, two hole replica BE emission [2] has been used to chemically identify Group II dopants. FB transitions are also extremely useful for dopant identification [3].

CHEMICAL IDENTIFICATION

Group II impurities form acceptor states by substituting for Ga on the Group III sublattice. The main transitions used for dopant identification are listed in the table.

Enhanced characterisation of acceptor states Zn (C and Ge) from excited state analysis are reviewed by Reynolds et al [6]. Several comprehensive studies of many shallow impurity transitions including Group II elements exist [3,4,5,7]. The upper concentration limit for precise chemical identification of the Group II elements is limited to about 10*16 cm*-3 by impurity screening. High resolution data for Group II elements, including acceptor excited state energies which can be obtained from two hole replica spectra for excitonic decay at neutral acceptors, and exciton lifetime can be found in [8].

GROUP II DOPING

The commonest Group II element used as a p-type dopant in LEC, LPE, VPE and MOCVD growth technologies is Zn. Zinc doping by diffusion is also of major importance. The luminescence characteristics of Zn substituted on a Ga site are well understood (table).

A rather broad emission band has been observed in Zn doped material (especially at relatively high Zn concentration) close to 1.4 eV [9]. This band has been associated with the zinc-arsenic vacancy Zn(Ga) - V(As) nearest neighbour pair. The defect has been investigated using high energy (2.2 MeV) electron irradiation, and annealing in order to probe defect reactions associated with this Zn related band [10]. Results indicate V(As) production leads to enhancement of the band (measured at 77 deg K with hf = 1.37 eV) through pairing with substitutional Zn. Anneal temperatures of 200 deg C (30 min) optimised the defect production as monitored by the PL intensity. At higher temperature mobile radiation induced defects interactions and thermal dissociation of the Zn(Ga) - V(As) pairs quench the luminescence.

The energy gap of GaAs can be made indirect by the application of hydrostatic pressure; the GAMMA(1) - GAMMA(8) gap converting to X1 - GAMMA(8) at around 40 kbar. All impurity related PL transitions reduce in efficiency at around this pressure [11]. The effect of hydrostatic pressure on heavily Zn doped GaAs PL spectra [12] showed the direct-indirect transition but with luminescence persisting up to 96 kbar in very heavily Zn doped (p = 9 X 10*19 cm*-3) material, due to the high free hole concentration [13]. At this pressure, the broad emission band peaked at approximately 2.4 eV at 120 deg K. The pressure coefficient for indirect gap (X1 GAMMA(8)) was found to be x = - (1.8 +- 0.6) X 10*-3 eV/kbar. Indirect luminescence was also observed.

Uniaxial stress has been shown to produce a short wavelength peak on the broad near edge emission measured in heavily Zn-doped GaAs (p above 5 X 10*18 cm*-3) . Polarisation studies [14] have demonstrated that this is due to valence band splitting.

Zinc related hot luminescence has been observed [15,16] with an emission threshold at 1.800 +- 0.3 eV. The observed emission was explained using hot hole and hot electron models.

Zinc has been used as a dopant in MBE growth, the low sticking coefficient associated with neutral Zn species being partially overcome by using a Zn+ ion source. Photoluminescence spectra indicate that p-type behaviour comparable with LPE can be achieved [17].

Beryllium doping is commonly used in MBE growth of GaAs. The increase of PL intensity within the Be emission bands has been observed to be sublinear with hole concentration [18]. The reason for this may be that a fraction of the incorporated Be atoms (possibly interstitial) play a role in forming non-radiative shunt paths [19].

Hot luminescence has been observed in Be doping MBE material [20]. A hot electron was used to explain the spectral emission and to estimate the L - GAMMA splitting of the conduction band to be 320 +- 4 meV.

TABLE: SPECTROSCOPIC FEATURES OF GROUP II IMPURITIES

ELEMENT	PL FEATURE (eV)			T	Ref
	FB	BE (1S)	THR(2S)	deg K	
Be	1.4923			2	[3]
	1.4915			5	[7]
		1.5124	1.4926	1.5	[7]
Mg	1.4911			5	[7]
		1.5124	1.4922	1.5	[7]

Ca doping experiments with VPE material

Sr	produces no extra PL emission lines				[7]
Zn	1.4888			5	
		1.5122	1.4904	1.5	[7]
Cd	1.4848			5	
		1.5123	1.4869	1.5	[7]

Note for the excitonic features measured at high resolution at
---- temperatures of below 2 deg K doublet centre energies are normally quoted.

REFERENCES
----

[1] P.J.Dean [Prog. Cryst. Growth & Charact. (GB) vol.5 (1982) p.89]

[2] P.J.Dean, D.J.Cuthbert, D.G.Thomas, R.T.Lynch [Phys. Rev. Lett. (USA) vol.18 (1967) p.548]

[3] K.Akimoto et al [Appl. Phys. Lett. (USA) vol.45 (1984) p.922]

[4] A.M.White, P.J.Dean, L.L.Taylor, R.C.Clarke, D.J.Ashen [J. Phys. C (GB) vol.5 (1972) p.1727]

[5] A.M.White, P.J.Dean, D.J.Ashen, J.B.Mullin, B.Day, P.D.Greene [J. Phys. C (GB) vol.6 (1973) p.L243]

[6] D.C.Reynolds, K.J.Bajaj, C.W.Litton [Solid State Comm. (USA) vol.53 no.12 (1985) p.1061]

[7] D.J.Ashen, P.J.Dean, D.J.T.Hurle, T.B.Mullins, A.M.White, P.D.Green [J. Phys. & Chem. Solids (GB) vol.36 (1975) p.1041]

[8] Landolt-Bornstein [vol.17 (1982) p.224-30]

[9] C.J.Hwang [Phys. Rev. (USA) vol.180 (1969) p.827]

[10] K.D.Glinchuck, A.V.Prokhorovich, N.S.Zayats [Phys. Status Solidi (Germany) (1984) p.625]

[11] P.Yu, B.Welber [Solid State Comm. (USA) vol.25 (1978) p.209]

[12] D.Dego, M.Cordona, H.Muller [J. Phys. Soc. Jpn. (Japan) vol.47 (1976) p.631]

[13] H.C.Casey, F.Stern [J. Appl. Phys. (USA) vol.47 (1976) p.631]

[14] N.S.Averkiev, S.A.Obukhov, A.A.Rogachev, N.A.Rud [Sov. Phys. Solid State (USA) vol.25 (1983) p.193]

[15] D.Olego, M.Cardona [Phys. Rev. B (USA) vol.22 (1980) p.1905]

[16] D.N.Mirlin, I.Y.Karlik, L.P.Nikitin, I.I.Reshina, V.F.Sapega [JETP Lett. (USA) vol.32 (1980) p.31]

[17] J.C.Bean, R.Dingle [Appl. Phys. Lett. (USA) vol.35 (1979) p.925]

[18] M.Illegems [J. Appl. Phys. (USA) vol.48 (1977) p.1278]

[19] N.Duhamal, P.Henoc, F.Alexandre, E.V.K.Rao [Appl. Phys. Lett. (USA) vol.39 (1981) p.49]

[20] E.A.Imhoff, M.I.Bell, R.A.Forman [Solid State Comm. (USA) vol.54 (1985) p.845]

12.9 PHOTOLUMINESCENCE SPECTRA OF GROUP IV DOPED GaAs

by B.Hamilton

Januaury 1986
EMIS Datareview RN=15604

Group IV elements can in principle form either donors or acceptors in GaAs by substitution into the Ga or As sublattices respectively (so called amphoteric behaviour). In practice the behaviour of Group IV dopants depends critically on the growth or doping conditions and thermodynamics so that generalisation can be difficult. For example, Si tends to exhibit donor activity Si(Ga) in most materials but under certain conditions shows acceptor activity in LPE GaAs [1]. Carbon favours acceptor activity [2] in LPE and MBE material and also in VPE material which is grown via the $AsCl_3$ route [3].

Donor related transitions are difficult to resolve in all but the highest purity samples, i.e. at the residual level. High resolution spectroscopy and magnetic field splitting are needed. Several Group IV donor states have been resolved using two electron transitions [4,5] and magnetic splitting [6].

Acceptor states, with larger binding energies and central cell corrections are more readily distinguishable [7] using excitonic or free to bound transitions.

The chief dopant role of the carbon acceptor is as a background contaminant in many major growth technologies [2,3] causing compensation in n-type material and degradation of mobility. The potentially complex role of C in photoluminescence transitions has been revealed by C ion implantation into pure GaAs grown by MBE [8]. These authors report seven transitions related to carbon including the `g´ line (1.5110 eV) which forms part of the sharp defect bound exciton luminescence commonly observed in MBE material [9].

High resolution data for the C neutral bound exciton (A(o)X) J = 3/2 and J = 5/2 transitions are reported in [6]. A transition resulting from recombination between a donor excited (n = 2) state and C (Zn and Si) acceptors has been observed between the normal free to bound (FB) and donor-acceptor pair (DAP) transitions [10].

The predominant electrical activity of Si in GaAs is that of a donor Si(Ga). High resolution measurements of transitions from magnetically split excited states of excitons bound to Si donors have been observed [6]. Such measurements are only possible at very low donor concentrations and normally Si is observed in the collective donor bound exciton peak in this spectral region.

In addition to Si(Ga), Si(As) and Si in a variety of complexed forms have been observed by photoluminescence.

Early work on Si doping of LPE material showed that Si could be made to favour the As site and that a variety of Si associated acceptor levels seemed to form [1], though at the relatively high doping levels produced, the luminescence bands were too broad to make definite assignments. Luminescence from what is thought to be the acceptor Si(As) at 1.485 eV has been reported for MBE material [11].

Silicon is an important n-type dopant in MBE growth. Recent work has shown extremely high quality spectra from lightly Si doped MBE layers [12]. These authors concluded that much residual dopant (Si and C) incorporation originated in the dopant sources. They further concluded that longer wavelength emission observed in MBE material (1.466-1.482 eV) was due to DAP recombination involving four rather deep, but unidentifed acceptors.

Silicon is also the major n-type dopant in ion implantation (into SI GaAs) technology. Photoluminescence has been used [13] to measure the ratio of Si(Ga):Si(As) and hence the degree of n-type activation for various rapid anneal schedules following implantation. The increase of FB luminescence, from the Si(As) acceptor, with anneal temperature demonstrates the siting preference of Si on the As site at higher temperatures.

Uniaxial stress splitting of free exciton luminescence and accompanying shift of donor bound exciton lines has been observed in Si implanted GaAs [14]. The complexes V(Ga)-Si(As) (1.35 eV) and V(Ga)-Si(Ga) (1.23 eV) together with Si(Ga) and Si(As) have been observed in annealed Si implanted SI material [15]. The ratio of the various defect species was shown to depend on the composition of capping material and its effectiveness as a barrier to Ga out-diffusion during anneal.

A gallium vacancy related band (probably V(Ga)-Si(Ga)) has been observed in Si doped MOCVD GaAs [15]. The intensity of the band varied with arsine molar ratio in the growth reactor. Similar bands have been observed in a variety of Si doped GaAs materials technologies, and similar luminescence is associated with other donor species [16].

In heavily Si doped material the role of conduction band tail states and compensation is important [1,17] and a recent analysis argues that the luminescence band shape and its excitation dependence can be explained by competition between valence band to tail transitions and transitions between tail states originating from both bands [18]. The complex nature of compensation caused by site competition between Si(Ga) and Si(As) and its influence on DAP transitions is discussed in [19]. Time resolved luminescence has been used to elucidate the dynamics of radiative recombination in heavily Si doped GaAs [20].

Neutron transmutation doping (NTD) is used to controllably dope material with Ge (Ga transmuted to Ge) and Se (As transmuted to Se). Photoluminescence has been used to study the formation of donor species and acceptor species caused by site switching e.g. Ge(Ga) to Ge(As) (acceptor). There is strong evidence for this acceptor incorporation in [2] monitored through the intensity of DAP and Ge A(o)X luminescence. Further evidence for site switching based on PL analysis is given in [21]. Changes in luminescence behaviour of NTD doped GaAs which relate to the properties of the neutron irradiation are discussed in [22].

Germanium doping in LPE Al(x)Ga(1-x)As with x varying from 0 to 0.32 is reported in [23]. The material remained p-type and the shift of Ge related peaks showed that the ionisation at the Ge acceptor depends strongly on x.

Luminescence from heavily Ge doped material is reported in [24] which demonstrates that compensation is important in determining band shape.

Ion implantation of Ge into GaAs has been reported in [25]. Luminescence spectra supported the view that significant damage remained after annealing, a peak at 1.360 eV (T = 6 deg K) being generally present after annealing and thought to be linked with Ga vacancies [26].

The epilayer substrate interface in Ge doped LPE material has been investigated in [27]. It was found that a defect line at 1.413 eV thought to be associated with carbon-arsenic vacancies [28] was reduced, probably by Ge occupation of the vacancy site.

Rather deep and broad bands are observed for Sn doped GaAs for various growth technologies. These bands have been associated with Ga vacancies both in the case of Sn and other donors [16]. The effect of excitation level, frequency response and thermal quenching have been used [29] to investigate the kinetics of a 1.39 eV band in Sn doped LPE material. This author discounted donor-acceptor transitions as a mechanism for this band, but invoked intra-centre or free to bound transitions.

Free to bound transitions were used to model the lineshape of Sn related luminescence in [30]. Luminescence decay and temperature evolution were used to show that conduction band to deep acceptor transitions fit the observed characteristics and to explain peak shift with doping. Group IV associated PL transitions are given in the table.

TABLE: Group IV PL transitions

element	peak energy eV	comments	ref
C	1.4935	FB transition measured at 5 deg K	[7]
	1.5124	A(o)X measured at 1.5 deg K 1S doublet centre	[7]
Si	1.4850	FB measured at 5 deg K	[7]
	1.5123	A(o)X measured at 1.5 deg K 1S doublet centre	[7]
Ge	1.4790	FB measured at 5 deg K	[7]
	1.5126	A(o)X measured at 1.5 deg K 1S doublet centre	[7]
Sn	1.349	FB measured at 5 deg K	[7]
	1.5067	A(o)X measured at 1.5 deg K 1S doublet centre	
Si(As)	1.4843	FB highly resolved in MBE material	[11]

complexes:

V(Ga)-Si(As)	1.35	measured at 12 deg K	[14]
V(Ga)-Si(Ga)	1.23	measured at 12 deg K	
C associated	1.5100	excitonic, measured at 2 deg K identical with g line in MBE GaAs [9]	[8]

REFERENCES

[1] H.Kressel, J.U.Dunse, H.Nelson, F.Z.Hawrylo [J. Appl. Phys. (USA) vol.39 (1968) p.2006]

[2] T.S.Low., M.H.Kim, B.Lee, B.J.Skromme, T.R.Lepkowski, C.E.Stillman [J. Electron. Mater. (USA) vol.14 (1985) p.477]

[3] M.Ozeki, K.Kitahara, K.Naki, K.Dazai, S.Okawe, O.Ruyzar [Jpn. J. Appl. Phys. (Japan) vol.16 (1977) p.1617]

[4] D.C.Reynolds, C.W.Litton, E.B.Smith, P.W.Yu, K.K.Bajaj [Solid State Commun. (USA) vol.42 (1982) p. 827]

[5] R.J.Almassy, D.C.Reynolds, C.W.Litton, K.K.Bajaj, C.L.Macoy [Solid State Commun. (USA) vol.38 (1981) p.1053]

[6] D.C.Reynolds, P.C.Colter, C.W.Litton, E.B.Smith [J. Appl. Phys. (USA) vol.55 (1984) p.1610]

[7] D.J.Ashen, P.J.Dean, D.T.J.Hurle, J.O.Mullin, A.M.White, P.D.Greene [J. Phys. & Chem. Solids (GB) vol.36 (1975) p.1041]

[8] Y.Makita et al [Nucl. Instrum. & Methods Phys. Res. (Netherlands) vol.B7/8 (1985) p.433]

[9] H.Kunzel, K.Ploog [Inst. Phys. Conf. Ser (GB) no.56 (1981) p.519]

[10] B.J.Skromme, C.E.Stillman [Phys. Rev. B (USA) vol.29 (1984) p. 1982]

[11] H.Kunzel, K.Ploog [Appl. Phys. Lett. (USA) vol.37 (1980) p.416]

[12] B.J.Skromme et al [J. Appl. Phys. (USA) vol.58 (1985) p.4685]

[13] S.Tiku, W.M.Duncan [J. Electrochem. Soc. (USA) vol.132 (1985) p.2237]

[14] Y.J.Chan, M.S.Lin, T.P.Chen [J. Appl. Phys. (USA) vol.58 (1985) p.545]

[15] K.Okamoto, S.Onozawa, T.Imai [J. Appl. Phys. (USA) vol.56 (1984) p.2293]

[16] E.W.Williams [Phys. Rev. (USA) vol.168 (1968) p.922]

[17] V.A.Vilkotskii, D.S.Domanevski, S.V.Zhokovets, M.V.Prokopenya [Sov. Phys. Semicond. (USA) vol.18 (1984) p.1368]

[18] V.L.Korolev, V.V.Rossin, V.G.Sidorov, Yu.K.Shalabutov [Sov. Phys. Semicond. (USA) vol.19 (1985) p.325]

[19] H.Takano, T.Kamijoh, M.Sakota [J. Lumin. (Netherlands) vol.31/32 (1984) p.442]

[20] E.O.Gobbel, W.Graudszus [Phys. Rev. Lett. (USA) vol.48 (1982) p.1277]

[21] J.Carrido, J.L.Castano, J.Piqueras [J. Appl. Phys. (USA) vol.57 (1985) p.2186]

[22] L.I.Kolenski et al [Sov. Phys. Semicond. (USA) vol.19 (1985) p.742]

[23] Y.R.Yuan, K.Mohammed, J.L.Merz [J. Appl. Phys. (USA) vol.57 (1985) p.2986]

[24] V.F.Kovalenko, V.A.Krasnov, Yu.E.Karanchuk [Sov. Phys. Semicond. (USA) vol.18 (1984) p.575]

[25] S.S.Chan, C.T.Marcyk, B.C.Streetmand [J. Electron. Mater. (USA) vol.10 (1981) p.213]

[26] P.K.Chatterjee, K.V.Vaidyanathan, M.S.Durschlag, B.C.Streetman [Solid State Commun. (USA) vol.17 (1975) p.1421]
[27] W.Y.Lum, A.R.Clawson, D.I.Elder, H.H.Wieder [J. Appl. Phys. (USA) vol.49 (1978) p.3333]
[28] W.H.Koschel et al [Inst. Phys. Conf. Ser. (GB) no.33a (1977) p.98]
[29] C.B.Norris [J. Appl. Phys. (USA) vol.48 (1977) p.4706]
[30] S.Zemon, M.O.Vasell, C.Lambert, R.H.Bartram [J. Appl. Phys. (USA) vol.53 (1982) p.3347]

12.10 PHOTOLUMINESCENCE SPECTRA OF GROUP VI DOPED GaAs

by B Hamilton

January 1986
EMIS Datareview RN=15605

Group VI elements form shallow donor states in GaAs by substitution for As on the As sublattice. Donor states in GaAs (and other low effective mass direct energy gap semiconductors) are extremely difficult to identify using photoluminescence because their bending energies are small (around 6 meV) as are the chemical shifts associated with individual impurities [1].

Attempts to chemically identify Group VI elements using sharp excitonic features are restricted to carefully doped high purity samples, with donor concentrations below approximately 10*15 cm*-3; above this concentration value impurity screening leads to line broadening, and the observation of a single donor bound exciton line.

Two exceptions to these comments are oxygen and chromium. The electrical role of O in GaAs is not yet fully understood but an excitonic line at 1.4889 eV [2] has been linked with O(As) in VPE material, though the defect involved may be more complex than a simple substitutional atom. Chromium shows many associated luminescence features including sharp 0.839 eV zero phonon line with associated sidebands that dominates in semi-insulating material. The band is thought to originate from an internal transition of a trigonally distorted Cr++ ion. This symmetry is thought to result from complexing between Cr(Ga) and some unknown defect [3,4]. The coupling constant of this band, S is approximately equal to 3 and associated phonon energies are approximately 10 meV (TA) and approximately 35 meV (LO) [5].

A variety of other Cr related bands have been reported in doped GaAs. These are generally broad and clearly involve vigorous phonon participation. These bands [6,7] are summarised in the table.

Donor species have been implicated in the appearance of broad bands near 1.2 eV in melt grown and epitaxial n-type GaAs [6]. These bands may come from donor-gallium vacancy (D-V(Ga)) complexes, and can be described by a multiphonon model on the strongly coupled limit with a coupling constant S approximately equal to 15. The bands are most readily observed in quite heavily doped material (n approx. 10*18 cm*-3). Details are given in the table.

The effect of doping level on the photoluminescence characteristics of S doped LPE material shows [12] that the emission intensity increases with S concentration up to about [S] = 3 X 10*18 cm*-3. The half width of the S donor luminescence also increases. Above this S concentration the luminescence efficiency drops and this was attributed to micro-precipitation of S. Sulphur associated deep level luminescence was also observed near one micron.

A comparison of luminescence efficiency for S and Se doping in VPE material shows [13] that the integrated intensity for both dopants increases linearly with electron concentration. However, Se doped material did not drop from linear variation until around n approx. 10*18 cm*-3, whereas the efficiency of S doped crystals began to degrade at about 5 X 10*17 cm*-3. Direct evidence for micro-precipitation of dopant was given in [13].

The influence of compensation (by Ge) on the band edge emission due to heavy Te doping indicates that a so-called `mobile' band is produced which shifts towards lower energies with increasing excitation power [14]. Recombination via tail states was rejected as a mechanism for this band, in favour of a donor-acceptor complex model. A more detailed model by the same authors is discussed [15].

Luminescence peaks have also been related to Te complexed with copper and with lattice vacancies. In [16] Te(As)-Cu(Ga) pairs were thought to be responsible for emission near 1.28 eV; and in [17,18] [V(Ga)-Te(As)]-V(As) for emission at 1.38 eV (all measurements at 77 deg K). Electron irradiation and annealing was used to study the growth and complexing kinetics of these defects.

Hydrostatic pressure was found to modify the properties of Te associated luminescence in a way which is dictated by the pressure dependence of the Fermi-level [17].

High resolution photoluminescence has been used [18] to resolve bound exciton luminescence from residual Te doping in VPE material. Luminescence transitions from excited (n=2) non-rigid stator states of the exciton were observed at 1.510216 eV.

Technique developments aimed at resolving donor states in GaAs using PL are reviewed in [19].

TABLE: Group VI related luminescence bands

element	peak energy eV	comments	refs
Cr (helium temperatures)	0.839 (plus sidebands)	Observed in SI and lightly n-type material. Internal transition of Cr++ charge state, trigonally distorted, possibly Cr++ associated with another point defect	[3,4,5,6]
	0.62	Observed in n-type GaAs, possibly due to electron capture at the Cr+++ ion	[7]
	0.57	Observed in SI and p-type GaAs, related hole capture into the Cr++ ion	[4,7]
	1 (approx.)	Bands near 1 eV have been observed in many n-type samples containing Cr but assignment is difficult	[6,8,9]

S	1.197	These bands are commonly observed in heavily doped	[11]
Se	1.224	material (e.g. [S] above 10*18 cm*-3) and have been	[11]
Te	1.216	linked to donor-gallium vacancy pairs. They are strongly phonon coupled. Peak energies quoted here correspond to T approx. 80 deg K	[11]
Te(As)Cu(Ga)	1.28	Measured at 77 deg K	[16]
[V(Ga)-Te(As)] -V(As)	1.35	Measured at 77 deg K	[17,18]

REFERENCES
----

[1] D.J.Ashen, P.J.Dean, D.T.J.Hurle, J.B.Mullin, A.M.White [J. Phys. & Chem. Solids (GB) vol.16 (1975) p.1041]

[2] A.M.White, P.J.Dean, D.Ashen, J.B.Mullin, B.Day [Proc. 12th Int. Conf. Phys. Semiconductors (Stuttgart) Tenbner p.381]

[3] F.Voillot, J.Barrau, M.Brousseau, J.C.Brabant [J. Phys. C (GB) vol.14 (1981) p.1855]

[4] G.Picoli, B.Deveaud, W.E Hagston [J. Phys. (France) vol.42 (1981) p.133]

[5] D.Xnan Tah, J.Barrau, J.C.Brabant, M.Brousseau [Semi Insulating III-V Compounds (Shiva) Evian (1982)]

[6] T.Instone, L.Eaves [J. Phys. C (GB) vol.11 (1978) p.L771]

[7] P.Leyral, F.Litty, S.Loualiche, A.Nouailhat, G.Guillot [Solid State Commun. (GB) vol.30 (1981) p.333]

[8] D.R.Wight, I.D.Blenkinsop, S.J.Bass [Semi Insulating III-V Compounds Conf. (Shiva) Nottingham (1980) p.174]

[9] G.A.Adegboyega, B.Tuck [Solid State Commun. (USA) vol.40 (1981) p.487]

[10] B.Deveaud, P.N.Tavennec [Solid State Commun. (USA) vol.43 (1982) p.953]

[11] E.W.Williams [Phys. Rev. (USA) vol.163 (1968) p.922]

[12] M.Ishi, T.Tanaka, W.Susaki [J. Cryst. Growth (Netherlands) vol.46 (1979) p.265]

[13] L.Hollan, M.Boulou and J.P.Chare [J. Electron. Mater. (USA) vol.10 (1981) p.193]

[14] B.G.Arnaudov et al [Sov. Phys. Semicond. (USA) vol.17 (1983) p.378]

[15] B.G.Arnaudov et al [J. Phys. C (GB) vol.17 (1984) p.331]

[16] K.D.Glinchuck, A.V.Prokorovich, N.S.Zayats [Phys. Status Solidi a (Germany) vol.72 (1982) p.715]

[17] D.Olego, M.Cardona, H.Muller [Phys. Rev. B (USA) vol.22 (1980) p.894]

[18] P.C.Colter, D.C.Reynolds, C.W.Litton, E.B.Smith [Solid State Commun. (USA) vol.45 (1983) p.375]

[19] D.C.Reynolds, K.K.Bajaj, C.W.Litton [Solid State Commun. (USA) vol.53 (1985) p.1061]

CHAPTER 13

DEFECTS AND INFRA-RED (IR) IMAGING OF DEFECTS IN GALLIUM ARSENIDE

13.1 Analysis of defects in undoped SI LEC GaAs by near-IR imaging

13.2 Defect densities in SI LEC GaAs

13.3 Structure of native defects in SI LEC GaAs

13.4 Defect energy levels in melt-grown GaAs

13.5 Defect energy levels related to the anion antisite defect in GaAs

13.6 Defect energy levels in LPE GaAs

13.7 Defect energy levels in VPE GaAs

13.8 Defect energy levels in MOVPE GaAs

13.9 Defect energy levels in MBE GaAs

13.10 Defect energy levels in electron irradiated GaAs

13.1 ANALYSIS OF DEFECTS IN UNDOPED SI LEC GaAs BY NEAR-IR IMAGING

by M.R.Brozel

November 1985
EMIS Datareview RN=15401

The major chemical contaminant in `undoped´, semi-insulating (SI), liquid encapsulated Czochralski (LEC) grown GaAs is carbon [1]. Carbon is a shallow acceptor and, as its concentration generally exceeds the total concentrations of chemical donor impurities in LEC GaAs, there must be a native compensating deep donor defect that produces semi-insulating (SI) behaviour and is not associated with an impurity. The identity of this defect is not known but it has been labelled EL2 (Electron Level 2).

The relative amounts of carbon and EL2 in LEC SI GaAs control the position of the Fermi level and thus the resistivity and other electrical properties [2,3]. It is therefore, important to measure the concentrations of carbon, [C], and EL2 centres, [EL2], within GaAs substrate wafers. Carbon concentrations are found by measuring the total absorption in the localised vibrational mode due to carbon acceptors that occurs at 582 cm*-1 [4]. By measuring this absorption across [110] diameters of thick GaAs slices, Holmes et al [2] have demonstrated that the distribution of carbon is rather variable and can vary by up to a factor of 2. However, more recent work by these authors [3] has reported that [C] is approximately uniform in some crystals. It must be emphasised that these data are macroscopic as any one measurement is an average over an area of up to 0.1 cm*2 representing the spatial extent of the beam in an infrared spectrophotometer when it is focused on the sample. To the author´s knowledge, nothing is known about inhomogeneities of carbon concentration on a microscopic scale.

Because it is well established that electrical properties are very variable across substrates and that they are probably determined by the relative values of [C] and [EL2] it follows that very non-uniform values of [EL2] may be expected. [EL2] is not easily measured directly in SI GaAs as the preferred method, Deep Level Transient Spectroscopy (DLTS), is not applicable to high resistivity material. However, Martin [5] has demonstrated that the magnitude of a broad infrared absorption band extending from 0.8 eV to the Band Edge at 1.5 eV correlates with the concentration of neutral EL2 centres [EL2(0)]. He also demonstrated that the Deep Level capacitance signal of EL2 could be `bleached´ by irradiating the sample with white light while it was held below 120 deg K and that this was also true for the infrared absorption band [5]. This is such an uncommon phenomenon that there is little doubt that the correlation is correct (but see note added in proof). Whilst EL2(0) centres are those which have not been ionised by the presence of the carbon acceptors, it follows that when [C] is approximately uniform, measurements of [EL2(0)] will reflect the distribution of total [EL2].

A calibration for [EL2(0)] as a function of absorption coefficient at a wavelength of 1 micron (1.24 eV) gives a cross section of 1.2 X 10*-16 cm*2, Martin [5]. This cross-section is rather small and in order to make quantitative estimates of [EL2(0)] several workers have found it necessary to perform the infrared absorption measurements on samples up to 5mm thick. Both Martin [5] and Holmes et al [2] found either W or U shape variations in [EL2] across [110] diameters of 3 inch (001) slices taken from the seeds and tails respectively of SI GaAs crystals and it was suggested that these reflect dislocation distributions. These smooth variations of [EL2(0)] were determined to be within +-20% of the mean. However, by using the focused beam of a spectrophotometer, Brozel et al [6] found considerable structure within the overall W and U patterns, indicating that the smooth variation of [EL2(0)] across wafers was an oversimplification. They also demonstrated that these `fine structures´ of [EL2(0)] could be imaged directly by viewing the light transmitted through the sample with a television camera fitted with an infrared sensitive tube such as a silicon vidicon. It was suggested that the origins of the fine structure were associated with dislocation arrays because infrared absorption images were similar to surface etching patterns which result from dislocations piercing the surface. The infrared imaging of 3 inch diameter slices [7] was accompanied by an attempt to describe all the features in terms of `cells´, `streamers´ and `sheets´. Brown et al [8] demonstrated that these features indeed correlated with complex dislocation arrays revealed by X-ray topography; the `cells´ were the images of polygonised dislocation arrays and the `streamers´ and `sheets´ represented lineage structures (very small angle grain boundaries, about 0.01 deg) running down the [001] growth axis. The formation of these infrared images and the orientations of the dislocation arrays that were required to satisfy these conditions were discussed by Brozel et al [9].

The correspondence between the images and dislocation arrays was confirmed by Stirland et al [10] who demonstrated, by comparing IR images and etch pit patterns, that a one-to-one correspondence between the images and complex groups of dislocation could be found.

Whilst it had been assumed that the images were produced entirely by EL2 centres, no direct evidence had been found. However, Skolnick et al [11,12] demonstrated that the images `bleached´ in the same way as the EL2 signature in photo-capacitance when the sample was correctly irradiated with white light. In addition, however, they showed that if the bleaching was performed at a temperature of about 20 deg K and the sample viewed with light of just below the band gap energy, a reverse image of the structure could be seen. Explanations for this extraordinary effect have been proposed but they remain speculative [13,14]. Whilst the correlation of the images with dislocation networks had been established, the question of whether high [EL2] was associated with individual dislocations remained open. This was because the dislocation density is high (10*4 - 10*5 cm*-2) in SI LEC GaAs and it is difficult to observe single dislocations. This limitation has recently been lifted by the availability of In-doped LEC GaAs samples of low and often zero dislocation density. Because dislocations are well separated in selected samples it was now possible to determine the validity of the previous correlations. Consequently infrared imaging of these samples was performed and this revealed discrete dark features [15] which were shown by A/B etching investigations to correlate exactly with single dislocations [16]. Thus the correlation of the infrared images with dislocations demonstrated that direct imaging of samples resulted in detailed mappings of dislocations.

Infrared absorbing features and therefore dislocations or dislocation arrays must be aligned roughly parallel to the beam direction to be observable because the absorption is cumulative and sufficient contrast is obtained only if absorbing features are viewed along their axis through samples over 3 mm thick. This restriction however, allows three-dimensional imaging to be performed, by taking photographs of images at slightly different angles and then viewing them in a stereoscopic viewer. This technique, `Infrared Absorption Topology´, has been demonstrated both for dislocation arrays in undoped GaAs [17] and also for single dislocations in indium-doped material [18].

Different approaches to infrared absorption mapping have been developed by many workers. Kaufmann et al [19] employed direct imaging by using infrared sensitive photographic film but later have turned to a method using detection by a linear diode array [20]. This latter technique is sensitive and quantitative, and can reveal much of the structure delineated using imaging. A point scanning technique is used by Dobrilla and Blakemore [21] where wafers of 300-500 microns can be assessed. This computer controlled system, using a cooled Ge photo-detector, is probably the most sensitive at present. Holmes and Chen [22] use a similar system with reduced spatial resolution, as do Gouteraux et al [23]. Castagne et al [26] have introduced electronic enhancers into their imaging equipment both to reveal further detail and also to investigate thinner samples. Very recently, sophisticated equipment has been developed by Silverberg et al [27] and Katsumata et al [28]. In addition to the analysis of `as-grown´ LEC GaAs, imaging systems have been used by Rumsby et al [24] and by Martin et al [25] to investigate the effects of annealing ingots after growth. They observe that ingots annealed at 850 deg C for several hours exhibit greatly reduced non-uniformities in [EL2] and, in some cases, [EL2] becomes completely uniform. It is expected that infrared mapping of LEC GaAs will be further exploited as an assessment tool both in the routine analysis of GaAs wafers and also as a research tool for SI GaAs substrates.

Note added in proof:

A very recent study by Ishida et al [29], who have repeated the work by Martin [5], has failed to substantiate the correlation between the near infrared absorption near 1 micron and [EL2(0)]. If this work is correct it follows that the infrared absorption imaging technique is detecting deep levels other than EL2 which aggregate around dislocations and exhibit `bleaching´.

REFERENCES

[1] D.E.Holmes, R.T.Chen, K.R.Elliott, C.G.Kirkpatrick [Appl. Phys. Lett. (USA) vol.40 (1982) p.46]

[2] D.E.Holmes, K.R.Elliott, R.T.Chen, C.G.Kirkpatrick [`Semi Insulating III-V Materials´ (Shiva publishing, 1982) p.19]

[3] R.T.Chen, D.E.Holmes, P.M.Asbeck [Appl. Phys. Lett. (USA) vol.45 (1984) p.459]

[4] M.R.Brozel, J.E.Clegg, R.C.Newman [J. Phys. D (GB) vol.11 (1978) p.133]

[5] G.M.Martin [Appl. Phys. Lett. (USA) vol.39 (1981) p.747]

[6] M.R.Brozel, I.Grant, R.M.Ware, D.J.Stirland [Appl. Phys. Lett. (USA) vol.42 (1983) p.610]

[7] M.S.Skolnick, M.R.Brozel, L.J.Reed, I.Grant, D.J.Stirland, R.M.Ware [J. Electron. Mater (USA) vol.13 (1984) p.107]

[8] G.T.Brown, M.S.Skolnick, G.R.Jones, B.K.Tanner, S.J.Barnett [`Semi-Insulating III-V Materials´ (Shiva publishing, 1984) p.76]

[9] M.R.Brozel, I.Grant, R.M.Ware, D.J.Stirland, M.S.Skolnick [J. Appl. Phys. (USA) vol.56 (1984) p.1109]

[10] D.J.Stirland, I.Grant, M.R.Brozel, R.M.Ware [Inst. Phys. Conf. Ser No.67 (1985) p.285]

[11] M.S.Skolnick, L.J.Reed, A.D.Pitt, [Appl. Phys. Lett. (USA) vol.44 (1984) p.447]

[12] M.S.Skolnick, D.A.O.Hope, B.Cockayne [`Semi-Insulating III-V Materials´ (Shiva Publishing, 1984) p.446]

[13] J.P.Fillard, J.Bonnafe, R.Castagne [J. Appl. Phys. (USA) vol.56 (1984) p.3020]

[14] H.J.Queisser [Appl. Phys. Lett. (USA) vol.46 (1984) p.757]

[15] E.J.Foulkes, M.R.Brozel, I.Grant, P.Singer, B.Waldock, R.M.Ware [`Semi-Insulating III-V Materials´ (Shiva publishing, 1984) p.160]

[16] D.J.Stirland, M.R.Brozel, I.Grant [Appl. Phys. Lett. (USA) vol.46 (1985) p.1066]

[17] M.R.Brozel, E.J.Foulkes, D.J.Stirland [Inst. Phys. Conf. No.74 (1984) p.59]

[18] S.Clark, M.R.Brozel, D.J.Stirland [Proceedings of `Defect Recognition and Image Processing in III-V compounds´ In press]

[19] U.Kaufmann, J.Windsheif, J.Schneider, F.Kohl [`Semi-Insulating III-V Materials´ (Shiva Publishing, 1984) p.246]

[20] J.Windscheif, M.Beaumler, U.Kaufmann [Appl. Phys. Lett. (USA) vol.46 (1985) p.447 and in Proceedings of `Defect Recognition and Image Processing in III-V Compounds´ (1985) In press.]

[21] P.Dobrilla, J.S.Blakemore [J. Appl. Phys. (USA) vol.58 (1985) p.208]

[22] D.E.Holmes, R.T.Chen [J. Appl. Phys. (USA) vol.55 (1984) p.3588]

[23] B.Gouteraux, N.Visentin, B.Lent, B.Bonnet [Inst. Phys. Conf. Ser. no.74 (1984) p.89]

[24] D.Rumsby, I.Grant, M.R.Brozel, E.J.Foulkes, R.M.Ware [`Semi-Insulating III-V Materials´ (Shiva Publishing, 1984) p.165]

[25] S.Martin, M.Duseaux, M.Erman [Inst. Phys. Conf. Ser. no.74 (1984) p.53]

[26] M.Castagne, Z.Fugui, J.Bonnafe, J.P.Fillard [Proceedings of `Defect Recognition and Image Processing in III-V Semi-conductors´ (1985) In press.]

[27] P.Silverberg, L.Neilsen, P.Omling, L.Samuelson [ibid]

[28] T.Katsumata, T.Obokata, M.Nakajima, T.Fukuda [ibid]

[29] K.Ishida, A.Yahata, T.Kikuta, [Jpn. J. Appl. Phys. (Japan) vol.24 (1985) p.L250]

13.2 DEFECT DENSITIES IN SI LEC GaAs

by M.R.Brozel

November 1985
EMIS Datareview RN=15400

Undoped LEC GaAs contains crystal defects that result in both the semi-insulating behaviour (that makes the material so attractive for many applications) and the instability and non-uniformity that have been the bane of the GaAs industry for many years.

Crystal defects can be divided into three groups:

A Line defects (dislocations)
B Volume defects such as precipitates
C Point defects

A. LINE DEFECTS

Dislocations in commercial 2 inch and 3 inch diameter Liquid Encapsulated Czochralski (LEC) GaAs crystals form well-defined structures. These result from polygonization and other interactions of slip dislocations which are produced during and soon after solidification. As a result, the dislocation density, D, is extremely non-uniform and ranges from 10*4 to 10*5 cm*-2 [1,2,3]. It is generally held that these values are unacceptably high and much work is in progress to reduce D considerably. These high values are a result of large thermal gradients being present when the crystal is grown and, as a result, several groups have attempted to reduce D by modifying these gradients. Methods that have been reported are listed below:

1. The Liquid Encapsulated Kyropoulos (LEK) process is similar to the LEC method except that the crystal is grown under the B2O3 encapsulant and is not withdrawn until growth is completed. As a result, thermal gradients are reduced. Values of D from 10*3 to 10*4 cm*-2 have been reported [4,5,6].

2. LEC growth is usually performed under a high ambient gas pressure. If this pressure is reduced, thermal gradients can be modified to produce crystals with low values of D (typically around 10*3 to 10*4 cm*-2) [7,8].

3. Reductions in thermal gradients can be achieved by inserting "thermal baffles" in the growth chamber to reduce convective currents and to reduce radiation losses. Values of D of around 10*3 cm*-2 have been reported [9].

4. Pulling the crystal into a thick B2O3 encapsulant instead of through it, Fully Encapsulated Czochralski (FEC), has resulted in values of D of 10*3 to 10*4 cm*-2 [10].

5. Large temperature gradients are produced by convective currents within the GaAs melt. These can be damped by the application of an external magnetic field and this has resulted in reported values of

D of about 10*3 cm*-2 [11].

6. The addition of isoelectronic impurities, such as indium, to the melt at high concentration (10*19 to 10*20 cm*-3) results in low values of D (below 10*2 cm*-2) and these can even be reduced to zero [2,12,13,14].

Clearly the latter is most effective in reducing the dislocation density but several workers have combined many of these methods to produce large LEC GaAs crystals that are effectively dislocation free throughout large volumes [15].

B. VOLUME DEFECTS

Volume defects are, in general, observed to be associated with dislocations and those that are large enough to be studied have been analysed by X-ray diffraction to be hexagonal arsenic precipitates [16,17]. However, the presence of microprecipitates consisting of GaAs particles has been reported by some workers [18] and amorphous precipitates by others [19]. The mechanisms by which these precipitates are produced are not known and their effects on material properties are also unknown.

C. POINT DEFECTS

Point defects can be separated into two major categories:

chemical impurities
native defects

There is also the possibility of complexing between such basic entities.

C.1 Chemical Impurities in LEC SI GaAs

An excellent review of both the analytical techniques and results from various samples of GaAs has been given by Clegg [20]. A table of typical results taken from Clegg is given below:

S:	mid 10*14 to mid 10*15 cm*-3
Se:	low 10*13 to low 10*15 cm*-3
Te:	10*12 to 10*14 cm*-3
Cr:	1 X 10*13 to mid 10*13 cm*-3
Fe:	around 10*14 cm*-3
Mn:	mid 10*13 cm*-3
Mg:	around 10*14 cm*-3
Si:	low 10*14 to mid 10*16 cm*-3 *
C:	low 10*15 to low 10*16 cm*-3 *
B:	mid 10*14 to mid 10*17 cm*-3 *

* these concentrations are a function of the crucible material (quartz or pyrolytic boron nitride, pBN) and the H_2O content of the boric oxide encapsulant [21]. High silicon concentrations are found in crystals grown from a quartz (SiO_2) crucible but only if the H_2O content of the B_2O_3 encapsulant is low. Carbon is observed at high

concentration in crystals grown from either crucible type but only if a "dry" B2O3 encapsulant is used. Boron is a troublesome impurity in crystals grown in a pBN crucible under dry B2O3.

C.2 Native Crystal Defects and "Deep Levels"

Generally, native defects can only be observed if they are associated with energy levels situated relatively deep in the band gap. They can be detected and their concentrations measured in SI material using thermally stimulated current (TSC) spectroscopy [22]. This technique can be extended to detect deeper traps by using light excitation; optical transient current spectroscopy (OTCS) [23]. Because of the difficulty in separating hole traps and electron traps using either TSC or OTCS another method, deep level transient spectroscopy (DLTS), has been developed [24]. Unfortunately DLTS requires that a rectifying junction be present and this is not possible when investigating semi-insulating material. Accordingly, many workers grow n-type LEC GaAs in order to perform DLTS measurements and then assume that defects present in n-type GaAs are also present in SI material [25].

All the techniques are very sensitive (detectivity about 10*12 cm*-3 in some cases) and it is often difficult to identify levels as being due to particular chemical impurities or native crystal defects. For this reason all these deep level defects are treated in the same section.

Deep levels are split into two groups: Deep donor or Electron Levels (EL group) and deep acceptor or Hole Levels (HL group) according to their signature in DLTS measurements (TSC and OTCS do not distinguish the two types). The table below is compiled mostly from data by Martin (TSC and OTCS) [26], Farges et al (DLTS) [21] and Iwata et et al (DLTS) [25], but other data have been included where necessary.

LABEL	ORIGIN	CONCENTRATION cm*-3	EMISSION PARAMETERS energy E(eV)	cross section cm*2	
EL11			Ec-0.17	3 X 10*-16	[26]
EL17			Ec-0.22	1.0 X 10*-14	[26]
EL14			Ec-0.215	5.2 X 10*-16	[26]
EL6	complex defect	10*14 - 10*16	Ec-0.35	1.5 X 10*-13	[26]
"		2 X 10*15	"	"	[21]
"		4 X 10*14	"	"	[25]
EL5		10*14 - 10*16	Ec-0.42	10*-13	[26]
"		below 2 X 10*14	"	"	[21]
EL3	Ni?	10*13 - 10*15	Ec-0.575	1.2 X 10*-13	[26]
"		below 2 X 10*14	"	"	[21]
"		4 X 10*13 - 1 X 10*14	"	"	[25]

EL2*	complex native defect	5 X 10*15 - 3 X 10*16	Ec-0.825	1.2 X 10*-13	[26]
"	"	2 X 10*15 - 6 X 10*15	"	"	[25]
HL1	Cr	0 - 3 X 10*17 (doping dependent)	Ev+0.886	10*-14	[26]
HL10			Ev+0.83	1.7 X 10*-13	[26]
"		below 2 X 10*14	"	"	[21]
HL9			Ev+0.69	1.1 X 10*-13	[25]
HL3	Fe		Ev+0.59	3 X 10*-15	[26]
HL4	Cu or Ni	below 10*15	Ev+0.42	3 X 10*-15	[26]
"		below 2 X 10*14	"	"	[21]
"		1 X 10*15	"	"	[25]
HL7		1.7 X 10*15	Ev+0.32		[25]
HL12	Zn-associated		Ev+0.27	1.3 X 10*-14	[26]
	Cu		Ev+0.14		[26]
	gallium antisite ** Ga(As)	3 X 10*15 to 3 X 10*16 (Ga-richness of melt, dependent)	Ev+0.077		[27,28]
	arsenic antisite As(Ga) (associated with EL2)	up to 10*16	Ec-0.75		[29]
-	boron antisite ** B(As)?	mid 10*16	Ev+0.2 (?)		[30]

* The DLTS signature of EL2 has been resolved into at least two components, the relative concentrations being a function of crystal growth parameters [31,32]. EL2, the most important of the native deep levels in SI, LEC GaAs gives rise to an infra red absorption band at 1 micron [33] and the measurement of this absorption is now the usual method of determining [EL2] [34,35,36]. For further discussion see Makram-Ebeid [37].

** p-type material grown from Ga-rich melts.

A NOTE

LEC GaAs improves in quality every month. For this reason, older analyses of defect concentrations in LEC GaAs have been omitted as they do not represent either current material or currently accepted accuracies of measurement. For more recent, but rather less comprehensive, discussions of chemical and native defect concentrations, the reader should consult Thomas et al [38], Stolte [7] and Kirkpatrick et al [39].

REFERENCES

[1] A.S.Jordan, R.Caruso, A.R.Von Neida [Bell Syst Tech J. (USA) vol.59 (1980) p.593] (theoretical discussions)

[2] A.S.Jordan, A.R.Von Neida, R.Caruso [J. Cryst. Growth (Netherlands) vol.70 (1984) p.155]

[3] D.Rumsby, R.M.Ware, M.Whittaker ["Semi-Insulating III-V Materials", Nottingham (Shiva Publishing, 1980) p.59]

[4] M.Duseaux [J. Cryst. Growth (Netherlands) vol.61 (1983) p.576]

[5] G.Jacob ["Semi-Insulating III-V Materials" (Shiva Publishing, 1982) p.2]

[6] G.Jacob [J. Cryst. Growth vol.58 (1982) p.455]

[7] C.A.Stolte ["Semiconductors and Semimetals" Ed. Willardson & Beer, Academic Press, vol.20 (1984) p.89]

[8] A.G.Elliott, C-L.Wei, R.Farraro, G.Woolhouse, M.Scott, R.Hiskes [J. Cryst. Growth (Netherlands) vol.70 (1984) p.169]

[9] T.Shimada, T.Obokata, T.Fukuda [Jpn. J. Appl. Phys. (Japan) vol.23 (1984) p.L441]

[10] See ref.8 and T.Shimada, K.Terashima, H.Nakajima, T.Fukuda [Jpn. J. Appl. Phys. (Japan) vol.23 (1984) p.L23]

[11] K.Terashima, T.Fukuda [J. Cryst. Growth (Netherlands) vol.63 (1983) p.423]

[12] G.Jacob, M.Duseaux, J.P.Farges, M.M.B.van den Boom, P.J.Roksnoer [J. Cryst. Growth (Netherlands) vol.61 (1983) p.417]

[13] M.Duseaux, S.Martin ["Semi-Insulating III-V Materials" (Shiva Publishing, 1984) p.118]

[14] H.Kimura, C.B.Afable, H.M.Oben, A.T.Hunter, H.V.Winston, [J. Cryst. Growth (Netherlands) vol.70 (9184) p.185]

[15] H.Kohda, K.Yamada, H.Nakanishi, T.Kobayasi, J.Osaka, K.Hosikawa [J. Cryst. Growth (Netherlands) vol.71 (1985) p.813] (an example of where several techniques have been applied)

[16] A.G.Cullis, P.D.Augustus, D.J.Stirland [J. Appl. Phys.. (USA) (1980) p.2556]

[17] D.J.Stirland, P.D.Augustus, M.R.Brozel, E.J.Foulkes ["Semi-Insulating III-V Materials" (Shiva Publishing, 1984) p.91]

[18] J.P.Cornier, M.Duseaux, J.P.Chevalier [Inst. Phys. Conf. Ser No.74 (1985) p.95]

[19] F.A.Ponce, F-C.Wang, R.Hiskes ["Semi-Insulating III-V Materials" (Shiva Publishing, 1984) p.68]

[20] J.B.Clegg ["Semi-Insulating III-V Materials" (Shiva Publishing, 1982) p.80]

[21] J.P.Farges, G.Jacob, C.Schemali, G.M.Martin, A.Mircea-Roussel, J.Halais [ibid p.45]

[22] M.G.Buehler [Solid State Electron. (GB) vol.15 (1972) p.69]

[23] C.Hurtes, M.Boulou, A.Mittonneau, D.Bois [Appl. Phys. Letts. (USA) vol.32 (1978) p.821]

[24] D.V.Lang [J. Appl. Phys. (USA) vol.45 (1974) p.3022]

[25] N.Iwata, F.Hasegawa, B.Yamamoto, Y.Nannichi [Jpn. J. Appl. Phys. (Japan) vol.223 (1983) Suppl.22-1 p.595] (an example of this technique where a p/n junction was produced after Zn diffusion into n-GaAs). See also, same authors [Jpn. J. Appl. Phys. (Japan) vol.22 (1983) p.L502

[26] G.M.Martin ["Semi-Insulating III-V Materials" (Shiva publishing, 1980) p.13]

[27] K.R.Elliott, D.E.Holmes, R.T.Chen, C.G.Kirkpatrick [Appl. Phys. Lett. (USA) vol.40 (1982) p.898]

[28] P.W.Yu, W.C.Mitchel, M.G.Mier, S.S.Li, W.L.Wang [Appl. Phys. Lett. (USA) vol.41 (1982) p.532] See also K.R.Elliott [Appl. Phys. Lett. (USA) vol.42 (1983) p.274]

[29] The identity of the As antisite and its relationship to EL2 is still unresolved. For discussions of these points see D.M.Hofmann, B.K.Meyer, F.Lohse, J.-M.Spaeth [Phys. Rev. Lett. (USA) vol.54 (1984) p.1187 and B.K.Meyer, J.-M.Spaeth, M.Scheffler [Phys. Rev. Lett. (USA) vol.52 (1984) p.851]

[30] J.Woodhead, R.C.Newman, I.Grant, D.Rumsby, R.M.Ware [J. Phys. C (Solid State Phys) (GB) vol.16 (1983) p.5523]

[31] M.Taniguchi, T.Ikoma [J. Appl. Phys. (USA) vol.54 (1983) p.6448]

[32] J.Lagowski, D.G.Lin, T.Aoyama, H.C.Gatos [Appl. Phys. Lett. (USA) vol.44 p.336]

[33] G.M.Martin [Appl. Phys. Lett. (USA) vol.39 (1981) p.747]

[34] M.R.Brozel, I.Grant, R.M.Ware, D.J.Stirland [Appl. Phys Lett. vol.42 (1983) p.610]

[35] D.E.Holmes, R.T.Chen, K.R.Elliott, C.G.Kirkpatrick [Appl. Phys. Lett (USA) vol.43 (1983) p.305]

[36] M.S.Skolnick, M.R.Brozel, L.J.Reed, I.Grant, D.J.Stirland, R.M.Ware [J.Electron. Mater (USA) vol.13 (1984) p.107]

[37] S.Makram-Ebeid, P.Langlade, G.M.Martin ["Semi-Insulating III-V Materials" (Shiva Publishing,1984) p.184]

[38] R.N.Thomas, H.M.Hobgood, G.W.Eldridge, D.L.Barrett, T.T.Braggins, L.B.Ta, S.K.Wang ["Semiconductors and Semimetals" Ed. Willardson & Beer, Academic Press, vol.20 (1984) p.1]

[39] C.G.Kirkpatrick, R.T.Chen, D.E.Holmes, P.M.Asbeck, K.R.Elliott, R.D.Fairman, J.R.Oliver [ibid p.159]

13.3 STRUCTURE OF NATIVE DEFECTS IN SI LEC GaAs

by S.Yasuami

December 1985
EMIS Datareview RN=15441

Native defects are classified into two categories: extended and point defects. In undoped SI LEC crystals, dislocations, lineage and precipitates are mainly extended defects and are observed by XRT, TEM etc. Microdefects such as dislocation loops with or without stacking faults in them are seldom observed.

A sole dislocation is rarely observed and usually bundles of dislocations form cellular walls or lineage. The former is predominant in the core region and the latter extends from the core to the peripheral region of a crystal. Since lineage is constituted of dislocations with similar directions and Burgers vectors, it forms a low angle tilt boundary. The cell wall is a complex agglomeration of dislocations with different directions and Burgers vectors [1].

Precipitates accompanying dislocations were observed for Cr-doped [2] and undoped [3] crystals by TEM. A possiblity was suggested that the precipitates formed directly on the dislocations by condensation of As point defects (interstitials) from the surrounding lattice. Amorphous precipitates with 2 to 3 nm dia. were also observed using lattice imaging technique of TEM [4].

Dislocations are considered to act as sinks for point defects and redistribute the point defects (non-radiative centres) [5]. Non-uniform luminescence efficiency associated with dislocations of cathodoluminescence [6,7] and photoluminescence [8] have been ascribed to their combined effects.

In undoped semi-insulating crystals, point defects that create deep levels in the band gap are crucial. Amongst them, studies are concentrated on the stoichiometry related [9] deep donor EL2, [10] which compensates carbon acceptor and makes the crystal semi-insulating. EL2 density is usually obtained from the near infrared 1.0 micron absorption coefficient, which is calibrated to the DLTS signals obtained for conductive crystals [11]. However, one-to-one correspondence between the EL2 signal and near infra-red absorption coefficient is still controversial [12,13].

The EPR spectrum for As(Ga) antisite defects was identified with EL2 [14]. However, there are two models proposed for EL2: As(Ga) [15,16] and its complexes [17].

See also Datareviews [18-20].

REFERENCES
----

[1] J.Matsui, T.Kitano, T.Kamejima, T.Ishikawa [11th Conf. GaAs and Related Compds., Biarritz, 1984]

[2] A.G.Cullis, P.D.Augustus, D.J.Stirland [J. Appl. Phys. (USA) vol.51 no.5 (May 1980) p.2556-60]

[3] T.Kamejima, J.Matsui [Unpublished work]

[4] F.A.Ponce, F.-C.Wang, R.Hiskes [Semi-Insulating III-V Materials (Shiva, Nantwich, 1984) p.68-75]

[5] A.H.Cottrel [Dislocations and Plastic Flow in Crystals (Clarendon Press, Oxford, 1953) Ch. II p.53]

[6] A.K.Chin, A.R.Von Neida, R.Caruso [J. Electrochem. Soc. (USA) vol.129 no.10 (Oct 1982) p.2386-8]

[7] T.Kamejima, F.Shimura, Y.Matsumoto, H.Watanabe, J.Matsui [Jpn. J. Appl. Phys. (Japan) vol.21 no.11 (Nov 1982) p.L721-3]

[8] K.Kitahara, M.Ozeki, A.Shibatomi [Jpn. J. Appl. Phys. (Japan) vol.23 no.2 (Feb 1984) p.207-11]

[9] D.E.Holmes, R.T.Chen, K.R.Elliot, C.G.Kirkpatrick [Appl. Phys. Lett. (USA) vol.40 no.1 (Jan 1982) p.46-8]

[10] G.M.Martin, A.Mittonneau, A.Mircea [Electron. Lett. (UK) vol.13 no.7 (Mar 1977) p.191-3]

[11] G.M.Martin [Appl. Phys. Lett. vol.39 no.9 (Nov 1981) p.747-8]

[12] M.Taniguchi, T.Ikoma [Appl. Phys. Lett. (USA) vol.45 no.1 (July 1984) p.69-71]

[13] K.Ishida, A.Yahata, T.Kikuta [Jpn. J. Appl. Phys. (Japan) vol.24 no.4 (April 1985) p.L250-2]

[14] K.Elliot, R.T.Chen, S.G.Greenbaum, R.J.Wagner [Appl. Phys. Lett. (USA) vol.44 no.9 (May 1984) p.907-9]

[15] M.Baeumler, U.Kaufmann, J.Windscheif [Appl. Phys. Lett. (USA) vol.46 no.8 (April 1985) p.781-3]

[16] N.Tsukada, T.Kikuta, K.Ishida [Jpn. J. Appl. Phys. (Japan) vol.24 no.9 (Sept 1985) p.L689-92]

[17] B.K.Meyer, J.-M.Spaeth [J. Phys. C. (UK) vol.18 no.4 (Feb 1985) p.L99-103]

[18] M.R.Brozel [EMIS Datareview RN=15401 (Nov 1985) "Analysis of defects in undoped SI LEC GaAs by near-IR imaging".

[19] M.R.Brozel [EMIS Datareview RN=15400 (Nov 1985) "Defect densities in SI LEC GaAs"]

[20] H.V.Winston [EMIS Datareview RN=13993 (Dec 1985) "Defect energy levels in melt-grown GaAs"]

13.4 DEFECT ENERGY LEVELS IN MELT-GROWN GaAs

by H.V.Winston

December 1985
EMIS Datareview RN=13993

The EMIS Datareview `Defect Densities in SI LEC GaAs´ [1] includes a table of deep energy levels and corresponding emission parameters based on reviews by Martin [2] and Hollan [3] of data on LEC and Bridgman GaAs. Neumark and Kosai [4] provide useful tables of energy levels and plots of the signatures (carrier emission rates divided by T*2 vs reciprocal T) for defects in various kinds of GaAs, including melt-grown. A major reason for the extensive work on deep levels is their technological importance in the preparation of semi-insulating GaAs substrates for microwave and high-speed digital devices and integrated circuits. Look [5] has reviewed deep levels in semi-insulating material and the methods for characterizing them. The present Datareview will concentrate on the behaviour of GaAs crystals grown from an approximately stoichiometric melt with or without intentional doping, and either in the as-grown condition or after a post-growth annealing. Energy levels related to radiation damage will not be considered. In comparing energy levels reported by different authors, it should be borne in mind that different experimental methods may produce different numerical values of energy. For example, the DLTS-derived activation energy for EL2 is 0.82 eV, in contrast to temperature dependent Hall values close to 0.75 eV; in this case the difference corresponds to the activation energy of the electron capture cross section. Another important point is that the conditions of crystal growth, such as stoichiometry and thermal gradients, and of annealing, may profoundly influence the nature of the defects and their spatial distribution in the crystal, and thus the electrical, optical, and crystallographic properties. An instructive demonstration of some of these issues is presented in the report of an exhaustive study of two typical semi-insulating ingots by J.Bonnafe et al [7].

The Cr deep acceptor level is technologically important because Cr doping was the first reliable way to obtain semi-insulating GaAs. The Cr level compensates the shallow donors that would otherwise dominate the conductivity of either Bridgman or LEC crystals grown using quartz crucibles or boats [3,6]. Difficulties caused by Cr redistribution near the surface during high-temperature processing and the resulting loss of the semi-insulating property made `undoped´ semi-insulating LEC GaAs grown from pyrolytic nitride (PBN) crucibles very attractive. In this material the residual impurities are predominantly acceptors, mainly C, and compensation is effected by a donor level about 0.75 eV below E(c) [2,3,8], corresponding to the EL2 level measured by DLTS. This level was once thought to be due to O at an As site, but Huber et al [9] showed that the crystals they studied did not contain enough O to account for the observed EL2 concentration.

The EL2 concentration increases with increasing As pressure over the melt in Bridgman growth [10], suggesting that it is related to native defects occurring in As-rich crystals. Lagowski et al [10] proposed that EL2 is the As antisite, As(Ga). In LEC growth, As-rich conditions also lead to high concentrations of EL2 [8,11,12].

The relation between EL2 and As(Ga) has been the subject of many investigations. A recent overview is provided by the Proceedings of the Thirteenth International Conference on Defects in Semiconductors, which includes a review of work presented at the conference [13]. As(Ga) has a paramagnetic singly ionized state that can be identified through its EPR spectrum [14]. Elliott et al [15] showed that the concentration of the singly positive state of EL2 corresponded to the spin density estimated from the EPR spectrum of the singly ionized As(Ga) in the same sample, thus establishing the close connection between the two entities.

However, neither EL2 nor As(Ga) behaves simply. Taniguci and Ikoma [16,17] showed that many melt-grown crystals contain a family of levels, all with energies and emission rates close to the accepted EL2 value, but differing slightly from each other. They suggested that the different members of the family might be different aggregates of excess As and other native defects. It should be noted that As(Ga) itself is a cluster of five As atoms. Moreover, optically detected electron nuclear double resonance experiments on the As antisite [18] reveal hyperfine structure corresponding to As(Ga) surrounded tetrahedrally by four equivalent As atoms as well as As(Ga) with different nearest neighbor interactions; that is, several kinds of atomic configuration are present. Another important similarity between EL2 and As(Ga) is that both undergo photoquenching and the formation of a metastable excited state at low temperatures [13]. The variety of EL2-like defects may account for the observation by Imamura and Osaka [19] of two kinds of deep traps in undoped semi-insulating LEC GaAs, at 0.72 and 0.86 eV below E(c). Lagowski et al [20] identified still another level with activation energy in the EL2 range as due to O; they designated it as ELO.

The correlation between the singly positive states of EL2 and As(Ga) eliminates the proposal [21] that EL2 is related to the second ionization of As(Ga) and that there is a shallow donor level related to the first ionization. This was also demonstrated by Hunter et al [22], who added donors to semi-insulating LEC GaAs by neutron transmutation doping and annealing, thus bringing the Fermi level into a region where the conductivity would be dominated by any supposed shallow first ionization of As(Ga). They found no levels in that region with as much as a tenth of the EL2 concentration.

Fine-scale inhomogeneities in EL2 concentration in as-grown LEC crystals are correlated with details of the dislocation network [23]. Large-scale variations of EL2 concentration across LEC wafers often correlate with the overall distribution of dislocation density, but the variation of EL2 concentration along the axis of an LEC crystal is controlled by melt stoichiometry, as shown by Holmes and Chen [24]. They suggested that the 10-20 per cent variations across a wafer are caused by the generation of EL2 by dislocation climb caused by thermal stresses.

The possible relationship of EL2 to dislocations led to the study of low-dislocation and dislocation-free GaAs. Kimura et al [25] found that in very lightly In-alloyed semi-insulating LEC undoped GaAs, which contains large dislocation-free regions, the major deep level has very similar electrical and optical properties to EL2 in unalloyed GaAs. Stirland et al, examining similar low-dislocation material, found enhanced concentrations of EL2 at dislocation sites [26]. Hope et al [27], studying small-diameter crystals of LEC GaAs, found that the EL2 distribution was much more uniform in dislocation-free than in dislocated regions. Alloying with In is thought to result in In on Ga sites, constituting an isoelectronic defect with no significant electrical effects. Mitchel and Yu [28] found that In-alloyed LEC semi-insulating GaAs was controlled electrically by an EL2-like level. However, LEC GaAs doped with Sb (an isoelectronic impurity on As sites) showed a temperature dependent Hall slope corresponding to 0.47 eV, which the authors suggest might be associated with Sb(Ga).

The distribution of EL2 in semi-insulating LEC GaAs can be changed remarkably by heat treatment. Holmes et al [29] found that EL2 distributions in ordinary dislocated semi-insulating LEC GaAs became very uniform after long-term annealing. Martin et al [30] made similar observations on samples that included In-alloyed material. Both groups found that the EL2 concentration tended toward a constant uniform value after annealing that was the same for all crystals whether the initial average value had been higher or lower.

Corresponding to the As(Ga) antisite appearing under As-rich growth conditions is the Ga(As) antisite expected in crystals grown from Ga-rich melts. Such crystals are p-type, containing shallow acceptors (principally C) and two deeper acceptors at about 0.077 eV [31,32,33,34] and 0.2 eV [35,36,37,38,39] above E(v) that have been assigned to the two ionizations of Ga(As). However, Ta et al [40] relate the 0.077 eV level (they determine it as 0.073 eV) to the presence of B in the LEC crystal and suggested that the level was due to complexes involving B(As). LEC material often contains B, but it usually has no electrical effects and is presumed to be occupying a Ga site. Dansas [41] found that annealed or Ga-implanted and annealed undoped LEC GaAs exhibits a 0.064 eV acceptor level. Material so treated would be expected to be Ga-rich. Only after B implantation does the 0.077 eV level appear. Dansas assigned the latter level to B(As) and the 0.064 eV level to Ga(As). Ga-rich conditions would be expected to favour the formation of As vacancies and hence the appearance of either B or Ga on As sites.

The deep levels related to impurities have an extensive literature that will not be reviewed here. However, it is worthwhile to point out that it is not always clear in particular cases if a known impurity actually introduces a deep level that might be inferred from the behavior of the material, or if the deep level is characteristic of a complex of native defects with the impurity, or if the level is related to native defects without any role for the impurity. The issue is discussed for V impurity by Gladkov and Ozanyan [42]. In connection with V, Kutt et al [43] showed that doping with V in a quartz crucible led to semi-insulating LEC GaAs with resistivity greater than 10*7 ohm cm and an acceptor level at 0.58 eV. However, Brandt et al [44] found no evidence for any level (besides EL2) other than an acceptor at 0.15 eV below E(c) in V-doped LEC crystals grown from PBN crucibles. They concluded that doping with V plays no direct role in the compensation process of semi-insulating V-doped GaAs.

REFERENCES

[1] M.R.Brozel [EMIS Datareview RN=15400 (Nov 1985) ` Defect Densities in SI LEC GaAs ´]

[2] G.M.Martin [Semi-Insulating III-V Materials Nottingham 1980 (Shiva, 1980) p.13-28]

[3] L.Hollan, J.C.Brice, J.Hallais [Current Topics in Materials Science vol.5 ch.1 (North Holland, 1980)]

[4] G.F.Neumark, K.Kosai [Semiconductors and Semimetals vol.19 ch.1 (Academic Press, 1983)]

[5] D.C.Look [Semiconductors and Semimetals vol.19 ch.2 (Academic Press, 1983)]

[6] A.M.White [Semi-Insulating III-V Materials, Nottingham 1980 (Shiva, 1980) p.3-12]

[7] J.Bonnafe et al [Mater. Res. Bull. (USA) vol.16 (1981) p.1193-212]

[8] D.E.Holmes, R.T.Chen, K.R.Elliott, C.G.Kirkpatrick [Appl. Phys. Lett. (USA) vol.40 no.1 (1982) p.46-8]

[9] A.M.Huber, N.T.Linh, J.C.Debrun, M.Valladon, G.M.Martin, A.Mittoneau, A.Mircea [J. Appl. Phys. (USA) vol.50 no.6 (1979) p.4022-5]

[10] J.Lagowski, H.C.Gatos, J.M.Parsey, K.Wada, M.Kaminska, W.Walukiewicz [Appl. Phys. Lett. (USA) vol.40 no.4 (1982) p.342-4]

[11] C.G.Kirkpatrick, R.T.Chen, D.E.Holmes, P.M.Asbeck, K.R.Elliott, R.D.Fairman, J.R.Oliver [Semiconductors and Semimetals vol.20 ch.3 (Academic Press, 1984)]

[12] R.N.Thomas, H.M.Hobgood, G.W.Eldridge, D.L.Barrett, T.T.Braggins, L.B.Ta, S.K.Wang [Semiconductors and Semimetals vol.20 ch.1 (Academic Press, 1984)]

[13] E.R.Weber [Thirteenth Int. Conf. on Defects in Semiconductors (Metallurgical Society of AIME, 1985) p.7-9]

[14] R.J.Wagner, J.J.Krebs, G.H.Stauss, A.M.White [Solid State Commun. (USA) vol.36 (1980) p.15-17]

[15] K.Elliott, R.T.Chen, S.G.Greenbaum, R.J.Wagner [Appl. Phys. Lett. (USA) vol.44 no.9 (1984) p.907-9]

[16] M.Taniguchi, T.Ikoma [J. Appl. Phys. (USA) vol.54 no.11 (1983) p.6448-51]

[17] M.Taniguchi, T.Ikoma [Appl. Phys. Lett. (USA) vol.45 no.1 (1984) p.69-71]

[18] D.M.Hoffman, B.K.Meyer, F.Lohse, J.-M.Spaeth [Phys. Rev. Lett. (USA) vol.53 no.2 (1984) p.1187-90]

[19] Y.Imamura, J.Osaka [Jpn. J. Appl. Phys. (Japan) vol.22 no.6 (1983) p.L333-5]

[20] J.Lagowski, D.G.Lin, T.Aoyama, H.C.Gatos [Appl. Phys. Lett. (USA) vol.44 no.3 (1984) p.336-8]

[21] W.Walukiewicz, J.Lagowski, H.C.Gatos [Appl. Phys. Lett. (USA) vol.43 no.1 (1983) p.112-4]

[22] A.T.Hunter, J.P.Baukus, O.J.Marsh, H.V.Winston, M.H.Young, R.R.Hart [Semi-Insulating III-V Materials, Kah-nee-tah 1984 (Shiva, 1984) p.429-36]

[23] M.R.Brozel, I.Grant, R.M.Ware, D.J.Stirland [Appl. Phys. Lett. (USA) vol.42 no.7 (1983) p.610-2]

[24] D.E.Holmes, R.T.Chen [J. Appl. Phys. (USA) vol.55 no.10 (1984) p.3588-94]

[25] H.Kimura, C.B.Afable, H.M.Olsen, A.T.Hunter, H.V.Winston [J. Cryst. Growth (Netherlands) vol.70 (1984) p.185-90]

[26] D.J.Stirland, M.R.Brozel, I.Grant [Appl. Phys. Lett. (USA) vol.46 no.11 (1985) p.1066-8]

[27] D.A.O.Hope, M.S.Skolnick, B.Cockayne, J.Woodhead, R.C.Newman [J. Cryst. Growth (Netherlands) vol.71 (1985) p.795-8]
[28] W.C.Mitchel, P.W.Yu [J. Appl. Phys. (USA) vol.57 no.2 (1985) p.623-5]
[29] D.E.Holmes, H.Kuwamoto, C.G.Kirkpatrick, R.T.Chen [Semi-Insulating III-V Materials, Kah-nee-tah 1984 (Shiva, 1984) p.204-13]
[30] S.Martin, M.Duseaux, M.Erman [Inst. Phys. Conf. Ser. no.74 ch.2 (1985) p.53-8]
[31] P.W.Yu, D.E.Holmes, R.T.Chen [Inst. Phys. Conf. Ser. no.63 ch.5 (1985) p.209-14]
[32] K.R.Elliott, D.E.Holmes, R.T.Chen, C.G.Kirkpatrick [Appl. Phys. Lett. (USA) vol.40 no.10 (1982) p.898-901]
[33] J.P.Fillard, M.Castagne, J.Bonnafe, M.deMurcia [J. Appl. Phys. (USA) vol.54 no.11 (1983) p.6767-70]
[34] S.R.Hetzler, T.C.McGill, A.T.Hunter [Appl. Phys. Lett. (USA) vol.44 no.8 (1984) p.793-5]
[35] P.W.Yu, W.C.Mitchel, M.G.Mier, S.S.Li, W.L.Wang [Appl. Phys. Lett. (USA) vol.41 no.6 (1982) p.532-4]
[36] K.R.Elliott [Appl. Phys. Lett. (USA) vol.42 no.3 (1983) p.274-6]
[37] A.T.Hunter, R.Baron, J.P.Baukus, H.Kimura, M.H.Young, H.Winston O.J.Marsh [Semi-Insulating III-V Materials, Evian 1982 (Shiva, 1982) p.396]
[38] S.G.Bishop, B.V.Shanabrook [Semi-Insulating III-V Materials, Kah-nee-tah 1984 (Shiva, 1984) p.302-7]
[39] W.J.Moore, B.V.Shanabrook, T.A.Kennedy [Semi-Insulating III-V Materials Kah-nee-tah 1984 (Shiva, 1984) p.453-9]
[40] L.B.Ta, H.M.Hobgood, R.N.Thomas [Appl. Phys. Lett. (USA) vol.41 no.11 (1982) p.1091-3]
[41] P.Dansas [J. Appl. Phys. (USA) vol.58 no.6 (1985) p.2212-6]
[42] P.S.Gladkov, K.B.Ozanyan [J. Phys. C (GB) vol.18 (1985) p.L915-20]
[43] W.Kutt, D.Bimberg, M.Maier, H.Krautle, F.Kohl, E.Bauser [Appl. Phys. Lett. (USA) vol.44 no.11 (1984) p.1078-80]
[44] C.D.Brandt, A.M.Hennel, L.M.Pawlowicz, F.P.Dabkowski, J.Lagowski, H.C.Gatos [Appl. Phys. Lett. (USA) vol.47 no.6 (1985) p.607-9]

13.5 DEFECT ENERGY LEVELS RELATED TO THE ANION ANTISITE DEFECT IN GaAs

by W.Potz

September 1985
EMIS Datareview RN=14410

The anion antisite defect As(Ga) has been identified as a double donor by Weber et al [1] by photo-EPR measurements on plastically deformed GaAs at a temperature of 6 deg K. The two donor levels (ionization energies) were obtained at 0.77 eV below the conduction-band edge and 0.52 eV above the valence-band edge. This finding is in qualitative agreement with most theoretical calculations. For a review see, for example, Potz and Ferry [2]. Meyer et al [3] determine two intra-defect transitions of As(Ga)+ by ESR-tagged magnetic circular dichroism measurements on semi-insulating GaAs at a temperature of 4.2 deg K. The two transitions were identified as transitions from the A1 ground state of As(Ga)+ to two excited T2 levels of the same defect. The measured transition energies are (1.05 +- 0.02) eV and (1.29 +- 0.02) eV. An alternative interpretation for these two transitions, proposed by Kaufmann [4], seems to be unlikely to be correct in view of recent results on the anion antisite defect in GaP [5]. Some experiments indicate that the anion antisite defect in GaAs may be involved in the defect that gives rise to EL2. As this point is not clear at the moment no EL2-associated levels are given here.

REFERENCES

[1] E.R.Weber, H.Ennen, U.Kaufmann, J.Windscheif, J.Schneider, T.Wosinski [J. Appl. Phys. (USA) vol.53 (1982) p.6140]

[2] W.Potz, D.K.Ferry [J. Phys. Chem. Solids (GB) vol.46 (1985) p.1101]

[3] B.K.Meyer, J.M.Spaeth, M.Scheffler [Phys. Rev. Lett. (USA) vol.52 (1984) p.851]. See also D.M.Hofmann, B.K.Meyer, F.Lohse, J.M.Spaeth [Phys. Rev. Lett. (USA) vol.53 (1984) p.1187]

[4] U.Kaufmann [Phys. Rev. Lett. (USA) vol.54 (1985) p.1332]

[5] B.K.Meyer, J.M.Spaeth [Phys. Rev. B (USA) vol.32 (1985) p.1409]

13.6 DEFECT ENERGY LEVELS IN LPE GaAs

by A.R.Peaker

August 1985
EMIS Datareview RN=13992

There is a consensus that LPE GaAs is relatively trap-free compared to GaAs grown by other methods. Indeed, workers looking specifically for electron traps in material doped only with shallow donors have found none. This means that the electron trap concentrations are certainly below 10*13/cm*3 [1,2] and probably as much as an order less than this. However, in p-type as-grown material two acceptor-like hole traps are commonly seen. These were first observed by Lang [3] and occur in roughly equal concentrations. He called them A and B and they have subsequently been investigated in considerable detail [4-11]. The states are in the 10*11 to 10*15/cm*3 concentration range in as-grown p-type although not in n-type material subsequently diffused with zinc [9]. It appears that this is due to some form of gettering action by the zinc because the states have been reported in undiffused n-type material [10] with a concentration given by:

[A] = [B] = 0.93 X 10*7 X n*0.5   per cm*3

where n is the free electron concentration (per cm*3) in material grown at 0.08 microns/min. A model is proposed in which the A centre is a gallium atom on the arsenic site paired with a gallium vacancy while the B centre is an arsenic atom on the gallium site paired with a gallium vacancy. However, Wang et al [6] present a persuasive case indicating that the A and B centres are merely different charge states of the same defect and propose that it is a gallium atom on an arsenic site. The concentration of the centres is also dependent on growth rate [10] and the presence of iso-valent impurities. Kalukhov has observed that the A and B concentrations decrease with the addition of either In or Sb [7]. However, Wang et al [6] observed that in their LPE system the concentrations of A and B were below 10*11/cm*3 with n = 2 X 10*16/cm*3. Only with fast cooling could measurable concentrations of A and B be produced. Part 1 in [11] has measured the capture cross-sections of the A and B centres at 300 deg K.

	Activation Energy (Holes) meV	Electron Cross-Section cm*2	Hole Cross-Section cm*2	Emission Rate At 300 deg K s*-1
A	400	1.4 X 10*-17	4.2 X 10*-15	1.2 X 10*-6
B	710	4.2 X 10*-20	2.1 X 10*-15	1.7

Comparable values have been obtained by Lang [3] while measurements at lower temperatures have been made by Wang et al [6].

Milnes [9] has calculated that the A centre is likely to dominate the recombination in n-type material at room temperature and so is of crucial importance in devices using holes as minority carriers. Photocapacitance measurements on the A and B centres have also been undertaken [6,8,12] .

Other workers have observed other hole traps in LPE GaAs with only shallow dopants or no dopants intentionally added. A number of deep states with different properties have been observed when additional impurities have been added to the melt. The table summarises both these data and those relating to inadvertant centres. Arrhenius plots of the emission rate of holes from these traps are given in the individual references or in a number of general reviews [8,12-16]

Name	Energy meV	Associated With	References
HB1	780	Cr	[3,17]
HL2	730	= B ?	[19]
HB2	710	= B	[3,6]
HS2	640	= B ?	[21]
HS1	580		[21]
HB3	520	Fe	[3,17]
HB4	440	Cu	[3,18]
HS3	440		[21]
HL5	410	= A ?	[19]
HB5	400	= A	[3,6]
HL12	270	Zn	[19]
-	140	Cu	[3]

All the above states have been detected by DLTS and its variants. In addition an electron trap has been associated with chromium doping with an activation energy for electron emission of 860 meV and designated EB1 [3]. Optical characterisation of the HB1 state is described in [20]. An important point is that in the cases where Fe, Cr or Cu were added to the melt the distribution coefficient was found to be very low (approx. 10*-16) in comparison to typical distribution coefficients from a stoichiometric melt of the order of 10*-3. It is this fortuitous situation which enables the LPE process to produce almost trap-free GaAs under slow cooling conditions.

REFERENCES

[1] R.Magno, M.Spencer, J.G.Giessner, E.R.Webber [Proc. 13th Intn. Conf. on Defects in Semicondcutors, Colorado USA 1984 (Metallurgical Soc. of AIME 1985) p.981-987]

[2] J.Baston, F.J.Tegude, K.Helme [Surf. Sci. (Netherlands) vol.132 (1983) p.465-8]

[3] D.V.Lang, R.A.Logan [J. Electron. Mat. (USA) vol.4 no.5 (1975) p.1053-66]

[4] D.V.Lang, R.A.Logan [J. Appl. Phys. (USA) vol.47 no.4 (1976) p.1533-37]

[5] T.Okumura, T.Ikoma [J. Cryst. Growth (Netherlands) vol.45 no.3 (1978) p.459]

[6] Zh.G.Wang, L.A.Ledebo, H.G.Grimmeiss [J. Phys. C (GB) vol.17 (1984) p.259-72]

[7] V.A.Kalukhov, S.I.Chikichev [Phys. Stat. Sol. (Germany) vol.88 no.1 (1985) p.K59-61]

[8] B.A.Bobylev, V.A.Kalukhov, Kh-M.Z.Torchinov, S.I.Chikichev [Sov. Phys. Semicond. (USA) vol.19 no.2 (1985) p.151-3]

[9] A.G.Milnes [Adv. Electronics & Electron Phys. vol.61 (1983) p.133-5]

[10] Zou Yuanxi [Proc. 13th Int. Conf. on Defects in Semiconductors, Colorado USA 1984, Metallurgical Soc. of AIME (1985) p.1021-7]
[11] D..Partin [J. Appl. Phys. (USA) vol.50 no.11 (1979) p.6845]
[12] M.M.Sobolev, S.G.Konnikov, M.N.Stepanova [Sov. Phys. Semicond. (USA) vol.18 no.2 (1984) p.240-2]
[13] G.M.Martin, A.Mitonneau, A.Mircea [Electron. Lett. (GB) vol.13 no.7 (1977) p.191-3]
[14] A.Mircea, D.Bois [Conf. Series Inst. of Physics vol.46 p.82-99]
[15] A.Mitonneau, G.M.Martin, A.Mircea [Elect. Lett. (GB) vol.13 no.22 p.666-7]
[16] G.F.Neumark, K.Kosai [Semicondcutors and Semimetal, Eds. Willardson & Beer (Academic Press 1983) vol.19 p.26-9]
[17] Yu B.Bolkovityanov, A.F.Kravchenko, S.I.Chikichev [Sov. Phys. J. (USA) vol.26 no.10 (1983) p.889-97]
[18] V.Kumai, L.A.Ledebo [J. Appl. Phys. (USA) vol.52 (1981) p.4866]
[19] G.M.Martin [Semi-Insulating III-V Materials Nottingham 1980 (Shiva Publishing 1980) p.13]
[20] P.Leyral, F.Lilly, S.Loualiche, N.Nouailhal, G.Guillot [Solid State Commun. (USA) vol.38 (1981) p.333]
[21] A.K.Saxena [J. Inst. Electron. & Telecommun. Eng. (Japan) vol.26 no.6 (1980) p.293-7]
[22] H.Hasegawa, A.Majerfeld [Electron. Lett. (GB) vol.11 (1975) p.286-9]

13.7 DEFECT ENERGY LEVELS IN VPE GaAs

by A.R.Peaker

November 1985
EMIS Datareview RN=13990

In GaAs material grown by the arsenic trichloride-gallium or arsine-gallium-hydrogen chloride processes both hole and electron traps are commonly seen.

ELECTRON TRAPS

The trap present in the highest concentration is EL2, a ubiquitous defect present in melt grown and VPE material [1]. A very substantial effort has been devoted to investigating this deep state which was originally associated with oxygen. Subsequent studies have shown conclusively that EL2 is quite unrelated to oxygen [2,3] as are all other commonly observed deep states in GaAs [4]. There is now considerable evidence that EL2 is associated with the arsenic antisite [5] and probably is a complex of arsenic on a gallium site and an arsenic interstitial [6]. EL2 is of great importance in VPE GaAs technology because of its high concentration (typically up to 3 X 10*14/cm*3) possibly resulting in semi-insulating material or in local variations of FET threshold voltages. The concentration of EL2 is influenced by dislocations [7] and by heat treatment after growth [8] but the major determining factor is the III/V ratio during growth. If a larger proportion of arsenic trichloride or arsine is used the concentration of EL2 increases [9-11]. However, the concentration is also a function of growth rate and substrate orientation [12]. Other commonly observed electron traps are EL5 (Nt = 10*12) traps characterised by emission behaviour similar to EL1 (Nt = 10*13). A comprehensive listing is given in the table. The trap nomenclature is based on that used by Martin et al [12]. However, considerable care must be exercised with the alternative names quoted. The A and B centres in VPE [13] are not the same as the A and B centres seen in LPE [14] while Ruby et al [4] have named the centres they observed in VPE GaAs E1, E2, E3 and E4 which appear to correspond to EL11, EL5, EL3 and EL2 respectively. This is particularly confusing as the E1 - E5 notation has been used previously for traps induced by radiation [15].

Name	Energy meV	Related to	Principal References
EL2 (A)	825	arsenic antisite complex	[12,1]
EB2 (EC1)	830	EL2	[14,17]
EL12 (A´)	780	EB4 ?	[12]
EB4 (E4)	710	gallium vacancy ?	[12,15]
EL3 (B)	575	reduces on anneal	[12]
EC2	480	Ni	[17]
EI1	430	EC2 ?	[18]
EL5 (C)	420	reduces on anneal	[12]
EL16	370		[12]
EL8 (D)	275		[13]

EL9 (E)	225		[13]
EI2	190		[18]
EI3	180		[18]
EL11 (F)	170	Cu?	[13]

Energies are apparent activation energies with T*2 correction measured in all cases by DLTS or its variants. Arrhenius plots of emission rate as a function of temperature are given in the references quoted.

HOLE TRAPS

The hole trap population of VPE GaAs is very variable. Only HL9 seems to occur in most layers, usually in concentrations of the order of 10*12/cm*3. HL2 is seen only in heat treated materials and is the same as the B state nearly always present in LPE GaAs. The other states are often undetectable. The association of HL1, HL3 and HL4 with Cr, Fe and Cu respectively is a result of diffusion and/or doping experiments.

Name	Energy	Notes	References
HL1	940	Cr	[20]
HL10	830		[20]
HL2	730	heat treatment	[7,8]
HL9	690	common	[20]
HL3	590	Fe	[23,20]
HT1	440		[19]
HL4	420	Cu	[21,22]
HL6	320		[20]

RECOMBINATION

In a study by Miller [9] it was shown that the minority carrier lifetime in n-type GaAs increased as the arsine to gallium chloride ratio was decreased over the range 3:1 to 1:3. The concentration of EL2 also decreased, so correlating with lifetime. Mitonneau rejects this hypothesis as he considers the hole cross-section of EL2 to be too small [24] and proposed HL10 as the important recombination centres. Partin et al [17] in a comprehensive study of VPE GaAs contaminated with nickel consider the state EC2 as the most likely lifetime killer in processed material. However, few of the observed states have reliable values for directly measured capture cross-sections and although it may be justified to make comments about recombination in specific layers, it is not yet possible to generalise about VPE GaAs.

REFERENCES

[1] A.E.Lin, E.Omekanovski, R.H.Bube [J. Appl. Phys. (USA) vol.47 (1976) p.1852]

[2] A.M.Huber, N.T.Link, J.C.Delsun, M.Vollandon, C.M.Martin, A.Mittonneau, A.Mircea [J. Appl. Phys. (USA) vol.50 no.6 (1979) p.4022]

[3] R.H.Wallis [Inst. Phys. Conf. Ser. vol.56 (1981) p.73]

[4] D.S.Ruby, K.Arai, G.E.Stillman [J. Appl. Phys. (USA) vol.51 (1976) p.3587-91]

[5] M.Baumler, U.Kaufmann, J.Windscheif [Appl. Phys. Lett. (USA) vol.46 (1985) p.781]
[6] H.J.von Bardelelen, D.Stievenard, J.C.Bourgoin, A.Huber [Appl. Phys. Lett. (USA) vol.47 no.9 (1985) p.970-2]
[7] N.Yamamoto, F.Hasegawa, M.Onomura, Y.Nannichi [Jpn. J. Appl. Phys. (Japan) vol.24 no.5 (1985) p.L326-8]
[8] K.Saxena, N.P.Singh [Indian J. Phys. vol.59A no.2 (1985) p.104-8]
[9] M.D.Miller, G.H.Olsen, M.Ettenberg [Appl. Phys. Lett. (USA) vol.311 (1977) p.538]
[10] M.Ozeki, J.Komeno, A.Shibatomi, S.Ohkawa [J. Appl. Phys. (USA) vol.50 (1979) p.4808]
[11] Zou Yuamxi [Inst. Phys. Conf. Series no.63 ch.4 (1982) p.185]
[12] G.M.Martin, A.Mittonneau, A.Mircea [Electronics Lett. (GB) vol.13 no.7 (1977) p.191-3]
[13] A.Mircea, A.Mittonneau [Appl. Phys. (Germany) vol.8 (1975) p.15-21]
[14] D.V.Lang, R.A.Logan [J. Electron. Mat. (USA) vol.4 (1975) p.1053-66]
[15] D.V.Lang, L.C.Kimerling, S.Y.Leung [J. Appl. Phys. (USA) vol.51 (1976) p.3587-91]
[16] A.Ashley, G.G.Roberts, D.J.Ashen, J.B.Mullin [Solid State Commun. (USA) vol.20 (1976) p.61-3]
[17] D.L.Partin, J.W.Chen, A.G.Milnes, L.F.Vassamillet [J. Appl. Phys. (USA) vol.50 no.11 (1979) p.6845-59]
[18] H.Lefevre, M.Schulz [Appl. Phys. (Germany) vol.12 (1977) p.45]
[19] K.Sakai, T.Ikoma [Appl. Phys. (Germany) vol.5 (1974) p.165-71]
[20] A.Mittonneau, C.M.Martin, A.Mircea [Electron Lett. (USA) vol.13 no.22 (1977) p.666-8]
[21] A.Mittonneau, A.Martin, A.Mircea [Inst. Phys. Conf. Series 33A (1977) p.73]
[22] A.Chantre [Phys. Rev. B (USA) vol.23 (1981) p.5335]
[23] M.Kleverman, P.Omling, L-A.Ledebo, H.G.Grimmeiss [J. Appl. Phys. (USA) vol.54 no.2 (1983) p.814-9]
[24] A.Mittonneau [Rev. Phys. Appl. (France) vol.14 (1979) p.853]

13.8 DEFECT ENERGY LEVELS IN MOVPE GaAs

by A.R.Peaker and J.Nicholas

December 1985
EMIS Datareview RN=15442

MOVPE gallium arsenide has a very variable deep state population but like conventional VPE and bulk grown material it always contains the deep electron trap EL2. This defect is associated with the arsenic antisite [1]. It is normally present in concentrations between 10*12 and 10*15/cm*3 dependent primarily on growth conditions [2]. Published work indicates that there are no other states which are ALWAYS detectable although many are quite common. It is apparent that the concentration of many of the states depends on the source materials [3] and so are probably related to chemical impurities rather than lattice defects. A broad generalisation is that higher purities (both deep and shallow levels) are achieved with triethyl gallium [4] rather than the trimethyl alkyl and that the kinetics of low pressure growth results in a lower concentration of impurities being incorporated [5].

ELECTRON TRAPS

Considerable effort has been expended in studying the effect of the III/V gas stream ratio on the trap population in MOVPE GaAs. Bhattacharya et al [6] observed a linear dependence of EL2 concentration on the vapour phase As/Ga ratio. However, Samuelson et al [7] observed a dependence on SQRT [AsH3]/[TMG] over the As/Ga range 5 to 50. Zhu et al [8] gathering together data from various sources prior to 1981 showed that there were obviously many other factors influencing the EL2 concentration apart from the III-V ratio. A very comprehensive study by Watanabe et al [2] shows a clear half power law at 720-740 deg C and a quarter power law at 630-660 deg C. However, other factors such as shallow doping level, growth rate and substrate, all have some effect on the EL2 concentration. Systematic differences in the deep state emission Arrhenius plots were seen between the higher and lower growth temperatures, suggesting that the structure of EL2 may well vary [2].

An electron trap which is often observed is labelled in the table EH5. This is usually seen in the 10*12 to 10*14/cm*3 concentration range [3] in GaAs but is present in much higher concentrations in GaAlAs [9]. Profiles of the state in GaAs near a heterojunction with GaAlAs have shown dramatically increasing profiles with concentrations reaching 10*16/cm*3 [10]. Another state, EH4, shows similar behaviour and appears as a shoulder on the EH5 DLTS peak. The energy quoted in the literature varies from 300 to 380 meV because of the problems of deconvolution. It is not clear whether these states are directly associated with aluminium, or with impurities in the trimethyl aluminium. Allsopp et al [10] describes another state at 250 meV which exhibits similar behaviour.

Zhu et al [8] have reported a number of defects which are detected near the interface region in homoepitaxy layers. Their origin in unknow but increased concentrations of inadertent impurities have been observed near MOCVD interfaces using SIMS [11].

Another state frequently seen [7] is similar to EL14 observed in bulk material. Wagner et al [9] associate it with tellurium and observe that in GaAs:Te it is present typically at 0.01 of the tellurium concentration.

In the table the nomenclature of [12] is followed wherever possible with additions for later work. The notation W1 etc. is from Wagner [9], EW2 etc. from Watanabe [2], ET1MO from Zhu [8] and EH1 etc. from Hashizume [3]. The energies quoted are apparent activation energies with T*2 correction. Arrhenius plots are available in the refrences quoted.

Name	Energy meV	Notes	References
EL2	785	EH1 ET1MO	[2,6,8]
ET3MO	520	EW2 (increase at interface	[2,8]
ET4MO	470	EL4 in [12]	[8]
EI1	440	low concentrations	[7]
EH5	380	W2 EW3 (Al)	[3,9,10]
ET6MO	370	interface only	[8]
ET5MO	350	interfae	[8]
EH4	300-380	W1 (Al)	[3,10]
EH3	220	EL8 ? in [12]	[3]
EL14	190	W4 (Te)	[7,9]
EH2	180	EI3 in [12]	[3]

HOLE TRAPS

Very little work has been published on hole traps in MOVPE GaAs. The only state which is seen consistently is HL3, a level which has been associated with iron because of diffusion experiments. It is present in as-grown MOVPE material in concentrations between 5 X 10*12 and 2 X 10*14/cm*3. The state HB4 which is related to copper has been seen near interfaces [8], while in the 10*13/cm*3 range a structural defect HL5 normally seen in LPE has been observed at the 10*12/cm*3 level. A state at 340 meV has been associated with chromium through radio tracer diffusions into MOVPE but is present only in low concentrations in as-grown material [13].

Name	Energy meV	Notes	Reference
-	310	HT4MO	[6]
HT5MO	330	HL5 HB5	[8]
-	340	Cr (diffused MOVPE)	[13]
-	350		[6]
HB4	370	HT4MO	[8]
HL3	570-590	HT3MO HB3 (Fe)	[2,8]

REFERENCES

[1] H.J.von Bardeleben, D.Stievenard, J.C.Bourgoin, A.Huber [Appl. Phys. Lett. (USA) vol.47 no.9 (1985) p.970-2]

[2] M.O.Watanabe, A.Yanaka, T.Udagawa, T.Nakanishi, Y.Zohta [Jpn.

J. Appl. Phys. (Japan) vol.22 no.6 (1985) p.923-9]
[3] T.Hashizume, E.Ikeda, Y.Akatsu, H.Ohno, H.Hasegawa [Jpn. J. Appl. Phys. (Japan) vol.23 no.5 (1984) p.L296-8]
[4] N.Kobayashi, T.Fukui [Electron Lett. (GB) vol.20 (1984) p.887]
[5] T.F.Kuech, R.Potemski [Appl. Phys. Lett. (USA) vol.47 no.8 (1985) p.821-3]
[6] P.K.Bhattacharya, J.W.Ku, S.J.T.Owen [Appl. Phys. Lett. (USA) vol.36 no.4 (1980) p.304]
[7] L.Samuelson, P.Omling, H.Titze, H.G.Grimmeiss [J. Cryst. Growth (Netherlands) vol.55 (1981) p.164-72]
[8] H.-Z.Zhu, Y.Adachi, T.Ikoma [J. Cryst. Growth (Netherlands) vol.55 no.1 (1981) p.154-63]
[9] E.E.Wagner, D.E.Mars, G.Horn, G.R.Stringfellow [J. Appl. Phys. (USA) vol.51 (1980) p.5434]
[10] D.Allsopp, A.R.Peaker, E.J.Thursh, G.Wale Evans [J. Cryst. Growth (Netherlands) vol.68 no.1 (1984) p.295-300]
[11] A.M.Huber, G.Mosillott, M.Bonnet [Appl. Phys. Lett. (USA) vol.41 no.7 (1982) p.638-40]
[12] G.W.Martin, A.Mitonneau, A.Mircea [Electron. Lett. (GB) vol.13 no.7 (1977) p.91-3]
[13] W.K.Ke, B.Hamilton, A.R.Peaker, M.Brozel, B.Tuck, D.R.Wight [Solid State Electron (GB) vol.28 no.6 (1985) p.611-5]

13.9 DEFECT ENERGY LEVELS IN MBE GaAs

by A.R.Peaker and J.Nicholas

August 1985
EMIS Datareview RN=13991

ELECTRON TRAPS

Using DLTS to study n-type MBE grown GaAs Lang [1] observed nine electron traps M0...M8. In subsequent work the levels M1...M4 have been observed consistently in MBE GaAs grown between 500 and 650deg C whether n-type or p-type. The states occur in the 10*12 to 10*14/cm*3 concentration range. However, M6...M8 do not appear reproducibly and it has been proposed that they may result from surface effects [2]. Martin [3] has reported electron traps in MBE layers (EL4, EL7 and EL10) which are similar to traps in the M series. Despite early associations with deep states observed in material grown by other techniques it is now apparent that M1, M3 and M4 are unique to MBE and are almost certainly related to lattice defects. Conversely, a number of levels often seen in GaAs grown by VPE techniques are not detected in MBE unless the GaAs is heat treated after growth (e.g. EL2) [4]. A listing of MBE states is given in the table.

Arrhenius plots of the electron emission rate as a function of temperature are given in the reference quoted and in reviews [3,7]. The activation energies vary in the literature, even between traps which are believed for other reasons to be physically the same. In the case of M1 the emission rate is known to be strongly field dependent [2]. In the table, wherever possible, the most reliabel low field value of the apparent activation energy is quoted with T*2 correction. The first reference is usually the one in which measurements have been done over the widest temperature range or in which comparisons with previous data appear.

Cross sections are only quoted if a direct measurement of the electron capture cross section has been made (not the Arrhenius plot intercept)

Name	Energy meV	Cross Section cm*-2	Notes (Tg in deg C)	Main Reference
EB3LT	858		Tg below 490	[5]
EL2A	825		annealed	[4-8]
E5A	759		annealed	[4]
EL3LT	566		Tg below 490	[5,3]
M5	565			[1,4,7]
EL4	510		M4 ?	[3]
M4 (EB4)	480	above 2 X 10*-16		[2,1,9]
EB6LT	405		Tg below 300	[2]
EL7	301		M3 ?	[3]
M3 (EB7)	300	1.1 X 10*-16		[2,1]
M2	220	above 4 X 10*-17	Tg below 600	[2,1,7]
M2 !	230	10*-20	Tg below 600	[2]
M1 (EB8)	189	above 2 X 10*-16		[2,1,9]

EL10	159	M1 ?	[3]
E10	165	annealed	[4]
M0	75		[1]
M00	30	low concn	[6]

In general, the concentrations of the deep states decrease with increasing substrate temperature during growth (Tg). The exception to this is M2 which is only present in material grown at around 650 deg C. At lower temperature a trap (M2 !) with similar emission but very different capture cross section is present [2]. However, at very low growth temperatures (300 deg C) the M series disappears, but insufficient data exist to quantify the changes between 300 and 500 deg C. The trap concentrations are also dependent on the III/V flux ratio [2,7] and on whether dimeric or tetrameric arsenic is used [9]. In particular, M1, M3 and M4 are found in much lower concentrations when the dimer is used. Heat treatment after growth increases the concentration of some traps and decreases others [8,10].

HOLE TRAPS

Studies of hole traps have not been as detailed as those of electron traps. However, it seems that, unlike the case of the electron traps, defects seen in LPE and VPE material are apparently indentical.

Name	Energy meV	Notes	Reference
900A	900	after 750 deg C anneal	[10]
HB1	780	associated with Cr in LPE	[5]
HL9	690	seen in VPE	[5]
HL3	590	seen in Fe diffused VPE	[2,10]
HL8	520	same as HL3?	[13]
HB4	440	seen in Cu doped LPE	[5]
HL7	350		[13]
HB6LT	290	seen for below 490 deg C	[5]

Although no capture cross sections have been determined in MBE material HL3, HL8, HB4 and HB1 have been measured in LPE and/or VPE [14]. All are classic acceptors with large hole cross sections and small electron cross sections.

Iron, manganese and chromium have been detected in MBE GaAs using Secondary Ion Mass Spectrometry (SIMS) and their origin ascribed to the effusion sources [15]. The concentration of these impurities was shown to decrease with increasing substrate temperature. SIMS measurements by Palmateer et al [16] indicate that substrates were a major source of manganese.

RECOMBINATION

Not enough measurements have been done to make definitive statements on recombination but Blood and Harris [2] suggest that the minority carrier lifetime in p-type MBE GaAs will be limited by M4 for growth temperarures below 625 deg C and by M2 above this temperature. In n-type material minority carrier lifetime at the low injection levels typical of photovoltaic devices may be limited by HB1.

REFERENCES

[1] D.V.Lang, A.Y.Cho, A.C.Gossard, M.Ilegems, W.Wiegmann [J. Appl. Phys. (USA) vol.47 no.6 (1976) .2558-64]

[2] P.Blood, J.T.Harris [J. Appl. Phys. (USA) vol.56 no.4 (1984) p.993-1007]

[3] C.M.Martin, A.Mittonneau, A.Mircea [Electronics Lett. (GB) vol.134 no.7 (1977) p.191-2]

[4] D.S.Day, J.D.Olerstav, T.J.Drummond, H.Markov, A.Y.Cho, B.G.Streetman [J. Electron. Mat. (USA) vol.10 no.13 (1981) p.445-53]

[5] R.A.Stall, C.E.C.Wood, P.D.Kirchner, L.F.Eastman [Electron. Lett. (GB) vol.16 no.5 (1980) p.171-2]

[6] R.Y.De Jule, M.A.Hause, G.E.Stillman, S.C.Palmateer, J.C.M.Hwang [J. Appl. Phys. (USA) vol.57 no.12 (1985) p.5287-9]

[7] C.Amano, A.Shiabakawa, K.Ando, M.Yamaguahi [Electron. Lett. (GB) vol.20 no.4 (1984) p.174-5]

[8] S.H.Xin, W.J.Schaff, C.E.C.Wood, L.F.Eastman [Appl. Phys. Lett. (USA) vol.41 no.8 (1982) p.742-4]

[9] O.Kumagi, K.Wimstrel, W.Jansche [Solid State Commun. (USA) vol.41 no.1 (1982) p.89-92]

[10] P.K.Bhattacharya, H.J.Buhlmann, M.Ilegems, J.1.Staehi [J. Appl. Phys. (USA) vol.53 no.9 (1982) p.6391-8]

[11] L.Hollan, J.C.Brice, J.Hallais [Current Topics in Materials Science vol.5 Ch.1 (North Holland 1980)]

[12] A.G.Milnes [Advances in Electronics and Electron Physics vol.61 (1983) p.76-81]

[13] A.Mittonneau, G.M.Martin, A.Mircea [Electron. Lett. (GB) vol.13 no.22 (1977) p.666-7]

[14] D.V.Lang, R.A.Logan [J. Electron. Mat. (USA) vol.4 no.5 (1975) p.1053-66]

[15] J.B.Clegg, C.T.Foxton, G.Weimann [J. Appl. Phys. (USA) vol.53 no.6 (1982) p.4518-20]

[16] S.C.Palmateer, W.J.Schaff, A.Galuska, J.D.Berry, L.F.Eastman [Appl. Phys. Lett. (USA) vol.42 no.2 (1983) p.183-5]

13.10 DEFECT ENERGY LEVELS IN ELECTRON IRRADIATED GaAs

by A.R.Peaker

December 1985
EMIS Datareview RN=15443

Extensive studies of electron damage in GaAs indicate that the introduction rate and the defect energies of many of the traps produced are essentially unchanged with the type of material (LPE, VPE or bulk) and doping. This indicates that the influence of impurities is negligible and strongly supports the view that the defects are intrinsic.

Lang [1] has reviewed much of the early work on carrier removal rates (compensation) in n-type material, most of which has been done using 1 MeV electrons. This shows a linear dependence on fluence and indicates the dominant species are acceptor-like, indeed p-type conversion has been observed and attributed to a level 100 meV from the valence band [2] although other acceptors are also produced in high concentrations.

ELECTRON TRAPS

Lang et al [3] studied the orientation dependence of the defect production rate using 1 MeV electrons. He concluded that the states he observed resulting from room temperature irradiation were associated with gallium displacement. These were EB10, EB9, EB6 with similar but less definitive conclusions for EB4 and EB3. These states all annealed out at 500 deg K. More recent work using electron energies nearer the displacement threshold leads to the conclusion that the defects result from a single arsenic displacement (not gallium) and not from the displacement of two or more neighbouring atoms [4,5]. The discrepancy with Lang´s results has been explained in terms of the recoil effects of the high energy electrons used by Lang [3] and it is almost certain that the defect reaction is indeed an arsenic displacement [6].

A summary of the electron traps created is given in the table.

Name	Energy meV	Irradiation Temp. degK	Introduction Rate cm*-1	Reference
EB3 (E5)	900	300	0.1	[3]
EL2 (I1)	720	570	1	[8]
P3 (EL2 ?)	720	after anneal	-	[7]
EB4 (E4)	710	300	0.08	[3]
I2	620	570	0.5	[8]
I3	500	570	below 0.01	[8]
PI (I3 ?)	500	after anneal	-	[7]
EB6 (E3)	410	300	0.7	[3]
EL15 (EB9 ?)	370	?	?	[9]
I4 (PI ?)	360	570	0.03	[8]
P1 (EL5)	260	after anneal	-	[7]
I5	230	570	0.03	[8]
E (0.23)	230	80	0.015	[10]

I6	190	570	below 0.01	[8]
EB9 (E2)	180	300	2.8	[3]
I7	160	570	0.03	[8]
EB10 (E1)	40-80	300K	1.8	[3]

The electron energy was 1 MeV in all cases but the defect production rate is only a rough guide as the rate is sometimes dependent on flux which was not the same in all cases. However EB9 (E2) has by far the highest introduction rate and appears to be acceptor-like so tending to compensate n-type material. The energy quoted is the apparent activation energy after T*2 correction. Defects EB6,EB9 and EB190 could be produced by 80 deg K as well as 300 deg K so are either primary defects or the result of interactions with species mobile at low temperature [10].

HOLE TRAPS

Less work has been done on hole traps and the results are much more variable particularly in respect of introduction rates.

Name	Energy meV	Introduction Rate cm-1	References
L13 (H5)	940-960	0.48	[11,12,13]
L16 (H3)	710-790	0.22	[11,12]
H2	370	-	[13]
HB6 (H1,L14)	250-290	0.07-0.7	[6,11,14,12]
H0	90-115	0.5-2.4	[11,6,14]
	60	0.8	[11]
	50	1.6	[11]

All irradiations were at 300 deg K with 1 MeV electrons. A number of other shallow traps have been seen associated with very high fluences. These tend to be sample dependent [11] and are not included above. The concentrate of H5 saturates with fluence at around 3 X 10*15 cm*-3 and is believed to be a complex with impurities in the starting material.

H1 and H0 appear consistently and H1 is believed to be a simple primary defect. Both H0 and H1 show the same anisotropic dependence as E1, E2 and E3 and so are also associated with the arsenic sublattice.

ANNEALING BEHAVIOUR

The annealing stages of defects induced by electron irradiation have been reviewed by Lang [1]. Essentially after room temperature irradiation levels E3, E5 and the hole trap H1 have identical annealing behaviour exhibiting first order kinetics with an activation energy of 1.4 eV while E1 and E2 have an activation energy for annealing of 1.75 eV [16] while E4 appeared stable. The annealing rate of the second pair of defects depends on Nd*2/3 where Nd is the pre-irradiation donor level. This implies that the donor atoms are acting as sinks for the defects during the anneal phase. However, both these groups of defects exhibit recombination enhanced annealing resulting in their removal at lower temperatures when minority carriers are generated by optical or electrical excitation and subsequent recombination [17,16].

Work by Pons et al [7] showed that E4 anneals out, generating the new states P1, P2 and P3 documented in the electron trap table. They also consider that the annealing of E2 proceeds along two concurrent paths and identify E2 as a vacancy.

REFERENCES
----

[1] D.V.Lang [Inst. Phys. Conf. Ser. (GB) vol.31 ch.1 (1977) p.70-94]

[2] J.W.Farmer, D.C.Look [J. Appl. Phys. (USA) vol.50 no.4 (1979) p.2970-2]

[3] D.V.Lang, R.A.Logan, L.C.Kimmerling [Phys. Rev. B (USA) vol.15 no.10 (1977) p.4874-82]

[4] D.Pons, J.C.Bougoin [Phys. Rev. Lett. (USA) vol.47 (1981) p.1293]

[5] D.Pons, P.M.Mooney, J.C.Bougoin [J. Appl. Phys. (USA) vol.51 no.4 (1980) p.2038-42]

[6] D.Pons [Physica B & C (Netherlands) vol.116B (1983) p.388-93]

[7] D.Pons, A.Mircea, J.C.Bougoin [J. Appl. Phys. (USA) vol.51 no.8 (1980) p.4510-4157]

[8] D.Stievenard, J.C.Bougoin, D.Pons [Physica B & C (Netherlands) vol.116B (1983) p.394-7]

[9] A.Mittonneau, G.M.Martin, A.Mircea [Electron Lett. (GB) vol.13 no.22 (1977) p.666-8]

[10] A.A.Rezazedeh, D.W.Palmer [J. Phys. C (GB) vol.18 no.1 (1985) p.43-54]

[11] S.Loualiche, A.Nouailhat, G.Guillot, M.Gavand, A.Laugier [J. Appl. Phys. (USA) vol.53 no.12 (1982) p.8691-8696]

[12] A.Mitonneau, A.Mircea, C.M.Martin, D.Pons [Rev. Phys. Appl. (France) vol.14 (1979) p.853]

[13] C.Guilot, S.Loualiche, A.Nouailhat, G.Martin [in "Defects and Radiation Effects in Semiconductors" Oiso 1980, Inst. Phys. Conf. Series 59 (1981) p.323]

[14] D.V.Lang, L.C.Kimmerling [in "Lattice Defects in Semiconductors" (1975) Inst. Phys. Conf. Series 23 p.581-8]

[15] A.G.Milnes [Adv Electron. & Electron Phys. (GB) vol.61 (1983) p.76-81]

[16] D.V.Lang, L.C.Kimmerling, S.Y.Leong [J. Appl. Phys. (USA) vol.47 (1976) p.3587-91]

[17] D.V.Lang, L.C.Kimmerling [Phys. Rev. Lett. (USA) vol.33 (1974) p.489-92]

CHAPTER 14

DIFFUSION OF IMPURITIES IN GALLIUM ARSENIDE

14.1 Diffusion of Cr in GaAs

14.2 Diffusion of Mn in GaAs

14.3 Diffusion of Zn in GaAs

14.4 Diffusion of Si in GaAs

14.5 Diffusion of Sn in GaAs

14.6 Diffusion of S in GaAs

14.1 DIFFUSION OF Cr IN GaAs

by B.Tuck

September 1985
EMIS Datareview RN=15429

There is a good deal of interest in this system, since Cr can be used to prepare GaAs in its high-resistivity (or `semi-insulating´) form. Despite this fact, there is relativity little systematic information on the diffusion of Cr in GaAs. The most extensive set of data is that due to Tuck and Adegboyega [1], who diffused radioactive Cr into GaAs over the temperature range 800-1100 deg C for times between 15 min and 10 hr, using a vapour source. The profiles showed a sharp peak at the surface, dropping by about an order of magnitude in the first 20 microns or so. This was followed by a long, slowly-decreasing tail, penetrating deep into the crystal. For the longer diffusion times, the tail became virtually horizontal and the Cr penetrated the whole of a slice a few hundred microns in thickness.

None of these profiles could be fitted to any of the well-known mathematical solutions to Fick´s law, and it is neither possible nor useful to try to ascribe a value of diffusion coefficient to them. A better idea of the profiles is given by the following table, which records concentrations (in cm*-3) at the surface and at depths of 20 microns and 150 microns for various diffusion conditions:

Temp deg C	Time h	Concn at x = 0	Concn at x = 20	Concn at x = 150
1100	4	3.0 X 10*17	1.0 X 10*17	5.0 X 10*16
1100	2	2.4 X 10*17	2.6 X 10*16	1.5 X 10*16
1100	1	1.1 X 10*17	1.4 X 10*16	8.0 X 10*15
1000	4	3.3 X 10*17	2.6 X 10*16	2.0 X 10*16
1000	2	1.9 X 10*17	1.3 X 10*16	5.0 X 10*15
1000	1	1.3 X 10*17	1.1 X 10*16	3.0 X 10*15
900	4	2.8 X 10*17	1.3 X 10*16	7.0 X 10*15
800	4	7.0 X 10*16	2.5 X 10*15	7.0 X 10*14

These results were taken at dissociation arsenic pressure. If the ambient arsenic pressure was increased, the Cr concentrations decreased while maintaining the shape of the profile. Note that the concentration in the tail can be very low, of the order of 10*15 in some cases. If the technique used to measure the profile is not sufficiently sensitive to detect the tail level, the method will give the result that the Cr has diffused only 10-20 microns into the GaAs. This gave rise to some misleading results in some of the earlier work and led to the mistaken belief that Cr diffused only very slowly in GaAs.

Some of the most interesting work in this area has been on out-diffusion from a Cr-doped substrate [2-4]. It was found, using radiotracers [2], that a peak developed at the surface if a homogeneously-doped GaAs slice was annealed at temperatures of 750 deg C and above. A similar result, using SIMS, was reported by Mizutani et al [3], who ignored the surface peaks and fitted the remainder of each out-diffusion profile to a standard solution of Fick´s law. They obtained values of `diffusion coefficient' which, in the range 800-900 deg C followed the law:

D = Do exp(-E/kT)

with Do = 8.53 X 10*4 cm*2/s and E = 3.53 eV. The values of D derived from this expression cannot be regarded as true diffusion coefficients, but in the absence of more rigorous data are useful if employed with care. Similar results were obtained by Kasahara et al [4] who found Do = 6.3 X 10*5 cm*2/s and E = 3.4 eV.

REFERENCES
----

[1] B.Tuck, G.A.Adegboyega [J. Phys. D (GB) vol.12 (1979) p.1875]

[2] B.Tuck, G.A.Adegboyega, P.R.Jay, M.J.Cardwell [Inst. Phys. Conf. Ser. no.45 (1979) p.114]

[3] T.Mizutani, T.Honda, S.Ishida, Y.Kawasaki [Solid State Electron. (GB) vol.25 (1982) p.885]

[4] J.Kasahara, N.Watanabe [Jpn. J. Appl. Phys. (Japan) vol.19 (1980) p.L151]

14.2 DIFFUSION OF Mn IN GaAs

by B.Tuck

July 1985
EMIS Datareview RN=15153

The only systematic study of the diffusion of Mn in GaAs appears to be that reported by Seltzer [1]. Radiotracer Mn was diffused into GaAs over the temperature range 850-1110 deg C and profiles were plotted. Solving Fick's law for the experimental conditions employed predicts profiles in the form of error function complement (erfc) curves. Some of Seltzer's profiles took this form and could therefore be assigned an unambiguous value of diffusion coefficient. Other profiles could not be described in this way but could, however, be fitted reasonably well by a pair of overlapping erfc curves. Using this purely empirical procedure, Seltzer assigned two diffusion coefficients to these anomalous profiles, labelling them 'fast' and 'slow'.

The fast diffusivity, D(F) and the slow diffusivity, D(S), were measured at 900 deg C for arsenic over-pressures, P(As), ranging from 10^{-3} to 10 atm. In both cases, the coefficients were found to vary inversely as the square root of the pressure:

$$\log D(F) = -11.377 - 0.47 \log P(As)$$
$$\log D(S) = -12.087 - 0.49 \log P(As)$$

i.e. the two coefficients differed by a factor of about five over the whole range. A graph was also presented showing the variation of D with temperature. All of the experiments contributing to this graph were taken at an ambient arsenic pressure of 8×10^{-3} atm, and none of the profiles were anomalous, so the problem of 'fast' and 'slow' diffusivities did not arise.
A relation of the form:

$$D = D_o \exp(-E/kT)$$

was found for 850-1100 deg C, with $D_o = 0.65\ cm^2/s$ and E = 2.49 eV. Seltzer proposed a diffusion mechanism wherebye the Mn diffuses by way of arsenic divacancies.

More recently, discussion in this area has centred on the vexed question of type conversion, i.e. the tendency of n-type and semi-insulating GaAs to turn p-type at the surface when subjected to heat-treatment. Good evidence has been produced showing that the onset of conversion correlates with the appearance of a characteristic 'Mn' photoluminescence line [2,3]. In addition, SIMS measurements have shown that most GaAs contains Mn and that after a heat treatment the element can pile up at the surface. Opinions differ on whether this pile-up is due to extra Mn entering from the outside [4] or to out-diffusion from the interior [5], but the above evidence seems to suggest that, one way or the other, Mn gives rise to thermal conversion. This simple conclusion was challenged by Look et al [6], however, who measured the concentration of the acceptor level using electrical techniques and also the concentration of Mn atoms, using SIMS. They came to the conclusion that there were not enough Mn atoms to give rise to the acceptors causing conversion.

REFERENCES
----

[1] M.S.Seltzer [J. Phys. & Chem. Solids vol.26 (1965) p.243]

[2] J.Hallais, A.Mircea-Roussel, J.P.Farges, G.Poiblaud [Inst. Phys. Conf. Ser. no.33b (1977) p.220]

[3] R.Zucca [Inst. Phys. Conf. Ser. no.33b (1977) p.228]

[4] J.B.Clegg, G.B.Scott, J.Hallais, A.Mircea-Roussel [J. Appl. Phys. vol.52 (1981) p.1110]

[5] P.E.Nordquist, P.B.Klein, S.G.Bishop, P.G.Siebrmann [Inst. Phys. Conf. Ser. no.56 (1981) p.569]

[6] D.C.Look, P.W.Yu, J.E.Ehret, Y.K.Keo, R.Kwor [In `Semi-insulating III-V Materials´ (S.Makram-Ebeid and B.Tuck eds): Shiva Publishing, Nantwich, (UK) (1982) p.372]

14.3 DIFFUSION OF Zn IN GaAs

by B.Tuck

July 1985
EMIS Datareview RN=15150

Early work in which diffusion profiles were determined [1,2] indicated that they did not correspond to mathematical solutions of Fick's law. The rates of diffusion appeared to depend on concentration, and a profile often showed a concave section. The diffusion front was invariably abrupt. It was proposed by Longini [3] that a substitutional-interstitial mechanism could be used to account for the results. This model assumes that Zn can exist in two different forms in the GaAs lattice: as an interstitial, in which case it acts as a donor, Zn(i), and also as a substitutional acceptor on the Ga site, Zn(Ga). Most of the impurity is in the substitutional form, but this species has a negligible diffusion coefficient. Diffusion therefore proceeds by movement of the highly mobile interstitial atom. Analysis of this mechanism gives a concentration-dependent diffusion coefficient,

$$D = \frac{3K\,D_i\,C^2}{C_v} \qquad (1)$$

where K is an equilibrium constant, C is the Zn concentration, Di is diffusion coefficient of interstitial atoms and Cv is the equilibrium Ga vacancy concentration. The diffusion coefficient is thus expected to be dependent both upon Zn concentration and the stoichiometry of the crystal.

The above relationship has been tested using the isoconcentration technique [4,5]. In this method, radiotracer Zn is diffused into GaAs which has previously been doped with Zn to a homogeneous level, C(I), say. Under these circumstances, a simple diffusion profile is obtained, to which a diffusion coefficient Di can be assigned. The coefficient Di corresponds to the concentration C(I). The experiment can be repeated for different values of C(I) and a plot of D v C made. A comprehensive set of results for a diffusion temperature of 1000 deg C was produced by Kadhim and Tuck [5]. They presented results for experiments carried out for two different values of ambient arsenic pressure, i.e. for two different values of C(v) over the concentration range 10^{18} - 10^{21} /cm^3. Power relationships of the form

$$D = A\,C^n \qquad (2)$$

were found in each case. For an ambient arsenic pressure of the order of 10^{-4} atm, values of 1.6×10^{-52} and 2.2 were found for A and n respectively (D measured in cm^2/s and C in cm^{-3}). For an arsenic pressure of about 3×10^{-2} atm, the values were 4.6×10^{-53} and 2.2. Thus the diffusion coefficient of Zn in GaAs can vary by five orders of magnitude at a single temperature. In the light of these results, papers which give a single value for diffusion coefficient at a given temperature should therefore be regarded with suspicion.

Unfortunately, substituting Eqn (2) into Fick's law and solving for

C(x) does not produce profiles corresponding to those determined in normal chemical diffusion experiments. A number of workers [6-8] have suggested that a further physical process becomes important in chemical diffusion. Zn interstitial atoms diffuse through the lattice very fast, but sooner or later need to join the lattice. This requires a vacancy to be lost, or a Ga atom to be displaced from its site. In either case, the transfer of Zn atoms to the lattice is impeded and the diffusion process slows down.

Values of D at a Zn concentration of 10*19 cm*-3 are given by Blum et al [9] over the temperature range 600-1000 deg C. An expression of the form

D = Do exp(-E/kT)

is followed, with E = 0.92 eV and Do = 2.7 X 10*-6 cm*2/s. No value of arsenic overpressure is given by these authors. A value for D at 700 deg C for a Zn concentration of 5 X 10*19 cm-3, at dissociation pressure of arsenic, is given as 4.1 X 10*-10 cm*2/s in ref [10].

When diffusion is taking place from the vapour phase, the effect of increasing the ambient vapour pressure is to enhance the Ga vacancy concentration, Cv. As noted above, this means that the zinc diffusion coefficient is dependent on the As vapour pressure as well as the zinc concentration. If the diffusion takes place inside a closed ampoule, as is often the case, it is possible to specify both the ambient Zn vapour pressure and the As pressure by using carefully weighed quantities of the two elements in the diffusion source. In this way, a satisfactory degree of control over the diffusion process can be achieved. Calculation of the vapour pressures of Zn and As created by given amounts of the two elements is not easy, however, since the Ga-As-Zn ternary phase diagram is quite complex. It would not be correct to assume, for instance, that all Zn and As added to a diffusion ampoule necessarily goes into the vapour phase. Discussion of the composition of diffusion sources and the use of the phase diagram can be found in refs [11,12].

REFERENCES

[1] F.A.Cunnell, C.H.Gooch [J. Phys. & Chem. Solids vol.15 (1960) p.127]

[2] B.Goldstein [Phys. Rev. vol.121 (1961) p.1305]

[3] R.L.Longini [Solid State Electron. vol.5 (1962) p.127]

[4] L.L.Chang, G.L.Pearson [J. Appl. Phys. vol.35 (1964) p.1960]

[5] M.A.Kadhim, B.Tuck [J. Mater. Sci. vol.7 (1972) p.68]

[6] B.Tuck, M.A.Kadhim [J. Mater. Sci. vol.7 (1972) p.585]

[7] U.Gosele, F.Morehead [J. Appl. Phys. vol.52 (1981) p.4617]

[8] A.H.van Ommen [J. Appl. Phys. vol.54 (1983) p.5055]

[9] S.E.Blum, M.B.Small, D.Gupta [Appl. Phys. Lett. vol.42 (1983) p.108]

[10] B.Tuck, A.J.Houghton [J Phys. D vol.14 (1981) p.2147]

[11] H.C.Casey, M.B.Panish [Trans Met. Soc. AIME vol.242 (1968) p.406]

[12] K.K.Shih, J.W.Allen, G.L.Pearson [J. Phys. & Chem. Solids vol.29 (1968) p.379]

14.4 DIFFUSION OF Si IN GaAs

by B.Tuck

July 1985
EMIS Datareview RN=15152

In early work [1,2] diffusion penetration of Si in GaAs was assessed by measuring junction depths. Two-hour diffusions at 1000 deg C gave junction depths of about 2 microns [1] while 1150 deg C for 90 min gave approximately 5 microns [2]. Vieland [2] carried out his experiments for different ambient arsenic pressures and found that the measured junction depth increased by a factor of about two with increasing pressure in the range 0.1-5.0 atm.

Diffusion of silicon into Cr-doped GaAs was studied by Greiner and Gibbons [3] in the temperature range 850-1050 deg C. The diffusion source used was a 100 A evaporated layer of Si, encapsulated by a layer of SiO2. Very short heat treatments, of the order of seconds, gave rise to shallow profiles which were determined by SIMS. Differential Hall measurements were also made and profiles of electron concentration plotted. Comparison of the two types of profile showed a large discrepancy, with Si concentrations exceeding electron concentrations by a least an order of magnitude. This suggests that the Si atoms were distributed between Ga and As sites so as to give almost complete compensation. They proposed that diffusion takes place by way of the donor-acceptor complex (Si(Ga)+Si(As)-), which can easily move by interaction with a vacancy of either type, e.g.

(Si(Ga)+Si(As)-) + V(Ga) = V(Ga) + (Si(As)+Si(Ga)-)

The effective diffusion coefficient of Si in GaAs, De, is given by

De = Dc [1 - (1 + 2[Si]/K)*-0.5]

where Dc is the diffusion coefficient of the complex, [Si] is the silicon concentration, and K is an equilibrium constant. Greiner and Gibbons used this expression to calculate theoretical profiles, which gave a good fit to the SIMS results. They also presented a graph of Dc against temperature, which followed the expression

Dc = Do exp (-E/kT)

over the range 850-1050 deg C, with values of 0.11 cm*2/s and 2.5 eV for Do and E, respectively.

REFERENCES
----

[1] G.R.Antell [Solid State Electron vol.8 (1965) p.943]
[2] L.J.Vieland [J. Phys. Chem. Solids vol.21 (1961) p.318]
[3] M.E.Greiner, J.F.Gibbons [Appl. Phys. Lett. vol.44 (1984) p.750]

14.5 DIFFUSION OF Sn IN GaAs

by B.Tuck

November 1985
EMIS Datareview RN=15410

There is very little information available on the diffusion of tin in GaAs. Work by Goldstein and Keller [1] and Fane and Goss [2] indicated that tin diffusion is a very slow process compared to that of acceptors in the same material. More recently it has been established that tin diffusions can be carried out from doped films of SiO_2 deposited on the GaAs surface [3,4,5].

The largest body of experimentation is that due to Tuck and Badawi [6], who diffused radioactive tin into n-type GaAs from the vapour phase over the temperature range 850-1100 deg C. Diffusion profiles were plotted and electrical measurements were taken so that the electron concentration could be compared with atomic doping levels. They found that the profiles could be fitted reasonably well by standard solutions to the diffusion equation, so that it was possible to fit values of diffusion coefficient, D, to the curves. An interesting finding, however, was that the value of D depended on the level of n-type doping in the original GaAs slice. Two plots of ln D vs 1/T were presented, one from a set of experiments using undoped starting material (group A) and the other experiments using n-type, previously doped in growth to a level 2×10^{18} cm^{-3} (group B). In both cases an expression of the form

$$D = D_o \exp(-Q/kT)$$

was followed with the values

group A $D_o = 3.2$ cm^2/s
$Q = 3.3$ eV

group B $D_o = 9.43 \times 10^{-8}$ cm^2/s
$Q = 1.9$ eV

These values can be compared with those of Goldstein and Keller [1]

$D_o = 6 \times 10^{-4}$ cm^2/s
$Q = 2.5$ eV

Unfortunately, these latter workers did not give any information on the original doping of their substrates. All of the experiments of Tuck and Badawi were carried out in the presence of an excess arsenic pressure. This was found necessary in order to maintain the integrity of the GaAs surface.

Arnold et al [5] diffused tin from spin-on sources at 800 deg C and 900 deg C. SIMS measurements were made on some of the diffused samples, and diffusion profiles were produced. They found that they could not fit standard solutions to the profiles and suggested that the curves corresponded to a diffusion process for which D was proportional to the square of tin concentration. The amount of data presented in the paper is small, however, and the SIMS profiles cover less than two orders of magnitude in tin concentration. They produced a plot of ln D vs 1/T corresponding to a tin concentration of approximately 10*18 cm*-3. An activation energy, Q, of 2 eV was estimated from this curve. Most of the experiments used gallium-rich substrates. A result using arsenic-rich substrate was also presented, and this indicated a value of diffusion coefficient about an order of magnitude smaller than the gallium-rich case.

Tuck and Badawi [6] found that for tin levels in excess of about 10*18 cm*-3, the atomic concentration was greater than that of electrons. Much the same result was reported by Von Munch [7]. In both cases this was assumed to be due to self-compensation by the tin, the group IV atom being able to exist on either the group III site as a donor or the group V site as an acceptor. Arnold et al [5] concluded that concentrations of tin and electrons were about the same in their experiments, but their diffusion profiles did not go much above 10*18 cm*-3, so their finding does not necessarily contradict the earlier results. The experiments described by Erkaya et al [8] are interesting in this context. They diffused tin into p-type GaAs from a solid SnS source at temperatures in the range 650-750 deg C, in an attempt to create n-p junctions. In fact no such junctions were produced, even though SIMS measurements indicated tin concentrations substantially in excess of the background p-type doping. It was concluded that tin introduced at a low processing temperature enters the GaAs lattice in an inactive state, such as an inactive interstial. Alternatively, complete self-compensation may remove all electrical activity.

REFERENCES

[1] B.Goldstein, H.Keller [J. Appl. Phys. (USA) vol.32 (1961) p.1180]

[2] R.W.Fane, A.J.Goss [Solid State Electron. (GB) vol.6 (1963) p.383]

[3] C.F.Gibbon, D.R.Ketchow [J. Electrochem. Soc. (USA) vol.118 (1971) p.975]

[4] H.Yamazaki, Y.Kawasaki, M.Fujimoto, K.Kudo [Jpn. J. Appl. Phys. (Japan) vol.14 (1975) p.717]

[5] N.Arnold, R.Schmitt, K.Heime [J. Phys. D (GB) vol.17 (1984) p.443]

[6] B.Tuck, M.H.Badawi [J. Phys. D (GB) vol.11 (1978) p.2541]

[7] W.Von Munch, [IBM J. Res. & Dev. (USA) vol.10 (1966) p.438]

[8] H.H.Erkaya, R.J.Roedel, R.Lareau, P.Williams, J.Leavitt, A.Von Neida [J. Electrochem. Soc. (USA) vol.132 (1985) p.2240]

14.6 DIFFUSION OF S in GaAs

by B.Tuck

July 1985
EMIS Datareview RN=15151

In the early work studying the diffusion of sulphur in GaAs, the process was mainly studied by measuring the depths of p-n junctions formed in p-type material [1-3] or by taking sheet resistivity measurements [4]. In only two pieces of recently published work have accurate diffusion profiles been ascertained, in both cases using radio-tracer sulphur, and the data presented in these papers [5,6] must be considered more reliable than the previous work. Refs [1,2,4,5] present values of diffusion coefficient D, as a function of temperature, and all of these authors find that results can be described by an expression of the form

$$D = D_o \exp(-E/kT)$$

although a good deal of variation exists in the quoted values of Do and E, as can be seen from the following Table

Do cm^2/s	E eV	temp. range deg C	ref.
1.2×10^{-4}	1.80	900-1050	[1]
2.6×10^{-5}	1.86	900-1100	[2]
1.6×10^{-5}	1.63	700-1000	[4]
1.85×10^{-2}	2.6	900-1200	[5]

It should be stressed, however, that the diffusion coefficient of an impurity in a compound is not normally a function of temperature alone. It also depends on the ambient vapour pressures of the host material, since these pressures influence the stoichiometry of the compound. Only refs [5,6] consider this factor in any detail. Ref [5] presents D as a function of ambient arsenic pressure for 1003 deg C and 1130 deg C. At the former temperature the value of D varies approximately as $\sqrt{P(As)}$, from 1.6×10^{-14} cm^2/s at a pressure of 10^{-4} atm. to 1.4×10^{-12} cm^2/s at 0.6 atm, above which D saturates. A variation of almost two orders of magnitude in D is therefore found at a single temperature. A similar result is found for 1130 deg C, with values of D a factor of 7 above those for 1003 deg C. The data given in the Table for reference 5 is for high ambient arsenic pressures.

A further complication arises from these results. The experimental conditions used in both refs [5,6] would normally be expected to give profile of the form

$$C = C_o \, \mathrm{erf}\left(\frac{x}{2\sqrt{Dt}}\right)$$

In fact, in neither case did the profiles fit well to this expression, and the values of D presented should therefore be treated with some caution.

Young and Pearson [5] attempted to explain their results by proposing that sulphur exists in GaAs in elemental form and also as part of the complex [V(Ga)S(As)V(Ga)]. Zahari and Tuck [7] proposed a rather different mechanism, in which the equilibrium of arsenic vacancies was taken into account. This second mechanism has the advantage that it accurately predicts the shape of the diffusion profile.

REFERENCES

[1] L.J.Vieland [J. Phys. & Chem. Solids vol.21 (1961) p.318]
[2] R.G.Frieser [J. Electrochem. Soc. vol.112 (1965) p.697]
[3] T.H.Yeh [J. Electrochem. Soc. vol.111 (1964) p.253]
[4] D.L.Kendall [Semiconductors and Semi-metals, vol.4 (1968), Eds R.K.Willardson, A.C.Beer (Academic Press, New York), p.163]
[5] A.B.Young, G.L.Pearson [J. Phys. & Chem. Solids vol.31 (1970) p.517]
[6] B.Tuck, R.G.Powell [J. Phys. D vol.14 (1981) p.317]
[7] M.D.Zahari, B.Tuck [J. Phys. D vol.15 (1982) p.1741]

CHAPTER 15

SURFACE STRUCTURE AND OXIDATION OF GALLIUM ARSENIDE

15.1 Surface structure of GaAs

15.2 Oxide layer structures on GaAs

15.3 Oxidation rates of GaAs

15.1 SURFACE STRUCTURE OF GaAs

by P.J.Dobson

January 1986
EMIS Datareview RN=15498

GENERAL COMMENTS

Most work has been channelled into studying the structure of the (110) and (001) surfaces. The former surface is non-polar and is the natural cleavage plane and has therefore attracted most of the fundamental surface science interest. The (001) plane is technologically important since most opto-electronic devices are grown on this plane. Recently there has been increased interest in some other low index surfaces, motivated partly by technological interest in seeking alternatives to the (001) plane. There is now substantial information concerning the (111) plane. This review will deal separately with the (110), (100) and (111) surfaces and briefly describe references to other low index surfaces.

1. GaAs (110)

This surface is one of the most understood in terms of surface crystallography. Details of this and other III-V (110) surfaces were given by Kahn [1] in a recent review and the historical development of this particular structure determination was outlined by Duke [2]. The structure determinations were made using elastic LEED intensity measurements [3-7] and result in a model in which the anions (As) rotated out of the surface by 27 deg i.e. producing a buckled atomic arrangement with the arsenic moving outwards by about 0.65 A relative to the gallium, leading to distortions in the next atomic layer. This arrangement is consistent with the observation that for carefully prepared surfaces there are no surface states present in the fundamental band gap [8-10]. Theoretical predictions for an unrelaxed surface suggest that surface states would be observed, but these would be removed by the 27 deg relaxation.

The 27 deg relaxation has been challenged by Gibson et al [11] who suggested that Rutherford Backscattering Spectroscopy (RBS) results are consistent with a much smaller, 7 deg angular relaxation. Froelich et al [12] reviewed the dependence of surface electronic state dispersion and could not distinguish between either model. Duke and Paton [13] have discussed this problem in detail with regard to the sensitivity of LEED to displacement of atoms parallel to the surface. Scanning tunnelling microscopy [14] has suggested that surface corrugation exists which may be consistent with a relaxation model. However the main finding in this work was evidence for surface point defects.

2. GaAs (001)

LEED studies on this surface are difficult to perform because the surface stoichiometry has to be maintained by constant exposure to a flux of arsenic molecules. For this reason the most reliable observations are those employing reflection high energy electron diffraction (RHEED) in molecular beam epitaxy (MBE) equipment. There is not a clear concensus of agreement between the sequence and specification of structures in going from gallium rich to arsenic rich surfaces which is broadly as follows: (4 X 2) or c(8 X 2); (4 X 6); c(6 X 4) or (3 X 1); (1 X 6); (2 X 4) or c(2 X 8); c(4 X 4), see refs [15-19].

The c(2 X 8) may be a special case of two dimensional surface fault arising from the (2 X 4) as pointed out by Neave et al [20,21]. They base their model on a suggestion by Dobson originally reported by Larsen et al [22] which gives a complex unit cell in terms of asymmetric tilted dimers similar to the original suggestion for the Si (001) (2 X 1) surface [23]. Daweritz has reported a substantially different model for this and other surfaces which is based on the concept of arrays of mini-facets or micro-steps [24,25]. These models have the merit of giving some explanation for the wide variations of arsenic coverage that have been reported and they may be consistent with the work function variations reported by Massies et al [26]. Any advance on these models probably awaits detailed RHEED intensity analyses. Larsen et al [19] gave evidence based on surface core-level shift that the c(4x4) structure is a chemisorbed phase of trigonally bonded arsenic. However, RBS channelling measurements [27] suggest that lattice distortion to a depth of several layers exists and this would be more consistent with an asymmetric dimer model.

The surface electronic structure of the (001) surface structures has been investigated by Larsen et al [22] and Salmon and Rhodin [28] using angle resolved photoemission.

3. GaAs (111)

There are two situations to consider i.e. the (111)Ga terminated and the (barl barl barl) As terminated surface. Both surfaces when prepared by MBE have shown a (2 X 2) reconstruction [29,30] in addition to more complex structures. Recent LEED studies [31,32] suggest that the (2 X 2) reconstruction is stable but that there are important differences between the (111) and (barl barl barl) with regard to electronic structure e.g. surface state dispersion [33], and local chemistry as indicated by surfaces core level shift [34]. The later measurements indicate the complete absence of gallium atoms at the surface which means that Chadi´s vacancy model [35] which is based on energy minimisation is not appropriate for the (barl barl barl) but it appears to fit very well with data for the (111). The vacancy formation on the (111) transforms the surface into a non-polar (110)-like surfaces with six-fold rings of Ga and As atoms. Further support for this model is given by photoemission [36]. Other reconstructions i.e. (3 X 3) [37,38] and (3 X 1) [39] have been noted for the As terminated (barl barl barl) surface.

4. OTHER GaAs SURFACES

Hren et al [40] have reported a LEED study of the (211) surfaces which showed a clear tendency to form (110) both the arsenic and gallium terminated surfaces, but the facet orientation differed on these surfaces. Stiles and Kahn [41] have reported LEED studies of the (221) and (311) surfaces. The (221) produced large (110) type facets but both arsenic and gallium terminated (311) surfaces gave (1x1) diffraction patterns and appeared to be stable up to 600 deg C. Electronic structures have been calculated for the (211) [42] and (311) [43] surfaces.

REFERENCES

[1] A.Kahn [Surf. Sci. Rep. (Netherlands) vol.3 (1983) p.193]

[2] C.B.Duke [Appl. Surf. Sci. (Netherlands) vol.11/12 (1982) p.1]

[3] S.Y.Tong, B.J.Mrstik, M.A.Van Hove [Phys. Rev. B (USA) vol.17 (1978) p.3303]

[4] A.Kahn, E.So, P.Mark, C.B.Duke [J. Vac. Sci. & Technol. (USA) vol.15 (1978) p.580]

[5] C.B.Duke, A.Paton, W.K.Ford, A.Kahn, J.Carelli [Phys. Rev. B (USA) vol.24 (1981) p.562]

[6] C.B.Duke, A.Paton, W.K.Ford, A.Kahn, G.Scott [Phys. Rev. B (USA) vol.24 (1981) p.3310]

[7] C.B.Duke, R.J.Meyer, P.Mark [J. Vac. Sci. & Technol. (USA) vol.17 (1980) p.971]

[8] A.Ebina, T.Unno, Y.Suda, H.Koinuma, T.Takahashi [J. Vac. Sci. & (USA) vol.19 (1981) p.301]

[9] J. Van Laar, J.J.Scheer [Surf. Sci. (Netherlands) vol.8 (1976) p.342]

[10] J. Van Laar, A.Huijser [J. Vac. Sci. & Technol. (USA) vol.13 (1976) p.769]

[11] W.M.Gibson, H.J.Gossmann [J. Vac. Sci. & Technol.B (USA) vol.2 (1984) p.343]

[12] D.V.Froelich, M.E.Lapeyre, J.D.Dow, R.E.Allen ["Superlattices and Microstructures" vol.1 (1985) p.87]

[13] C.B.Duke, A.Paton [J. Vac. Sci. & Technol. B (USA) vol.2 (1984) p.327]

[14] R.M.Feenstra, A.P.Fein [Phys. Rev. B (USA) vol.32 (1985) p.1394]

[15] A.Y.Cho, [J. Appl. Phys. (USA) vol.47 (1976) p.2841]

[16] P.Drathen, W.Ranke, K.Jacobi [Surface Sci. (Netherlands) vol.77 (1978) p.L162]

[17] R.Z.Bachrach, R.S.Bauer, P.Chiaradia, G.V.Hansson [J. Vac. Sci. & Technol. (USA) vol.18 (1981) p.797]

[18] T.C.Chiang, R.Ludeke, M.Aonom G,Landgren, F.J.Himpsel, D.E.Eastman [Phys. Rev. B (USA) vol.27 (1983) p.4770]

[19] P.K.Larsen, J.H.Neave, J.F. van der Veen, P.J.Dobson, B.A.Joyce [Phys. Rev. B (USA) vol.27 (1983) p.4966]

[20] J.H.Neave, B.A.Joyce, P.J.Dobson, N.Norton [Appl. Phys. A (Germany) vol.31 (1983) p.1]

[21] B.A.Joyce, J.H.Neave, P.J.Dobson, P.K.Larsen [Phys. Rev. B (USA) vol.29 (1984) p.814]

[22] P.K.Larsen, J.F. van der Veen, A.Mazur, J.Pollmann, J.H.Neave,

B.A.Joyce [Phys. Rev. B vol.26 (1982) p.3222]
[23] W.S.Yang, F.Jona, P.M.Marcus [Phys. Rev. B (USA) vol.28 (1983) p.2049]
[24] L.Daweritz [Surface Sci. (Netherlands) vol.118 (1982) p.585]
[25] L.Daweritz [Surface Sci. (Netherlands) vol.160 (1985) p.171]
[26] J.Massies, P.Etiene, F.Dezaly, N.T.Linh [Surf. Sci. (Netherlands) vol.99 (1980) p.121]
[27] T.Narusawa, K.L.I.Kobayashi, H.Nakashima [Jpn. J. Appl. Phys. Part 2 (Japan) vol.24 81985) p.L98]
[28] L.G.Salmon, T.N.Rhodin [J. Vac. Sci. Technol. B (USA) vol.1 (1983) p.736]
[29] J.R.Arthur [Surface Sci. (Netherlands) vol.453 (1974) p.449]
[30] A.Y.Cho [J. Appl. Phys. (USA) vol.41 (1970) p.2780]
[31] S.Y.Tong, G.Xu, W.N.Mei [Phys. Rev. Lett. (USA) vol.52 (1984) p.1693]
[32] M.Alonso, F.Soria, J.l.Sacedon [J. Vac. Sci. & Technol. A (USA) vol.3 (1985) p.1598]
[33] R.D.Bringans, R.Z.Bachrach [Phys. Rev. Lett. (USA) vol.53 (1984) p.1954]
[34] A.D.Katnani, H.W.Sang, Jr., P.Chiaradia, R.S.Bauer [J. Vac. Sci. & Technol. B (USA) vol.3 (1985) p.608]
[35] D.J.Chadi [Phys. Rev. Lett. (USA) vol.52 (1984) p.1911]
[36] A.D.Katnani, D.J.Chadi [Phys. Rev. (USA) vol.31 (1985) p.2554]
[37] A.U.MacRae [Surface Sci. (Netherlands) vol.4 (1966) p.247]
[38] W.Ranke, K.Jacobi [Surface Sci. (Netherlands) vol.63 (1977) p.33]
[39] D.J.Frankel, J.Anderson, G.J.Lapeyre, H.H.Farrell [J. Vac. Sci. & Technol. B (USA) vol.3 (1985) p.1093]
[40] P.Hren, D.W.Tu, A.Kahn [Surface Sci. (Netherlands) vol.146 (1984) p.69]
[41] K.Stiles, A.Kahn [J. Vac. Sci. & Technol. B (USA) vol.3 (1985) p.1089]
[42] A.Mazur, J.Pollman [Phys. Rev. B (USA) vol.30 (1984) p.2084]
[43] D.J.Chadi [J. Vac. Sci. & Technol. B (USA) vol.3 (1985) p.1187]

15.2 OXIDE LAYER STRUCTURES ON GaAs

by H.L.Hartnagel

December 1985
EMIS Datareview RN=15472

Native oxide films produced by electrolytic or gas-plasma anodization are usually amorphous, when produced with current densities less than 2 mA/cm*2 at temperatures only close to room temperature; they become partly crystalline and contain beta-Ga2O3 when annealing temperatures above 600 deg C are applied [1]. The initial stoichiometric mixture of Ga2O3 and As2O3 [2] with a possibly very small accumulation of elemental As at the oxide-semiconductor interface (see review paper [3]) is changed into pure Ga2O3 after exposure to temperatures above 450 deg C by evaporation of the volatile As2O3 [3]. After anodization with ideal growth conditions, the oxide-semiconductor transition region has a width which is comparable to or less than the one found in thermally oxidized Si/SiO2 systems (less than 10 A) [4].

Thermal oxidation, even below 530 deg C, gives primarily Ga2O3 [5]. It is possible to produce a sequence of different oxides by first evaporating various metal films on GaAs and then oxidizing such a structure by anodization. Depending on the type of material transport during oxidation, various oxide compositional profiles result [2,6,13]. Thin native oxides formed by the room-temperature exposure of GaAs surfaces to various oxidizing environments such as air, water, etch solutions etc can exhibit a composition of chemically adsorbed oxygen (called beta-type) and (primarily?) physically adsorbed oxygen (called alpha-type) [7], whose ratio depends on type of environment [8] and orientation of GaAs [7]. XPS and other surface analytical techniques have shown that the oxide layer composition after long exposure to air (12 months) varies with depth over the oxide film thickness of less than 50 A [9]. In fact, both XPS and ISS have shown that the top layer of a few atomic distances consists of a mixture of Ga2O3, As2O3 or some other forms of oxides of Ga and As with a Ga concentration which is larger than that of As. Deeper down, where a similar mixture of Ga oxides and As oxides exists, an excess of As is detected [10,11,12]. Finally, after some 25 A (this value varies strongly depending on the surface treatment to which the GaAs was exposed), the interface with the semiconductor occurs, where an accumulation of elemental As is probable [12].

REFERENCES

[1] B.L.Weiss, H.L.Hartnagel [Thin Solid Films (Switzerland) vol.56 (1979) p.143]
[2] C.W.Fischer, J.D.Canaday [J. Electrochem. Soc. (USA) vol.130 (1983) p.1740]
[3] C.W.Wilmsen [J. Vac. Sci. & Technol. (USA) vol.19 (1981) p.279]
[4] D.E.Aspnes et al [J. Electrochem. Soc. (USA) vol.128 (1981) p.590]
[5] Y.Mizokawa et al [Jpn. J. Appl. Phys. Suppl. (Japan) vol.17 suppl. 17-1 (1978) p.327]
[6] B.Bayraktaroglu, H.L.Hartnagel [Int. J. Electron. (GB) vol.45

(1978) p.449]
[7] W.Ranke et al [Surf. Sci. (Netherlands) vol.122 (1982) p.256]
[8] E.Huber, H.L.Hartnagel [Solid State Electron. (GB) vol.27 (1984) p.589]
[9] S.Kohiki et al [Jpn. J. Appl. Phys. (Japan) vol.23 (1984) p.L15]
[10] C.M.Demanet, M.A.Marais [Surf. & Interface Anal. (GB) vol.7 (1985) p.13]
[11] K.Rohkel, H.L.Hartnagel [unpublished data, so far initially reported only in grant report to European Research Office of US Army, London, 1984 (Grant Number DAJA, 45-83-C-0012), Hartnagel et al]
[12] G.P.Schwartz [Thin Solid Films (Switzerland) vol.103 (1983) p.3]
[13] J.T.Grant et al [Int. J. Electron. (GB) vol.46 (1979) p.209]

15.3 OXIDATION RATES OF GaAs

by H.L.Hartnagel

December 1985
EMIS Datareview RN=15471

A) NATURAL OXIDES:

a) Powdered GaAs exposed to oxygen at 0.40 torr at a temperature of 26.2 deg C yields an oxygen sorption rate which is initially extremely high (giving 10*15 oxygen atoms per cm*2 after the first 2 minutes of exposure) and reduces then to a rate of 0.3 X 10*12 oxygen atoms per cm*2 per minute [1].

b) Cleaved GaAs, i.e. the (110) single crystal surface, exposed to oxygen at room temperature, gives an oxide film thickness (as obtained by ellipsometry) with a straight-line increase from 7 to 43 A (linear scale) for a logarithmic scale of time of exposure from 1 min to 10*6 min [2]; more recently a growth rate of less than 10 A/h was found (referred to in [3]).

c) A cylindrical GaAs sample allows the measurement of the orientation dependent oxide growth rate, which is strongest for [bar1 bar1 bar1] [4]. Here AES, XPS and UPS reveal the two types of oxides, called alpha and beta, (distinguished by the chemical shift of electron emission and the distinct difference in the UPS spectra); the former is less tightly bound and desorbs at increased temperatures up to 300 deg C whereas the latter is more stable and desorbs at 420-480 deg C. The ratio of alpha/beta is only weakly dependent on exposure and orientation [4].

d) Exposure to other environments, like aqueous solutions of various types, gives an environment-dependent oxidation rate of strongly different alpha/beta ratio [4]. alpha-type oxygen could be mainly physically adsorbed whereas beta-type would be chemically adsorbed. A growth rate for exposure to stationary deionized water was found to be 10-20 A/h [3]. The vacuum ultraviolet photochemical oxidation of GaAs by N2O depends on the surface treatment before insertion into the reactor (Br2 etching gives equal amounts of As2O3 and Ga2O3, whereas HCl etching gives predominantly AsO), and depends strongly on whether island growth is involved (to be measured by platinum-carbon secondary replicas [6]). Using XPS with its depth of analysis, the percentage change of oxide increases at room temperature from 40 to 50% for the time from 30 to 60 minutes after starting of the reaction [7].

B) THICK NATIVE OXIDES:

a) When thermally oxidised in pure oxygen or water-vapour containing oxygen, the film consists mainly of crystalline beta-Ga2O3 (beta represents here the crystalline structure and not the type of thin native oxygen) for temperatures above 600 deg C [8,9] and of amorphous Ga_2O_3 for temperatures from 400 deg C to 525 deg C [10]. Using a mixture of As_2O_3 and O_2 as ambient, not only Ga_2O_3 but also $GaAsO_4$ can be found (for a review, see [11]).

b) Wet anodic oxidation has a growth constant which depends linearly on the current density and is proportional to the voltage Vox, appearing across the oxide. This proportionality ranges from 19 to to 24 A/V for a current density from 10^0 to 10^{-2} mA/cm^2 for an electrolyte consisting of glycol and an aqueous solution (AGW) [12]. The value of Vox increases proportionally with time, for a current density of 100 $microA/cm^2$ at a rate of 2.8 V/min [13,14] with AGW.

c) Plasma anodic oxidation exhibits very similar growth behaviour to that of the wet process. Using a multipole plasma, growth rates of 200-250 A/min were obtained [15].

REFERENCES

[1] A.J.Rosenberg [J. Phys. & Chem. Solids (GB) vol.14 (1960) p.175]

[2] F.Lukes [Surf. Sci. (Netherlands) vol.30 (1972) p.90]

[3] J.Massies, J.P.Contour [Appl. Phys. Lett. (USA) vol.46 (1985) p.1150]

[4] W.Ranke, Y.R.Xing, G.D.Shen [Surf. Sci. (Netherlands) vol.122 (1982) p.256]

[5] E.Huber, H.L.Hartnagel [Solid-State Electron. (GB) vol.27 (1984) p.589]

[6] W.H.Makky, K.M.Geib, C.W.Wilmsen [J. Vac. Sci. & Technol. (USA) vol.21 (1982) p.417]

[7] P.A.Bertrand [J. Electrochem. Soc. (USA) vol.132 (1985) p.973]

[8] K.H.Zaininger, A.G.Revesz [J. Phys. (France) vol.25 (1964) p.208]

[9] B.J.Sealy, P.L.F.Hemment [Thin Solid Films (Switzerland) vol.22 (1974) p.539]

[10] K.Navratil [Czech. J. Phys. (Czechoslovakia) vol.B18 (1968) p.266]

[11] C.W.Wilmsen [J. Vac. Sci. & Technol. (USA) vol.19 (1981) p.279]

[12] H.Hasegawa, H.L.Hartnagel [J. Electrochem. Soc. (USA) vol.123 (1976) p.713]

[13] H.L.Hartnagel [Oxides and Oxide Films, vol.6, Ed. A.K.Vijh (Marcel Dekker, New York & Basel, 1981) ch.2 p.85]

[14] B.Bayraktaroglu, H.L.Hartnagel [Int. J. Electron. (GB) vol.45 (1978) p.337]

[15] S.Gourrier, A.Mircea, M.Bacal [Thin Solid Films (Switzerland) vol.65 (1980) p.315]

CHAPTER 16

ETCHING RATES OF GALLIUM ARSENIDE

16.1 Etching rates of wet etched GaAs

16.2 Etching rates of plasma etched GaAs

16.3 Etching rates of ion-beam milled and sputter etched GaAs

16.4 Etching rates of reactive ion and reactive ion-beam etched GaAs

16.5 Etching rates of GaAs etched by laser-assisted methods

16.1 ETCHING RATES OF WET ETCHED GaAs

by C.I.H.Ashby

December 1985
EMIS Datareview RN=15420

Wet etching processes have three general applications: pattern formation, polishing, and defect or damage visualization. This Datareview is restricted to wet etches suitable for pattern formation for device fabrication. Chemo-mechanical methods suitable for polishing and defect visualization have been reviewed [1,2,3]. Some polishing etches which are `free´ etches, i.e., not chemo-mechanical, are useful for pattern formation. Some general characteristics of wet etches are:

1. Preferential etching based on crystallographic orientation with the (111)Ga face generally etching a factor of 2 to 5 slower than the (100), (110), or (111)As faces.

2. Greater tendency for etch pits to form on (111) faces.

3. Etching reactions generally based on oxidation of Ga.

There are two categories of wet etches: non-electrolytic [3-15,21] (Table I) and electrolytic [3,16-20] (Table II).

Non-electrolytic etching rates can be either diffusion limited or chemical reaction limited. For diffusion-limited etches, the rate is controlled by the mass transport of reactants to the surface or of products from the surface. These etches tend to be isotropic and relatively insensitive to temperature but highly sensitive to changes in the nature and degree of agitation. Diffusion-limited etches are especially good for polishing wafers. For chemical-reaction-limited etches, the rate is controlled by the chemical reactions at the GaAs surface. These etches tend to be anisotropic with respect to certain crystallographic orientations, quite sensitive to temperature, and relatively insensitive to agitation. They are well suited for etching geometric shapes along crystallographic planes. Whether a given etching rate is diffusion or chemical-reaction controlled is often determined by the relative proportions of the constituents of the etching solution. An increase in the viscosity of the solution increases the relative importance of diffusion.

Electrolytic etches can provide accurate depth control but require ohmic contact to the wafer. They are based on anodic oxidation of GaAs followed by dissolution of products. Etching of n-type material is much slower than p-type unless it is illuminated to photogenerate holes in the surface region or unless sufficient voltage is applied to produce breakdown of the surface barrier.

Ion bombardment to amorphize the surface can greatly increase the etching rate for both electrolytic [20] and non-electrolytic [21] etching.

Table I: Nonelectrolytic Wet Etches

etchant	etch rate micron/min	T deg K	(111)Ga / (111)As	Ref.	Comments
0.3N NH4OH: 0.1N H2O2	0.18-0.2	300	1/6-1/4	[4]	n-type
	0.12-0.14	300	1/6-1/4	[4]	p-type
3N NH4OH:1N H2O2	1.4	300	below 1	[4]	little mask attack
NH4OH:H2O2:H2O	0.5-5	293	composition selectable	[5]	very smooth
NaOH:H2O2	0.2	303		[6]	
	1	328		[6]	
H2SO4:H2O2:H2O	3.5	273	up to 1	[7]	orientation dependence f(H2SO4:H2O2)
1H2SO4:8H2O2:1H2O	14	300	1/4	[8]	
H3PO4:H2O2:H2O	0.01-4	303	composition selectable	[9]	reproducible to 10's of A for below 0.1 micron/min
H3PO4:H2O2:3CH3OH	2	300	1/2	[10]	extremely smooth
Citric acid:H2O2	0.6	297	1/2	[11]	no mask attack
HCl:4H2O2:H2O	0.22	300	below 1	[8]	
40HCl:4H2O2:H2O	5	300	1	[8]	isotropic
K2Cr2O7:H2SO4: HCl:CH3COOH	0.01-1	293		[12]	no mask attack no pits
K2Cr2O7:H2SO4:HCl	2.5	298		[13]	no mask attack
	20	333		[13]	no pits
KI:I2	1	300	below 1	[14]	
5%Br2-CH3OH	5-7	300	1/5	[15]	lower Br2, more preferential

Table II: Electrolytic Wet Etches

electrolyte	etch rate micron/min	current density mA/cm*2	T deg K	ref.	comments
3M NaOH	4	100	300	[16]	p faster than n
0.5N KOH	1	20	298	[17]	p-type
1N HC104	1	20	298	[17]	p-type
1HC104:4CH3COOH	1	20	300	[18]	+V required for n-type
0.1N HCl	0.2	10	300	[19]	surface residue if above 40 mA/cm*2
0.1N HNO3	0.12	10	300	[19]	surface residue if above 40 mA/cm*2

REFERENCES

[1] B.Tuck [J Mater. Sci. (GB) vol.10 (1975) p.321]

[2] D.J.Stirland, B.W.Straughan [Thin Solid Films (Switzerland) vol.31 (1976) p.139]

[3] W.Kern [RCA Rev. (USA) vol.39 (1978) p.278]

[4] J.J.Gannon, C.J.Nuese [J. Electrochem. Soc. (USA) vol.121 (1974) p.1215]

[5] L.P.Molodyakova, V.M.Andreev [Inorg. Mater. (USA) vol.12 (1977) p.956]

[6] I.Shiota, K.Motoya, T.Ohmi, N.Miyamoto, J.Nishizawa [J. Electrochem. Soc. (USA) vol.124 (1977) p.155]

[7] S.Iida, K.Ito [J. Electrochem. Soc. (USA) vol.118 (1971) p.118]

[8] D.W.Shaw [J. Electrochem. Soc. (USA) vol.128 (1981) p.874]

[9] Y.Mori, N.Watanabe [J. Electrochem. Soc. (USA) vol.125 (1978) p.1510]

[10] J.L.Mertz, R.A.Logan [J. Appl. Phys. (USA) vol.47 (1976) p.3503]

[11] M.Otsubo, T.Oda, H.Kumabe, H.Miki [J. Electrochem. Soc. (USA) vol.123 (1976) p.676]

[12] S.Adachi, K.Oe [J. Electrochem. Soc. (USA) vol.131 (1984) p.126]

[13] S.Adachi, H.Kawaguchi, G.Iwane [J. Mater. Sci. (GB) vol.16 (1981) p.2449]

[14] R.P.Tijburg, T.van Dongen [J. Electrochem. Soc. (USA) vol.123 (1976) p.687]

[15] Y.Tarui, Y.Komiya, Y.Harada [J. Electrochem. Soc. (USA) vol.118 (1971) p.118]

[16] C.J.Nuese, J.J.Gannon [J. Electrochem. Soc. (USA) vol.117 (1970) p.1094]

[17] W.W.Harvey [J. Electrochem. Soc. (USA) vol.114 (1967) p.472]

[18] N.A.Pakhanov, A.S.Terekhov [Instrum. & Expt. Tech. (USA) vol.18

(1975) p.62]

[19] B.Schwartz, F.Ermanis, M.H.Brastad [J. Electrochem. Soc. (USA) vol.123 (1976) p.1089]

[20] Y.Yamamoto, S.Yano [J. Electrochem. Soc. (USA) vol.122 (1975) p.260]

[21] T.Inada, K.Kodama, M.Kitahara [Nucl. Instrum & Meth. Phys. Res. B (Netherlands) vol.7/8 (1978) p.278]

16.2 ETCHING RATES OF PLASMA-ETCHED GaAs

by C.I.H.Ashby

December 1985
EMIS Datareview RN=15421

Plasma etching is based on chemical reaction of plasma-generated species such as halogen atoms or H atoms. Conditions are selected to minimize contributions from ion bombardment. Plasma etching is faster than reactive ion etching due to higher pressure and consequent higher number densities of reactants. The greater contribution of chemical versus ion-assisted mechanisms results in aspect ratios less than 2:1. Rates are generally increased by increasing power density, RF frequency, or pressure and by decreasing flow rates. The increase in rate is due to a higher number density of etchant species such as atomic Cl, Br, or H. Higher etch rates tend to increase surface roughness. The etch rate often decreases with time from its initial value [2].

Preferential crystallographic etching is observed with (111)As greater than (100) greater than (110) greater than (111)Ga [2]. No difference between n-GaAs and p-GaAs has been reported. Chlorine atoms are the most common active species but Br and H have also been used (Table I). Sources of Cl include Cl2, chlorocarbons, PCl3, HCl, and COCl2 [7,9]. Chlorocarbons deposit surface films which usually diminish etch rates. Cl-based etches all produce surfaces rich in Ga [3]. The etch rate with Cl2 increases with temperature with an Arrhenius-type dependence at a given power density [3]. The measured activation energy depends on power density, however. In contrast, etch rates with chlorocarbons such as CCl4 decrease with increasing temperature; this has been attributed to a temperature-dependent morphology of the deposited chlorocarbon film, which is believed to control rates by controlling reactant and/or product diffusion [6].

Cl2 provides the highest etch rate, but a factor of 2-3 variation in rate from run to run has been reported. The higher rates can produce surface roughness and pitting [6]. Etch rates with CCl4 are very reproducible (+-10%) but polymer deposition occurs [6]. Etch rates with HCl are slower but surfaces are smooth with no polymer deposits [5,7]. Rates can be increased by adding O2 to Cl2, CCl4, or CCl3F [5,9]. Addition of O2 increases Cl number density. Polymer formation is also decreased. Maximum rates are achieved with O2 fractions between 40 and 60%; above this optimum concentration oxide formation reduces the etch rate.

GaAs generally etches faster than its oxide. Addition of H2 to HCl or CCl2F2 decreases absolute etch rates while increasing the oxide:GaAs etch rate ratio [7]. Pure H2 plasmas etch oxide faster than GaAs and do not produce Ga-rich surfaces [11]. However, oxide growth replaces etching if base pressures exceed 10*-6 Torr or H2O exceeds 10*-7 torr [10,11].

Electrode material influences film deposition. Graphite, Al or Al2O3-coated electrodes are preferred over stainless steel, which produces deposits of Fe and Ni chlorides on the etched surface [5].

Table I: Plasma Etching Rates

============================

etchant	rate micron/ min	press. torr	flow sccm	temp. deg K	freq. MHz	power W/cm*2	ref	comments
Cl2	2.5	0.3	40	343	13	0.25	[3]	
	over 80	0.3	40	373	13	0.25	[3]	
Br2	37-60	0.3	30	373	14	0.10	[2]	depends on time
	20-70	0.15-0.3	30	373	0.1-14	0.1-0.5	[2]	
40%O2/Cl2	7.3	0.05	6(O2) 10(Cl2)	573	0.055	0.22	[4]	
Cl2	5.8	0.05	10	573	0.055	0.22	[4]	
40%O2/CCl4	2.5	0.05	6(O2) 10(CCl4)	573	0.055	0.22	[4]	
CCl4	1.4	0.05	10	573	0.055	0.22	[4]	
40%O2/CCl4	12.3-14	0.16	10	573	0.055	0.7	[5]	slows as polymer deposits
CCl4	3-0.4	0.16	10	573	0.055	0.7	[5]	slows as polymer deposits
60%Cl2/ CCl4	34.5-3.3	0.16	10	573	0.055	0.7	[5]	slows as polymer deposits
CCl4	1.7	0.16	10	440	0.055	0.7	[5]	
	0.4	0.16	10	700	0.055	1.8	[6]	stainless steel electrode
Cl2	5	0.5	20	335	13.56		[7]	pitting, 5 W total to plasma
HCl	0.06	0.5	50	335	13.56		[7]	smooth, 50 W total to plasma
CCl2F2:He	1.1	0.04	--	--	13.56	0.35	[8]	
	0.5	0.04	--	--	13.56	0.18	[8]	

CCl3F	2.2-0.7	0.16	150	573	0.055	0.22	[9]	f(residence time)
58% O2/ CCl3F	9.9-5.5	0.16	150	573	0.055	0.22	[9]	f(residence time)
H2	0.12	0.15	--	below 423	30	2-10 W/cm*3	[11]	tube reactor

REFERENCES

[1] D.E.Ibbotson, D.L.Flamm, V.M.Donnelly, B.S.Duncan [J. Vac Sci. & Technol. (USA) vol.20 (1982) p.489]

[2] D.E.Ibbotson, D.L.Flamm, V.M.Donnelly [J. Appl. Phys. (USA) vol.54 (1983) p.5974]

[3] V.M.Donnelly, D.L.Flamm, C.W.Tu, D.E.Ibbottson [J. Electrochem. Soc. (USA) vol.129 (1982) p.2533]

[4] R.H.Burton, G.Smolinsky [J. Electrochem. Soc. (USA) vol.129 (1982) p.1599]

[5] G.Smolinsky, R.A.Gottscho, S.M.Abys [J. Appl. Phys. (USA) vol.54 (1983) p.3518]

[6] R.A.Gottscho, G.Smolinsky, R.H.Burton [J. Appl. Phys. (USA) vol.53 (1982) p.5908]

[7] G.Smolinsky, R.P.Chang, T.M.Mayer [J. Vac. Sci. & Technol. (USA) vol.18 (1981) p.12]

[8] K.Hikosaka, T.Mimura, K.Joshin [Jpn. J. Appl. Phys. (Japan) vol.20 (1981) p.L847]

[9] R.H.Burton, C.L.Hollien, L.Marchut, S.M.Abys, G.Smolinsky, R.A.Gottscho [J. Appl. Phys. (USA) vol.54 (1983) p.6663]

[10] C.W.Tu, R.P.H.Chang, A.R.Schlier [J. Vac. Sci. & Technol. A (USA) vol.1 (1983) p.637]

[11] R.P.H.Chang, C.C.Chang, S.Darack [J. Vac. Sci. & Technol. (USA) vol.20 (1982) p.45]

16.3 ETCHING RATES OF ION BEAM MILLED AND SPUTTER ETCHED GaAs

by C.I.H.Ashby

December 1985
EMIS Datareview RN=15422

Ion beam etching, sputter etching, or ion beam milling are purely physical processes based on momentum transfer from an energetic ion or neutral to surface atoms. Atom ejection occurs when the transferred momentum exceeds the atomic binding energy (10 eV). A pronounced dependence of etch rate on ion/neutral incidence angle is seen, with maximum etch rates near an incidence angle of 60 degrees, where surface atoms acquire a large momentum component directed away from the surface [1,2]. This angular dependence is responsible for mask facetting, substrate trenching, and nonequal etch rates for sloped or bilevel surfaces [1]. Redeposition of sputtered material decreases the accuracy of pattern reproduction. Steeper sidewalls increase the problem and use of thin metal masks and sample rotation decreases the problem [1]. There is little selectivity between materials so significant mask erosion occurs. Consequently, ion etching is not suitable for high aspect ratio structures. Rates are generally linearly dependent on ion current density. Significant degradation of electrical characteristics occurs under ion bombardment. For n-GaAs the surface barrier, breakdown voltage, and photoluminescence intensity decrease while effective doping concentrations increase [3,6]. These effects are more severe at higher ion energies. Thermal annealing can remove damage from low energy ions but damage from ions with energies of 500 eV or more is not removed [3,6]. Electrical properties are degraded due to amorphization of the surface and due to diffusion of beam-induced damage to depths up to 10 times deeper than the amorphous region [3]. The degradation of electrical properties can generally be explained by formation of donor-like traps [6].

Table I: Ion Beam Etching Rates
==============================

etchant	rate micron/min/ mA/cm*2	energy (eV)	incidence angle deg	ref
Ar	0.10	500	0	[2]
	0.15	500	30	[2]
	0.11	500	60	[2]
	0.04	500	75	[2]
Ar	0.01	100	0	[3]
	0.08	500	0	[3]
	0.10	1000	0	[3]
	0.09	1500	0	[3]
	0.09	500	60	[3]
Ar+	0.18	400	0	[4]
Ar+	0.25	500	0	[5]

REFERENCES

[1] R.E.Lee [J. Vac. Sci. & Technol. (USA) vol.16 (1979) p.164]

[2] S.Somekh, H.C.Casey,Jr [Appl. Opt. (USA) vol.16 (1977) p.126]

[3] M.Kawabe, N.Kanzaki, K.Masuda, S.Nambe [Appl. Opt. (USA) vol.17 (1978) p.2556]

[4] R.A.Powell [Jpn. J. Appl. Phys. (Japan) vol.21 (1982) p.L170]

[5] M.W.Geis, G.A.Lincoln, N.Efremow, W.J.Piacentini [J. Vac. Sci. & Technol. (USA) vol.19 (1981) p.1390]

[6] S.W.Pang, W.J.Piacentini [J. Vac. Sci. & Technol. B (USA) vol.1 (1983) p.1334]

16.4 ETCHING RATES OF REACTIVE ION AND REACTIVE ION-BEAM ETCHED GaAs

by C.I.H.Ashby

December 1985
EMIS Datareview RN=15423

Reactive ion etching (RIE) is plasma-assisted etching using ions whose energy is greater than the positive space-charge potential of the plasma. This is typically achieved by placing the material to be etched on the driven electrode of an RF parallel plate reactor. Reactive ion beam etching (RIBE) involves generation of reactive chemical species such as Cl^+ and (Cl_2^+) in an ion source using source gases such as Cl_2 and CCl_4; a beam of these ions is used to etch GaAs. Ion-beam-assisted etching (IBAE) employs an ambient reactive neutral gas, such as Cl_2, and a beam of ions, usually Ar^+. In all three processes, GaAs is etched through the combined effect of chemical reaction and physical sputtering. Rates are intermediate between those of physical sputtering and plasma etching. Appropriate reaction conditions can give both anisotropic etching and reasonable etching rates. Ion energies are generally in excess of 100 eV; consequently the ion-induced degradation of the electronic properties of GaAs is an important consideration in selecting etching conditions [7,14,15,19]. IBAE permits easiest control of conditions to minimize ion damage.

Etch rates for RIE under a variety of conditions are listed in Table I. In general, slower etch rates give greater anisotropy and better surface morphology. For a fixed power density, the rate increases with pressure, reaches a maximum between 5 and 20 mtorr, and then decreases with further increase in pressure [7,9,10]. This is due to the interplay between ion energy, which decreases with increasing pressure due to decreasing self-bias voltage, and reactant concentrations, which increase with increasing pressure. At constant bias voltage, rates increase linearly with pressure [10]. Rates increase with increasing RF power due to the increase in self-bias voltage [5]. Pronounced dependence on self-bias voltage indicates the importance of ion-assisted mechanisms. Rates first increase, then reach a plateau with increasing flow rate [7]. More vertical walls are achieved by increasing the importance of ion-assisted processes by increasing bias voltage or RF power density and by decreasing pressure. Etching profiles can be changed from overcut to vertical to undercut by increasing pressure [9,11]. No significant differences in etch rates have been reported for different impurity doping [5] or for bulk or epitaxially grown (MBE or LPE) materials [8]. Highest rates are achieved with Cl_2 as the Cl source, but Cl_2 produces surface roughness and exhibits low reproducibility in etch rates from run to run. Reproducibility can be improved by an initial short etch in H_2 or CCl_2F_2 [1]. Chlorocarbons such as CCl_4 and CCl_2F_2 produce smoother surfaces but can deposit chloropolymer films. Etching with BCl_3 is slower but highly reproducible [3]. Addition of O_2 to CCl_4 or CCl_2F_2 increases etch rates by increasing Cl atom densities [9]. Addition of Ar to Cl_2, CCl_4, or CCl_2F_2 increases anisotropy [1,2,7]. Addition of H_2 to CCl_4, CCl_3F, or CCl_2F_2 increases anisotropy, decreases damage, and gives a residue-free surface [5,14]. Crystallographic orientation dependence is seen with $SiCl_4$ at intermediate pressures (20 mtorr) and low power density. Addition of Ar removes the orientation dependence [11].

In contrast to RIE, RIBE and IBAE permit independent control of the fluxes of chemically reactive neutrals and ions since ions are generated in an ion source rather than in a plasma in contact with the sample to be etched. Etching rates for RIBE and IBAE are listed in Table II. RIBE and IBAE have certain characteristics in common due to their beam-dependent character. Etch profiles can be changed by changing beam incidence angle. Normal incidence produces overcutting; increasing the incidence angle will produce vertical walls [17,19]. Sample rotation also increases verticality [17]. Higher etch rates result at larger incidence angles due to the occurrence of proportionately more collision events near the surface. This produces a greater number of excited surface atoms to undergo chemical reaction. Maximum rates occur for angles between 60 and 70 degrees [20,21]. Increasing neutral flux, which increases chemical contributions to etching, can both improve electronic properties [18,19] and decrease overcutting [22]. Normalizing rates measured at lower ion current densities to 1 mA/cm*2, as is usually done, can be misleading. The increased sample heating resulting from a higher ion flux can increase thermally-activated chemical reactions. Consequently, a higher ion flux can produce a higher normalized etch rate. Thermally-activated chemical reactions can also reduce the degree of anisotropy at high ion flux [18,19].

IBAE generally produces less degradation of electrical properties than RIE [19]. Damage can be minimized by minimizing ion penetration by reducing ion energy or using heavier ions [15]. Undercutting can be minimized by improving beam collimation [18]. Some beam spread due to scattering from neutrals occurs for ion energies below 500 eV [18].

Table I: Reactive Ion Etching Rates
====================================

etchant	rate micron/min	press. mtorr	flow sccm	power density W/cm*2	self bias V	ref	comments
Cl2	1.4	5			270	[1]	vertical walls
80%Cl2:20%Ar	0.80	5			200	[1]	severe undercut
20%Cl2:80%Ar	0.27	5			200	[1]	vertical walls
75%Cl2:25%Ar	3.4	15	12	0.75	330	[2]	isotropic
17%Cl2:83%Ar	0.9	15	12	0.75	330	[2]	anisotropic
17%Cl2: 83%BCl3	0.6	60		0.42	270	[3]	high reproduciblity
17%Cl2: 83%BCl3	1.1	150		0.42		[3]	high reproducibility

BC13	0.017	11	20	0.20		[4]	very smooth
CC14	1.1	11	10	0.16		[5]	isotropic, rough
90%CC14:10%02	1.7	11	10	0.16		[5]	isotropic, rough
10%CC14:90%H2	0.45	11	10	0.16		[5]	anisotropic, smooth
CC12F2	3.3	230	30	0.33	below 100	[6]	
	0.12	10	32	0.85		[7]	
40%CC12F2: 60%Ar	0.076	100	20	0.6		[7]	max. rate at 60% Ar
50%CC12F2: 50%He	1.1	38		0.35	200	[8]	
1 CC12F2: 1 02:10 Ar	0.35	5	20	0.5	650	[9]	smooth
SiC14	0.129	10	10	0.81	600	[10]	no polymer film
	0.067	10	10	0.27	275	[10]	no polymer films, max. rate at 10 mTorr
	4.3	20	3-15	0.10	90	[11]	(100) exceeds (111) exceeds (011)
	5.8	10	3-15	0.10	120	[11]	orientation independent
H2	0.15	5			600	[1]	
CC12F2	0.8	40			3000	[12], [13]	DC plasma
10%CC12F2: 90%Ar	5.8	40			3000	[12]	DC plasma
10%CC14: 90%Ar	0.65	40			3000	[12]	DC plasma
10%CF4:90%Ar	0.50	40			3000	[12]	DC plasma
CF4	0.31	40			3000	[12]	DC plasma

Table II: Reactive Ion-Beam and Ion-Beam-Assisted Etching Rates

etchant	rate micron/min/ (1 mA/cm*2)	press. mtorr	ion energy keV	current density mA/cm*2	incidence angle deg	ref	comments
RIBE							
(Cl+)/(Cl2)+ (Cl2)	0.067	0.2	0.2	0.05	30	[16]	no loss of As
	0.230	0.07	0.5	0.22	30	[16]	no loss of As
	0.405	0.3	0.5	0.21	30	[16]	no loss of As
(Cl+)/(Cl2)+ (CCl4)	0.78	0.01	0.75		0	[17]	rate depends on sample-source distance
IBAE							
Cl2 + Ar+	5-10	10	1-2	1	0	[18]	+-5% uniformity
	0.085	10	0.5	0.02	0	[18]	
	5	0.08	0.5	0.25	0	[19]	
	3	0.08	0.5	0.02	0	[19]	
Cl2 + Ga+	1.4	20	35	80	0	[20]	
	3.6	20	35	80	70	[20]	

REFERENCES

[1] E.L.Hu, R.E.Howard [J. Vac. Sci. & Technol. B (USA) vol.2 (1984) p.85]

[2] H.Yamada, H.Ito, H.Inaba [J. Vac. Sci. & Technol. B (USA) vol.3 (1985) p.884]

[3] H.Tamura, H.Kurihara [Jpn. J. Appl. Phys. (Japan) vol.23 (1984) p.L731]

[4] G.J.Sonek, J.M.Ballantyne [J. Vac. Sci. & Technol. B (USA) vol.2 (1984) p.653]

[5] S.Semura, H.Saitoh, K.Asakawa [J. Appl. Phys. (USA) vol.55 (1984) p.3131]

[6] L.G.Hipwood, P.N.Wood [J. Vac. Sci. & Technol. B (USA) vol.3 (1985) p.395]

[7] J.Chaplart, B.Fay, N.T.Linh [J. Vac. Sci. & Technol. B (USA) vol.1 (1983) p.1950]

[8] K.Hikosaka, T.Mimura, K.Joshin [Jpn. J. Appl. Phys. (Japan) vol.20 (1981) p.L847]

[9] E.L.Hu, R.E.Howard [Appl. Phys. Lett. (USA) vol.37 (1980) p.1022]

[10] M.B.Stern, P.F.Liao [J. Vac. Sci. & Technol. B (USA) vol.1 (1983) p.1053]
[11] J.Z.Li, I.Adesida, E.D.Wolf [J. Vac. Sci. & Technol. B (USA) vol.3 (1985) p.406]
[12] R.E.Klinger, J.E.Greene [J. Appl. Phys. (USA) vol.54 (1983) p.1595]
[13] R.E.Klinger, J.E.Greene [Appl. Phys. Lett. (USA) vol.38 (1981) p.620]
[14] S.Semura, H.Saitoh [J. Vac. Sci & Technol. A (USA) vol.2 (1984) p.474]
[15] S.W.Pang, M.W.Geis, N.N.Efremow, G.A.Lincoln [J. Vac. Sci. & Technol. B (USA) vol.3 (1985) p.398]
[16] R.A.Barker, T.M.Mayer, R.H.Burton [Appl. Phys. Lett. (USA) vol.40 (1982) p.583]
[17] R.A.Powell [Jpn. J. Appl. Phys. (Japan) vol.21 (1982) p.L170]
[18] G.A.Lincoln, M.W.Geis, S.Pang, N.N.Efremow [J. Vac. Sci. & Technol. B (USA) vol.1 (1983) p.1043]
[19] M.W.Geis, G.A.Lincoln, N.Efremow, W.J.Piacentini [J. Vac. Sci. & Technol. (USA) vol.19 (1981) p.1390]
[20] Y.Ochiai, K.Gamo, S.Namba [J. Vac. Sci. & Technol. B (USA) vol.3 (1985) p.67]
[21] Y.Ochiai, K.Gamo, S.Namba [J. Vac. Sci. & Technol. B (USA) vol.1 (1983) p.1047]
[22] J.D.Chinn, A.Fernandez, I.Adesida, E.D.Wolf [J. Vac. Sci. & Technol. A (USA) vol.1 (1983) p.701]

16.5 ETCHING RATES OF GaAs ETCHED BY LASER-ASSISTED METHODS

by C.I.H.Ashby

December 1985
EMIS Datareview RN=15424

Lasers or other high intensity photon sources can produce enhanced etch rates by thermal heating, photochemical generation of reactants, photogeneration of electrons and holes, or some combination of these processes.

Laser-assisted wet etching (Table I) usually involves direct participation of photogenerated carriers in the etching reaction. Photoelectrolytic etching requires ohmic contacts to the GaAs and an external power supply. Etch rates are controlled by the applied current density while photons are used to create a plentiful supply of holes in n-type GaAs so anodic oxidation and oxide dissolution can occur readily [1]. The dependences of the dissolution potential on carrier concentration and surface orientation seen in dark electrolytic etching diminish as incident photon flux is increased [2]. In contrast to dark etching, defects and ion-bombardment-damaged regions etch more slowly than undamaged areas [2]. Therefore, defect visualization is possible [2,9]. No crystallographic orientation dependence is observed [2]. Uniform thicknesses of n-GaAs on semi-insulating GaAs can be achieved by selecting depletion layer thickness [4] and light intensity [5].

Photovoltaic wet processes require a low-resistance contact between light-exposed regions of different conductivity type [6] but no external power supply is required. Where photons are absorbed, holes are etched on n-type regions and mesas on p-type, with depths or heights determined by the ratio of light and dark reaction rates [7]. Other wet processes require no power supply or junction. The higher etch rates from UV versus visible light have been attributed to the shallower absorption depth of UV photons, which permits more photogenerated carriers to reach the surface [10,11]. Etching with UV light has been used to drill 1 micron VIAs through wafers and has produced features with a resolution of 100 nm [11]. The high peak power available with pulsed lasers produces high etch rates during the pulse, but the low duty cycle results in lower overall rates than can be achieved with CW photon sources [12].

Laser-assisted dry etching methods (Table II) are based on heating effects (photothermal/pyrolytic), photochemical generation of reactants (gas-phase photochemical), or photogeneration of carriers (solid-phase photochemical). The highest etch rates are achieved at high pulsed power densities (exceeding 10*5 W/cm*2) [13-19] when samples may be heated to the melt [13]. These high-peak-power etching processes using pulsed lasers can produce high rates and narrow lines, if the beam is scanned [14-16], but the high temperatures involved may affect GaAs properties adversely [16]. Patterns may be etched into GaAs using substrate contact masking, projection etching, and direct writing with scanned beams [14-16,24]. Even at laser powers which can produce ablation in the absence of reactive gases, laser-assisted etching with CF3Br proceeds at 100 times the physical ablation rate [18]. Ablation becomes significant at power densities above 2-4 X 10*6 W/cm*2

[18,19]. Etching can also occur when the laser beam passes through the reactant gas parallel to the surface, but is slower and less localized than when perpendicular beams are used [19]. With HBr or CF_3Br, orientation dependent etch rates of the order (111)As exceeding (100) equal to or exceeding (110) exceeding (111)Ga result [19]. Etch rates with CW lasers using much lower power densities [18,20-23], which produce negligible heating, are 10 to 1000 times slower than the higher power density reactions which have major photothermal contributions. Solid phase photochemical etching is based on the reaction of GaAs with Cl atoms that is driven by photogeneration of electrons and holes in the semiconductor surface using photons of band-gap or greater energy [21-23]. The requirement for photogenerated carriers at the surface permits very selective wavelength-controlled etching of GaAs and GaAs-related ternary materials [22] and bias-voltage-controlled selective etching based on dopant concentrations [23]. Materials of nearly equal composition or of slightly different dopant levels can be selectively etched.

Table I: Laser-Assisted Wet Etching Rates

etchant	etch rate micron/min	laser power density W/cm*2	wavelength nm	ref	comments
Electrolytic					
0.1M H_2SO_4 + 0.1M NaSCN	0.13	0.75	633	[3]	p-GaAs, bias V oscillation
1M KOH	0.012	0.0026	white	[4]	n-GaAs, rate controlled by bias V
Photovoltaic					
$Fe_2(SO_4)_3$ + EDTA	0.4	0.130	white	[6]	n- greater than n+ much greater than p
Other					
I_2 + MI + H_2O	0.01-0.08	0.10	413	[8]	n-GaAs (100), heavily doped; photodissociation of I_2
CrO_3 + HF + H_2O	1.3	0.12	white	[9]	n-GaAs(100)
10% HNO_3	1.4	5 X 10*4	334-364	[10]	Cr-doped GaAs (100)

10% HNO3	0.35	5 X 10*4	514	[10]	Cr-doped GaAs (100)
1:1:100 H2SO4:H2O2:H2O	9.2	0.20	257	[11]	n-GaAs(100), heavily doped
	0.18	0.20	514	[11]	n-GaAs(100), heavily doped
5% HNO3	0.175	10*4 (peak)	249	[12]	30Hz pulse, 0.5 W/cm*2 (average)
	0.875	0.5	257	[12]	n-GaAs

Table II: Laser-Assisted Dry Etching Rates

etchant	etch rate micron/min	laser power density W/cm*2	wavelength nm	ref	comments
Photothermal/Pyrolytic					
5 torr C12	2 X 10*3	2.5 X 10*6	488	[13]	also gas-phase photochemical
0.07 torr C12	37-75	1.5 X 10*5	514	[14]	f(scan speed)
80 torr CC14	9-300	2.3 X 10*5	514	[15]	higher rate for slower scan
30 torr CC14	375	2.6 X 10*6	514	[16]	
3 CC14:7 H2	8.6	2-6 X 10*5	532	[17]	laser-enhanced RIE, rough
Gas-Phase Photochemical					
5 torr CF3Br	0.16	1.2-2.5 X 10*6	193	[18]	also photothermal
5 torr HBr	0.25	0.5-1 X 10*6	193	[19]	normal beam rate = parallel beam rate
5 torr HBr	0.44,2.9	1-2 X 10*6	193	[19]	also photothermal, normal exceeds parallel
5 torr HBr	0.75,8.2	1.7-3.5 X 10*6	193	[19]	also photothermal, normal exceeds parallel

750 torr CH3Br	0.03	100	257	[20]	not photothermal
25 torr CF3I	0.0006	100	257	[20]	not photothermal
Solid-Phase Photochemical					
3.3 torr 0.2% HCl/He	0.03	1750	514	[21]	depends on Cl atom concn, linear in laser power

REFERENCES

[1] C.R.Elliott, J.C.Regnault [J. Electrochem. Soc. (USA) vol.127 (1980) p.1557]

[2] A.Yamamoto, S.Yano [J. Electrochem. Soc. (USA) vol.122 (1975) p.260]

[3] F.W.Ostermayer, P.A.Kohl [Appl. Phys. Lett. (USA) vol.39 (1981) p.76]

[4] H.J.Hoffmann, J.M.Woodall, T.I.Chappell [Appl. Phys. Lett. (USA) vol.38 (1981) p.564]

[5] A.Shimano, H.Takagi, G.Kano [IEEE Trans. Electron Devices (USA) vol.ED-26 (1979) p.1690]

[6] P.D.Greene [Proc. 6th Int. Symp. on Gallium Arsenide and Related Compounds, Edinburgh, 20-22 Sept, 1976) p.141]

[7] E.Kuhn-Kuhnenfeld [J. Electrochem. Soc. (USA) vol.119 (1972) p.1063]

[8] R.W.Haynes, G.M.Metze, V.G.Kreismanis, L.F.Eastman [Appl. Phys. Lett. (USA) vol.37 (1980) p.344]

[9] J.Weyher, J. van de Ven [J. Crystal Growth (Netherlands) vol.63 (1983) p.285]

[10] G.C.Tisone, A.W.Johnson [Appl. Phys. Lett. (USA) vol.42 (1983) p.530]

[11] D.V.Podlesnik, H.H.Gilgen, R.M.Osgood,Jr [MRS Proc. `Laser-Controlled Chemical Processing of Surfaces´, eds. A.W.Johnson, D.J.Ehrlich, H.R.Schlossberg vol.29 (1984) p.161]

[12] D.V.Podlesnik, H.H.Gilgen, P.D.Brewer, D.M.McClure, R.M.Osgood,Jr [Proc. of Symp. on Laser Chemical Processing of Semiconductor Devices, Materials Research Society Mtg, Boston, Nov 1984 p.109]

[13] A.W.Tucker, M.Birnbaum [SPIE Proc. `Laser Processing of Semiconductor Devices´, vol.385 (1983) p.131]

[14] M.Takai, J.Tokuda, H.Nakai, K.Gamo, S.Namba [see ref. 11 p.211]

[15] M.Takai, J.Tokuda, H.Nakai, K.Gamo, S.Namba [Jpn. J. Appl. Phys. (Japan) vol.22 (1983) p.L759]

[16] M.Takai, J.Tsuchimoto, H.Nakai, K.Gamo, S.Namba [Jpn. J. Appl. Phys. (Japan) vol.23 (1984) p.L852]

[17] N.Tsukada, S.Semura, H.Saito, S.Sugata, K.Asakawa, Y.Mita [J. Appl. Phys. (USA) vol.55 (1984) p.3417]

[18] P.Brewer, S.Halle, R.M.Osgood,Jr [Appl. Phys. Lett. (USA) vol.45 (1984) p.475]

[19] P.D.Brewer, D.McClure, R.M.Osgood,Jr [Appl. Phys. Lett. (USA) vol.47 (1985) p.310]

[20] D.J.Ehrlich, R.M.Osgood,Jr, T.F.Deutsch [Appl. Phys. Lett. (USA) vol.36 (1980) p.698]

[21] C.I.H.Ashby [Appl. Phys. Lett. (USA) vol.45 (1984) p.892]
[22] C.I.H.Ashby, R.M.Biefeld [Appl. Phys. Lett. (USA) vol.47 (1985) p.62]
[23] C.I.H.Ashby [Appl. Phys. Lett. (USA) vol.46 (1985) p.752]
[24] P.D.Brewer, G.M.Reksten, R.M.Osgood,Jr [Solid State Technol. (USA) vol.28 no.4 (1985) p.273]

CHAPTER 17

INTERFACES AND CONTACTS WITH GALLIUM ARSENIDE

17.1 Structure of the GaAs/Ag interface

17.2 Structure of the GaAs/Al interface

17.3 Structure of the GaAs/Au interface

17.4 Structure of the GaAs/Au-Ge interface

17.5 Structure of the GaAs/Cu interface

17.6 Barrier height at the GaAs/Ag interface

17.7 Barrier height at the GaAs/Al interface

17.8 Barrier height at the GaAs/Au interface

17.9 Barrier height at the GaAs/Pt interface

17.10 Barrier height at the GaAs/W interface

17.11 Band bending at GaAs/dielectric interfaces

17.12 Conduction band and valence band offset at the GaAs/AlGaAs heterostructure interface

17.1 STRUCTURE OF THE GaAs/Ag INTERFACE

by EMIS Group

April 1984
EMIS Datareview RN=11124

Nucleated growth of Ag on MBE GaAs(100) has been observed at room temperature [1] on both the As-stabilized c(2 X 8) surface and the Ga stabilized (4 X 6) surface. The Ag film was not laminar, but island growth occurred. Interaction between the Ag and the substrate was relatively weak. The interface was abrupt, with no interdiffusion or segregation occurring.

Behaviour of Ag on a cleaved GaAs(110) surface showed similar tendencies [2]. Clustered growth proceeded much earlier than for (100) surfaces (0.005 A rather than 0.5 A). The interface was abrupt, with no detectable film-substrate interaction.
For Ag deposited in-situ onto MBE grown GaAs(100) surfaces, crystallographic relationships were found to be independent of reconstruction, stoichometry of the substrate surface or metal growth rate, but were dependent only on substrate temperature during Ag deposition. A 2D layer-by-layer growth mode was consistent with experimental data for low temperature growth (below 100 deg C), while at higher temperatures 3D crystallites develop [3,4].

When various metal overlayer systems on (100) n/n+ GaAs epitaxial structures were subjected to heat treatment up to 700 deg C, it was found that Ag-based metallizations exhibited only small As evaporation losses. Codeposition of Ga with the contact metal decreased As loss and inhibited metallization/substrate interaction [5].

In a study of Schottky barrier formation for noble metals deposited on GaAs (110) at room temperature it was found that Ag showed less tendency than Au or Cu to intermix with the substrate [6].

Some earlier work on this interface is briefly reviewed in a paper by Ludeke [7].

REFERENCES
----

[1] R.Ludeke, T.-C.Chiang, D.E.Eastman [J. Vac. Sci. & Technol (USA) vol.21 no.2 (1982) p.599]
[2] R.Ludeke, T.-C.Chiang, T.Miller [J. Vac. Sci. & Technol. B (USA) vol.1 no.3 (July/September 1983) p.581-7]
[3] J.Massies, N.T.Linh [Surf. Sci. (Netherlands) vol.114 (1982) p.147-60]
[4] J.Massies, N.T.Linh [J. Cryst. Growth (Netherlands) vol.56 (1982) p.25-38]
[5] I.Mojzes, T.Sebestyen, D.Szigethy [Solid-State Electron. (GB) vol.25 no.6 (June 1982) p.449-60]
[6] S.H.Pan, D.Mo, W.G.Petro, I.Lindau, W.E.Spicer [J. Vac. Sci. & Technol. B (USA) vol.1 no.3 (July/Sept 1983) p.593-7]
[7] R.Ludeke [Surf. Sci. (Netherlands) vol.132 (Sept 1983) 143-6]

17.2 STRUCTURE OF THE GaAs/Al INTERFACE

by G.P.Schwartz

April 1985
EMIS Datareview RN=11123

This Datareview concentrates on the ordered interface. Interfaces subject to high temperature processing or sputter annealing are not considered.

The Al/GaAs interface is the most widely studied system consisting of a metal on a binary semiconductor. The best characterized interfaces are of two general types: vacuum cleaved (110) and MBE grown (100). The exact structural nature of the Al/GaAs interface is still a matter of debate; useful lists of experimental references are given in a review by Ludeke [1]; theoretical perspectives have been presented by Ihm and Joannopoulos [2].

Al/GaAs (110)

The growth mode for Al deposition onto a room temperature substrate remains controversial in the submonolayer coverage regime. Huijser et al [3] contend that for submonolayer coverage the Al forms an ordered overlayer, whereas other LEED [4,5,6] and photoemission [7] studies suggest that the initial overlayer is dispersed and may even represent very small clusters. For Al deposition near liquid nitrogen temperatures a more uniform dispersed overlayer is observed which is believed to consist primarily of isolated atoms or very small clusters [8]. A preliminary analysis of LEED intensity profiles indicates that the surface relaxation which results on the (110) face of GaAs after cleaving is lifted by the deposition of isolated Al atoms [8]. A similar lifting of the surface relaxation has been reported in room temperature studies of Sb on (110) GaAs [9]. As the coverage regime increases to 1-3 monolayers evidence is seen for the formation of nucleated Al islands rather than layer-by-layer growth [10]. The question of whether Al-Ga exchange reactions occur for room temperature deposition remains open. Reference [3] did not see evidence for an Al-Ga exchange reaction whereas most other authors have [4,10,7]. Ludeke [1] has suggested that the (110) surface should represent a ´passivated´ version of the (100) face and that exchange reactions at low coverage may be related to individual experimenter´s substrates via the quality of the cleave and the number of resulting steps and terraces. It is difficult to address this question using MBE because the (110) surface has proven difficult for homoepitaxial growth. Recent mass-resolved, atom-probe studies by Nishikawa et al [11,12] have followed the course of Al-Ga replacement reactions on both (110) and (100) field ion GaAs tips as a function of substrate temperature and bulk doping for initial Al coverage of a few monolayers. On heavily Si-doped wafers, atomically abrupt interfaces were obtained at 200 deg K. As the substrate temperature was increased, a mixed layer of Al+AlAs was observed, until at 900 deg K for 70 seconds, only AlAs was detected. The studies were inconclusive concerning the detection of Ga on the surface of the reacted layer due to what the authors believe to be the rapid diffusion of that species away from the emitting region of the field ion tip. On heavily Zn-doped substrates, the Al-GaAs reaction was promoted so that interdiffusion occurred at substantially lower temperatures than experienced on Si-doped substrates. The reaction products were similar, however. Ihm and Joannopoulos [2] have used a pseudopotential energy minimization calculation to consider chemisorption, clustering, and exchange reactions for Al on (110) GaAs in the absence of steps or other defects. At low coverages they predict that chemisorption occurs at two-fold sites. At higher coverages their calculations predict that clustering occurs. The most stable calculated configuration is one in which Al atoms replace second layer (or deeper) Ga atoms. The epitaxial relations between Al and GaAs (110) have been reported by Prinz et al [14] for heat cleaned substrates. The (001) direction in GaAs becomes the (110) direction in the Al film, and the (001) direction in the film is aligned along the (110) substrate direction.

Al/GaAs (100)

MBE grown (100) GaAs/Al interfaces have been examined for a number of different surface reconstructions. At high temperatures, strong interactions with compound formation invariably occur [1]. At room temperature most of the data have focused on three surfaces: the As-rich c(4 X 4), the As-stabilized c(2 X 8), and the Ga-rich (4 X 6). Ludeke has recently reviewed the available data [1]. For Al adsorption at low coverage, RHEED data indicate that the adsorbed Al atoms are dispersed rather than forming an ordered overlayer. With increasing coverage, nucleated growth and island formation are observed [15]. An exchange reaction has been observed on the Ga-rich (4 X 6) reconstruction following the nucleation step [16]. Initially, no exchange reaction was observed on the As-stabilized c(2 X 8) [16], but a recent report by Andersson et al [17] purports that an Al-Ga exchange reaction does occur both before and after the nucleation step. The epitaxial Al overlayer-substrate relations on (100) GaAs are quite complex, involving coexisting (100), (110) and (110)R Al domains at intermediate coverages. For thick Al films, the As-rich reconstructions tend to yield (110) oriented films or a mixture of (110) and (100) depending on deposition conditions. On Ga-rich surfaces (100) oriented films are favoured [18].

REFERENCES

[1] R.Ludeke, [Surf. Sci. (Netherlands) vol.132 (1983) p.143]

[2] J.Ihm, J.D.Joannopoulos [J. Vac. Sci. & Technol. (USA) vol. 21 (1982) p.340]

[3] A.Huijser, J.van Laar, T.L.van Rooy [Surf. Sci. (Netherlands) vol.102 (1981) p.264]

[4] A.Kahn, J.Carelli, D.Kanani, C.B.Duke, A.Paton, L.Brillson [J. Vac. Sci. & Technol. (USA) vol.19 (1981) p.331]

[5] C.B.Duke et al [Phys. Rev. Lett. (USA) vol. 46 (1981) p.440]

[6] A.Kahn [J. Vac. Sci. Technol. A (USA) vol.1 (1983) p.684]

[7] R.R.Daniels, A.D.Katnani, T.X.Zhao, G.Margaritondo [Phys. Rev Lett. (USA) vol.49 (1982) p.895]

[8] C.R.Bonapace, K.Li, A.Kahn, [J. Phys. Colloq. (France) vol.45 (1984) p.C5-409]

[9] J.R.Myron, G.J.Lapeyre [17th Intern. Conf. Phys. Semicond. (San Francisco, 1984) paper FrA-A3]
[10] N.G.Stoffel, M.K.Kelley, G.Margaritondo [Phys. Rev. B (USA) vol.27 (1983) p.6561]
[11] O.Nishikawa, O.Kaneda, M.Shibata, E.Nomura [J. Phys. Colloq. (France) vol.45 (1984) p.C9-459]
[12] O.Nishikawa, O.Kaneda, M.Shibata, E.Nomura [Phys. Rev. Lett. (USA) vol.53 (1984) p.1252]
[13] R.Ludeke, G.Landgren [J. Vac. Sci. & Technol. (USA) vol.19 (1981) p.667]
[14] G.A.Prinz, J.M.Ferrari, M.Goldenberg [Appl. Phys. Lett. (USA) vol.40 (1982) p.155]
[15] G.Landgren, S.P.Svensson, T.G.Andersson [Surf. Sci. (Netherlands) vol.122 (1982) p.55]
[16] G.Landgren, R.Ludeke [Solid State Commun. (USA) vol.37 (1981) p.127]
[17] T.G.Andersson, S.P.Svensson, G.Landgren [J. Phys. C (GB) vol.15 (1982) p.6673]
[18] P.M.Petroff, L.C.Feldman, A.Y.Cho, R.S. Williams [J. Appl. Phys. (USA) vol.52 (1981) p.7317]

17.3 STRUCTURE OF THE GaAs/Au INTERFACE

by C.R.M.Grovenor

November 1985
EMIS Datareview RN=11126

In a review paper Ludeke [1] has pointed out that the Au/GaAs interface is generally believed to be interdiffused or strongly reacted.

Wang and Holloway [2] have analyzed the effect of Ar+ surface bombardment prior to Au deposition and correlated the surface composition changes with variations in the Schottky barrier height. This work clearly demonstrates the crucial importance of local stoichiometry variations on contact properties.

Prelimary experiments on the structure of Au/GaAs interfaces by ion channelling techniques [3] have shown that a limited amount of atomic displacement occurs at the interface, but no conclusive measurements on the extent of Ga and As replacement by Au atoms have been made.

Brillson et al [4] have investigated the use of metal interlayers to modify microscopic bonding at III-V semiconductor/Au interfaces. For Au/GaAs(110) with no interlayer present more As than Ga diffuses through the Au to the vacuum interface. An Al interlayer (10 A thickness) reverses this change in stoichiometry.

An Al interlayer of submonolayer thickness does not strongly affect the diffusion process and has been used as an immobile marker layer to study diffusion at the Au/GaAs(110) interface. It was found that at room temperature Au atoms first diffuse into the GaAs and this process is followed by nonstoichiometric outdiffusion of Ga and As into the Au overlayers [5].

Caprile et al [6] and Pandelisev and Yang [7] have shown that 1-2 monolayers of Al or Cr strongly enhance the reaction between Au and GaAs. This work suggests that there are 3 regimes of interlayer thickness (submonolayer, 1-2 monolayers and a relatively thick layer) each with a characteristic effect on the chemical behaviour at the Au/GaAs interface.

When various metal overlayer systems on (100) n/n+ GaAs epitaxial structures were subjected to heat treatment at up to about 600 deg C it was found that Au-based metallisations exhibited high As evaporation losses at relatively low temperatures. Codeposition of Ga with the contact metal decreased As loss and inhibited metallisation-substrate interaction [8].

In a study of Schottky barrier formation for noble metals deposited on GaAs(110) at room temperature it was found that Au showed less tendency than Cu to intermix with the substrate, but was more reactive than Ag [9].

SXPS experiments on Au (thickness 0.3-80 monolayers) deposited on cleaved n-GaAs(110) surfaces showed that at room temperature an Au-Ga-As intermixed layer was formed. At the highest Au coverages preferential As segregation to the surface was observed. On heat treatment increased intermixing and alloy formation was observed. It was likely that clustering of the Au and Ga took place, exposing areas of the GaAs surface [10].

UPS and XPS studies of GaAs(110) surfaces exposed to controlled amounts of oxygen before Au deposition have shown that O-GaAs interactions inhibit Au intermixing with GaAs and cause a more abrupt GaAs/Au interface than that formed on unoxygenated GaAs [11].

Newman et al [12] have used photoemission spectroscopy to measure chemical reactions at Au/GaAs interfaces, and propose that the decrease in Schottky barrier height on annealing is a result of the formation of Au/Ga layers of indeterminate nature under the contacts.

TEM studies of monocrystalline (100)-type GaAs wafers with Au overlayers deposited at 280 deg C showed that the as-deposited metal films were polycrystalline. Annealing at 330 deg C led to formation of square or rectangular patches of nearly pure Au aligned with edges parallel to [100] directions of GaAs. Evidence was obtained for the existence of an intermediate hexagonal Au-Ga phase (approx. composn Au7Ga2) separating the Au and the GaAs substrate [13].

The rectangular etch pit morphology has been investigated by Narusawa et al [14], who claim that it is not necessary to have Au/Ga phases present at these features as has been previously suggested by Yoshiie and Bauer [13].

These studies have been extended to other deposition and anneal conditions [15]. For a film deposited at ambient temperature, a stable orthorhombic Au2Ga phase was formed during annealing up to about 320 deg C. At higher temperatures this phase redissolved, promoting regrowth of GaAs.

In situ X-ray diffaction analysis of the reactions of Au films with GaAs [16] has shown that AuGa is the first phase formed on heating above 450 deg C and only on subsequent heating are the frequently reported Au7Ga2 and Au2Ga phases observed.

Rai et al [17] have studied the metallurgy of the Au/GaAs interface after annealing at 350-450 deg C, and their findings are roughly in agreement with earlier reports except in the identification of beta-Ga2O3 at higher temperatures.

REFERENCES
----

[1] R.Ludeke [Surf. Sci. (Netherlands) vol.132 (Sept 1983) p.143-68]

[2] Y.X.Wang, P.M.Holloway [J. Vac. Sci. & Technol. B (USA) vol.2 no.4 (Oct-Dec 1984) p.613-9]

[3] H.J.Gossman, W.M.Gibson [Surf. Sci. (Netherlands) vol.139 no.1 (April 1984) p.239-59]

[4] L.J.Brillson, C.F.Brucker, G.Margaritondo, J.Slowik, N.G.Stoffel [J. Phys. Soc. Jpn. (Japan) vol.49 (1980) Suppl. A p.1089-92]

[5] L.J.Brillson, R.S.Bauer, R.Z.Bachrach, G.Hansson [Phys. Rev. B (USA) vol.23 no.12 (15 June 1981) p.6204-5]

[6] C.Caprile, S.Chang, P.Philip, A.Franciosi [17th Users meeting of

Synchrotron Radiation Center, Univ. of Wisconsin Madison 1984]
[7] K.A.Pandelisev, E.Y.Wang [Thin Solid Films (Switzerland) vol.115 no.1 (May 1984) p.43-50]
[8] I.Mojzes, T.Sebestyen, D.Szigethy [Solid-State Electron. (GB) vol.25 no.6 (June 1982) p.449-60]
[9] S.H.Pan, D.Mo, W.G.Petro, I.Lindau, W.E.Spicer [J. Vac. Sci. & Technol. B (USA) vol.1 no.3 (July-Sept 1983) p.593-7]
[10] W.G.Petro, I.A.Babalola, T.Kendelewicz, I.Lindau, W.E.Spicer [J. Vac. Sci. & Technol. A (USA) vol.1 no.2 pt.II (Apr-June 1983) p.1181-4]
[11] Z.M.Lu, W.G.Petro, P.H.Mahowald, M.Oshima, I.Lindau, W.E.Spicer [J. Vac. Sci. & Technol. B (USA) vol.1 no.3 (July-Sept 1983) p.598-601]
[12] N.Newman, W.G.Petro, T.Kendelewicz, S.H.Pan, S.J.Eglash, W.E.Spicer [J. Appl. Phys. (USA) vol.57 no.4 (15 Feb 1985) p.1247-51]
[13] T.Yoshiie, C.L.Bauer [J. Vac. Sci. & Technol. A (USA) vol.1 no.2 pt 1 (Apr-June 1983) p. 554-7]
[14] T.Narusawa, N.Watanabe, K.L.I.Kobayashi, H.Nakashima [J. Vac. Sci. & Technol. A (USA) vol.2 no.2 pt 1 (Apr-June 1984) p.538-41]
[15] T.Yoshiie, C.L.Bauer, A.G.Milnes [Thin Solid Films (Switzerland) vol.111 no.2 (13 Jan 1984) p.149-66]
[16] X.-F.Zeng, D.D.L.Chung [Solid State Electron. (GB) vol. 27 no.4 (Apr 1984) p.339-45]
[17] A.K.Rai, R.S.Bhattacharya, Y.S.Park [Thin Sol. Films (Switzerland) vol.114 no.4 (Apr 1984) p.379-98]

17.4 STRUCTURE OF THE GaAs/Au-Ge INTERFACE

by C.R.M.Grovenor

November 1985
EMIS Datareview RN=11125

The Au-Ge alloy containing 12 at% Ge is widely used in forming ohmic contacts to n-GaAs. However, understanding of this metallisation system is far from complete. Contact formation is thought to involve Ge incorporation onto Ga sites leading to an n+ surface region.

Studies of the Au/Au-Ge/GaAs systems [1] have shown that for sintering temperatures above 250 deg C inhomogeneous Ge diffusion into GaAs occurred. For sintering temperatures below the Au-Ge eutectic point (363 deg C) there was no major interaction between Au and GaAs. Sintering at 450 deg C led to complete mixing of all the elements, the surface showing three distinct areas - agglomerated areas consisting of Au with some trace of Ga, a matrix area consisting of GaAs with some small trace of Au and rectangular particles containing about 72-85 at% Au, 13 at% Ga and 2-15 at% As.

High resolution TEM and X-ray analysis of AuGeNi contacts to GaAs have shown that the interface is very rough after annealing [2,3] and that the extent of this roughness is dependent on the evaporation sequence of the layered contact [2]. Au/Ga, Ni/As and Au rich phases have been observed at the interface [3], but although it has been suggested that high Ge and Ni concentrations in the GaAs surface create the ohmic contact, no identification of the phases responsible for the localised concentration of these elements has been made.

Nebauer and Trapp [4] have shown that there is a high concentration of Ge under AuGe contacts annealed at 420 deg C [4] and have estimated an indiffusion coefficient of 10*-14 cm*2/s at this temperature.

In situ SEM observations have identified a Au/Ge/As phase forming very rapidly at 360 deg C and forming deep pits into the GaAs surface [5]. At 450 deg C this ternary phase, and Ge and Au/Ge, crystals are formed at a very rough interface. The lowest contact resistivity has been correlated with the dissolution of the ternary phase at temperatures above 400 deg C. (A detailed analysis of the structure of this metastable ternary phase has been presented by Rackham and Steeds [6].)

Kim and Chung have studied the structure of Au/Ge/Au contacts after annealing at 300-500 deg C. alpha´-Au/Ga is formed epitaxially at 400 deg C and Au2Ga at higher temperatures [7]. The 3 layer contact structure seems to improve the morphology of the contact, lessening the balling up seen in Au alloy contacts.

The thickness of an Au/Ge/Ni contact has been shown to have a strong influence on the contact resistivity. After annealing at 470 deg C a minimum resistivity was achieved with 750 A of contact, and this has been correlated with a maximum interface density of a Ni+Ge rich phase of undetermined character [8].

Experiments on n-GaAs with Au-Ge based ohmic contact formed by firing treatment showed the existence of a variety of precipitate phases at the contact/GaAs interface. Phases identified included As2Ge, Au2Ge, Au7Ga2 and Au/10% Ga [9]. For samples fired at a relatively high temperature (giving good ohmic character) the GaAs interfaces contained at least 1% Au in solid solution and traces of Ge. It was suggested that the in-diffused Au and Ge concns, rather than microscopic chemical and structural properties of the interface, were crucial in determining the ohmicity of Au-Ge based ohmic contacts.

When various metal overlayer systems on (100) n/n+ GaAs epitaxial structures were subjected to heat treatment at up to about 600 deg C it was found that Au-based metallisations exhibited high As evaporation losses at relatively low temperatures. Codeposition of Ga with the contact metal decreased As loss and inhibited metallisation/substrate interaction [10].

Elemental distributions in electron beam alloyed Pt/AuGe/GaAs ohmic contacts were similar to those in as-evaporated contacts. Thermal ageing at 250 deg C led to interdiffusion and compound formation which resulted in contact degradation [11]. After ageing it was found that Pt had moved inwards and Ge had moved outwards (probably forming an intermetallic compound with Pt).

Alloying reactions between an Au/Ni/(Au-Ge) composite film and a (100)GaAs substrate have been studied by TEM and STEM [12]. During the initial stage of alloying (below 390 deg C) NiGe was formed within the metal layers and NiAs was formed underneath the GaAs substrate surface. During the middle stage of alloying (390-410 deg C) Ge diffused out of the NiGe phase into the NiAs phase to form a Ni2GeAs ternary phase. On alloying at upto 450 deg C, Au, NiGe and Ni2GeAs grains larger than 200 nm across were formed.

A comparative study on ageing of alloyed Au-Ge, Ni/(Au-Ge) and Au/Ni/(Au-Ge) ohmic contacts to epitaxial n-GaAs has confirmed the importance of Ni in providing good ohmic contacts [13]. SEM observations showed the presence of nodular or cluster form microstructures under both Au/Ni/(Au-Ge) and Ni/(Au-Ge) metallisations, the nodules being less well defined and of low concn in the latter case.

For Au-Ge ohmic contact to (100) oriented n-GaAs:Si complete intermixing of metallisation and GaAs phases occurred on thermal alloying at 450 deg C. Intermixing also occurred if samples were laser annealed at energy densities above 0.7 J/cm*2. Laser annealing resulted in a much smoother contact than thermal annealing [14]. Laser annealed Au-Ge ohmic contacts (with and without thin Ni overlayer) to n+ GaAs were also found to have better surface morphology than furnace annealed contacts, the Ni overlayer giving better contact properties [15]. Beyond an optimum laser energy density (0.356 J/cm*2 for Au-Ge films and 0.445 J/cm*2 for (Au-Ge)/Ni films) the metallisation began to evaporate, resulting in contact degradation.

Nd-glass laser irradiation of eutectic Au/Ge alloy contacts at between 0.5-2.4 J/cm*2 laser power with 30 ns pulse width has been used to produce alloyed contacts with low contact resistivities. In the energy range 0.8-1.3 J/cm*2 contacts with resistivities of the order of 10*-5 ohm cm*2 were produced [16].

Irradiation by UV lasers (0.5 J/cm*2) with pulse widths of 40 ns also gave good ohmic contacts [17]. In both these cases the morphology of the contact layer was very much smoother than can be achieved by thermal annealing in the conventional manner.

Kabarov et al [18] have also shown that annealing with xenon discharge lamps (15-20 J/cm*2 and 600 microsec pulse width) gives good ohmic contacts with Au-Ge alloys.

REFERENCES
----

[1] A.Iliadis, K.E.Singer [Solid State Electron. (GB) vol.26 no.1 (Jan 1983) p.7-14]

[2] D.Fathy, O.L.Krivanek, J.C.M.Spence, W.M.Paulson [Thin Films and Interfaces II. Eds J.E.E Baglin et al (North Holland, New York, 1984) p.557-62]

[3] Z.Liliental, R.W.Carpenter, J.Escher [Ultramicroscopy (Netherlands) vol.14 nos 1,2 (1984) p.135-43]

[4] E.Nebauer, M.Trapp [Phys. Status Solidi a (Germany) vol.84 no.1 (July 1984) p.K39-42]

[5] A.A.Lokhani, D.R.Urech [Thin Films and Interfaces II. Eds J.E.E. Baglin et al (North Holland, New York 1984) p.551-6]

[6] G.M.Rackham and J.W.Steeds [Developments in Electron Microscopy & Analysis, Inst. Phys. Conf Ser. 36 (1977) p.297]

[7] T.I.Kim, D.D.L.Chung [Thin Solid Films and Interfaces II. Eds J.E.E.Baglin et al (North-Holland, New York, 1984) p.69-74]

[8] T.S.Kalkur, J.Dell, A.G.Nassibian [Int. J. Electron. (GB) vol.57 no.5 (Nov 1984) p.729-36]

[9] C.R.M.Grovenor [Thin Solid Films (Switzerland) vol.104 nos 3-4 (1983) p.409]

[10] I.Mojzes, T.Sebestyen, D.Szigethy [Solid State Electron (GB) vol.25 no.6 (June 1982) p.449-60]

[11] C.P.Lee, B.M.Welch, J.L.Tandon [Appl. Phys. Lett. (USA) vol.39 no.7 (1 Oct 1981) p.556-8]

[12] T.S.Kuan, P.E.Batson, T.N.Jackson, H.Rupprecht, E.L.Wilkie [J. Appl. Phys. (USA) vol.54 no.12 (Dec 1983) p.6952-7]

[13] G.S.Marlow, M.B.Das, L.Tongson [Solid State Electron (GB) vol.26 no.4 (1983) p.259-66]

[14] A.H.Oraby, K.Murakami, Y.Yuba, K.Gamo, S.Namba, Y.Masuda [Appl. Phys. Lett. (USA) vol.38 no.7 (1 Apr 1981) p.562-4]

[15] P.Sircar [Can. J. Phys. vol.61 (1983) p.1218-21]

[16] E.Sandow, W.Wesch, E.Nebauer [Phys. Status Solidi a (Germany) vol.85 no.2 (Oct 1984) p.K169-73]

[17] P.Sircar, M.Aubin [Phys. Status Solidi a (Germany) vol.85 no.2 (Oct 1984) p.649-54]

[18] A.N.Kabanov, A.B.Kamnev, B.A.Lapshinov [Phys. Chem. & Mech. Surf. (GB) vol.2 no.6 (1985) p.1750-5]

17.5 STRUCTURE OF THE GaAs/Cu INTERFACE

by EMIS Group

April 1984
EMIS Datareview RN=11127

GaAs/Cu is the least studied of the GaAs/noble metal interfaces. In a study of Schottky barrier formation for noble metals deposited on GaAs(110) at room temperature it was found that Cu showed a greater tendency than Au or Ag to intermix with the substrate [1]. (Current understanding of the GaAs/Au and GaAs/Ag interfaces is covered in Ludeke's review on formation of interfaces on GaAs and related semiconductors [2]).

Bolmont et al [3] have shown that deposition of Cu in UHV on clean cleaved GaAs(110) leads to formation of a uniform non-metallic double layer made of a two dimensional compound estimated to consist of one Cu atom per GaAs surface unit cell plus at least one monolayer of Cu atoms. The structure of the two dimensional compound is thought to be such that the Ga atom remains on the outside of the system while the Cu atom goes inside, the As atom remaining in between. Growth of the two-dimensional compound is completed below 2 monolayers and metallic Cu islands then start to form and progressively cover the substrate.

REFERENCES

[1] S.H.Pan, D.Mo, W.G.Petro, I.Lindau, W.E.Spicer [J. Vac. Sci. & Technol. B (USA) vol.1 no.3 (July-Sept 1983) p.593-7]

[2] R.Ludeke [Surf. Sci. (Netherlands) vol.132 (Sept 1983) p.143-68]

[3] D.Bolmont, V.Mercier, P.Chen, H.Luth, C.A.Sebenne [Surf. Sci. (Netherlands) vol.126 nos 1-3 (Mar 1983) p.509-17]

17.6 BARRIER HEIGHT AT THE GaAs/Ag INTERFACE

by EMIS Group

September 1985
EMIS Datareview RN=15272

Ag forms a relatively abrupt Schottky barrier with GaAs because of its inertness. The barrier height depends largely on the conductivity-type of the semiconductor and to a lesser extent on residual oxides on the semiconductor, the surface structure of the semiconductor and any chemical interaction with the metal.

Waldrop [1,2] measured barrier heights of Ag and 13 other metals at a clean oxide-free interface with (100) plane oriented GaAs of both conductivity-types. X-ray photoemission and LEED analysis showed an ordered, contamination-free surface prior to ultra high vacuum evaporation of a 2000 A (or more) thick Ag layer onto the room temperature surface. The samples were n-GaAs:Se doped to 6 X 10*16/cm*3 (LEC), and p-GaAs:Cd doped to 8 X 10*16/cm*3 (Bridgman) and the barrier heights obtained were:

I-V method		C-V method
n-type	p-type	n-type
0.90 eV	0.50 eV	0.89 eV

The current-voltage values incorporate an image force correction of + 0.04 eV (n-type) and + 0.03 eV (p-type). Note that at a doping level of 6 X 10*16 /cm*3 there may be noticeable thermionic field emission which would reduce the BH for the n-type samples.

Ludeke et al [3] studied the effect of surface structure on barrier heights by means of photoemission spectra. Ag was evaporated on to smooth room temperature, uncontaminated, undoped (100) oriented MBE GaAs films. The values obtained were:

0.83 (+- 0.03) eV for an As-stabilised c(2 X 8) surface
0.97 (+- 0.03) eV for a Ga-rich (4 X 6) surface

Later, Ludeke [4] used photoemission spectroscopy to investigate the mechanism of GaAs/Ag Schottky barrier formation. Various thicknesses of Ag (up to 20 monolayers) were deposited on n-type and p-type (110) cleaved GaAs. Clustered Ag growth and undetectable chemical interactions with the substrate were indicated, and the surface Fermi level changed with the thickness of the Ag layer. For the thickest layer the barrier heights were 0.89 eV (n-type) and 0.35 eV (p-type).

OTHER WORK

Electron trapping and interfacial charges in the oxide layer of MIS-type n-GaAs/Ag Schottky diodes have been studied [5,6]. The effect of ion bombardment cleaning of the GaAs, thermal annealing and residual oxides on the GaAs/Ag barrier height are given in ref [7].

REFERENCES

[1] J.R.Waldrop [J. Vac. Sci. & Technol. B (USA) vol.2 no.3 (July-Sept 1984) p.445-8]

[2] J.R.Waldrop [Appl. Phys. Lett. (USA) vol.44 no.10 (May 1984) p.1002-4]

[3] R.Ludeke, D.E.Eastman, T.C.Chiang [J. Vac. Sci. & Technol. (USA) vol.21 no.2 (July/Aug 1982) p.599-606]

[4] R.Ludeke [J. Vac. Sci. & Technol. B (USA) vol.1 no.3 (July-Sept 1983) p.581-7]

[5] W.H.Laflere, R.L.Van Meirhaeghe, F.Cardon [Solid-State Electron. (GB) vol.25 no.5 (May 1982) p.389-93]

[6] R.L.Van Meirhaeghe, W.H.Laflere, F.Cardon [Solid-State Electron. (GB) vol.26 no.12 p.1189-92]

[7] J.M.Palau, E.Testemale, A.Ismail, L.Lassabatere [Solid-State Electron. (GB) vol.25 no.4 (April 1982) p.285-94]

17.7 BARRIER HEIGHT AT THE GaAs/Al INTERFACE

by EMIS Group

September 1985
EMIS Datareview RN=15271

The GaAs/Al Schottky barrier is of considerable interest largely because Al is often used to metallise GaAs structures. For example, it serves as a gate electrode in FET devices. Most of the research has been on the technologically important (100) GaAs/Al barrier.

The mechanism by which the Schottky barrier is formed is not clearly understood. Spicer et al [1] propose that the barrier height (BH) results from metal-induced defect states in the GaAs which pin the Fermi level at the interface, while Brillson et al [2] suggest that the interface dipole is a major determining factor. No fully experimentally tested model exists. It is, however, generally accepted that thermionic emission dominates the conduction process in Schottky diodes.

The BH depends on the conductivity-type, ranging from 0.7-0.8 eV for n-type to 0.6 eV for p-type. Residual semiconductor oxides and other interfacial layers (e.g. AlGaAs) are also important.

BH AT CLEAN INTERFACES

Barrier heights at clean oxide-free Al (and 13 other metal) interfaces with (100) n-GaAs and p-GaAs were measured by Waldrop [3,4]. X-ray photoemission and LEED analysis verified that the GaAs surface was clean and ordered. At least 2000 A of Al was UHV (ultra high vacuum) evaporated onto this surface at room temperature. The samples were:

n-GaAs:Se doped to 6 X 10*16/cm*3, LEC
p-GaAs:Cd doped to 8 X 10*16/cm*3, Bridgman

and the BH values were:

I-V method		C-V method
n-type	p-type	n-type
0.85 eV	0.61 eV	0.84 eV

The I-V values incorporate image force corrections of +0.04 eV (n-type) and +0.03 eV (p-type). Note that at doping levels of 6 X 10*16/cm*3 there may be noticeable thermionic-field emission which would reduce the BH for the n-type samples.

Another measurement on a clean (100) GaAs/Al interface was made by Eglash et al [5].Up to 30 monolayers of Al were slowly UHV evaporated on an n-type MBE GaAs layer doped with Si to 5 X 10*16/cm*3. There was no significant heating of the substrates during deposition. The following BH values were obtained:

Photoemission spectroscopy	0.70	(+- 0.10) eV
I-V method	0.77	(+- 0.02) eV
C-V method	0.99	(+- 0.10) eV

EFFECT OF INTERFACIAL LAYERS ON BH

Eglash et al [6] studied the effect of interfacial doped layers on the BH of (110) MBE Al on (100) MBE n-GaAs:Si doped to 5 X 10*16/cm*3. A heavily doped interfacial layer of p-GaAs:Be was grown on the n-type substrate prior to Al deposition. The growth coditions gave abrupt, ´ideal` interfaces. The following table gives BH as a function of acceptor concentration (Na) and thickness (d) of the interfacial layer:

Na	d	BH (eV)	
(10*18/cm*3)	(A)	I-V meas.	C-V meas.
1	0	0.79	0.75
1	120	0.80	0.76
1	209	0.90	1.04
1	269	0.98	1.19
1	318	1.08	1.38
1	361	1.12	1.46
20	50	0.98	1.02
60	50	1.24	1.20

The maximum BH of 1.24 eV compares with a room temperature band gap of 1.42 eV. These data show the possibility of controlling the BH of Schottky diodes for engineering purposes.

Schwartz and Gualtieri have measured BH at Al interfaces with (100) n-GaAs:Te wafers doped to about 1 X 10*17/cm*3 in the presence of native oxide [7] and graded AlGaAs [8] interfacial layers:

oxide layer present	0.59 eV	(I-V meas.)
oxide layer absent	0.53 eV	(I-V meas.)
abrupt interface	0.70 eV	(C-V meas.)
graded interface	0.70 eV	(C-V meas.)
AlAs-rich interface	0.80 eV	(C-V meas.)

Sakaki et al [9] also observed a decrease in BH of MBE (100) GaAs/Al when an interfacial oxide was removed and assumed this to be due to the potential drop in the oxide layer caused by negatively charged states at the GaAs/oxide interface. The ideality factor, deduced from the slope of the I-V characteristics was not affected by the oxide layer (in contrast to the findings of Oda and Sugano [10]). This would be expected if thermionic emission processes dominate transport.

Svensson et al [11] annealed MBE (001) n-GaAs:Si at 573 deg K after Al deposition and found that this caused a rapid increase in BH which they attribute to the creation of a graded AlGaAs-like compound by interdiffusion and the formation of defect-related electronic states pinning the Fermi level. Annealing-induced BH increases (e.g. from 0.80 to 0.90 eV for annealing temperatures of 630 to 840 deg K) were also measured by Newman et al [21] at an evaporated Al interface with UHV-cleaved (110) n-GaAs:Si doped to 10*17/cm*3. Similar reasons were attributed for the BH increase. Aina and Pande [12] formed an AlGaAs layer by ion implantation of Al into GaAs prior to metal deposition. This increased the BH from 0.82 to 0.99 eV. The Schottky contacts had lower leakage currents and higher breakdown voltages than for unimplanted samples.

Hydrogen sulphide chemisorbed onto MBE n-GaAs prior to MBE growth of an Al film considerably reduced the BH in an experiment by Massie et al [13]. The (100) GaAs layers were Sn doped to 10*16 - 10*19/cm*3 and of thickness 1-2 microns. The presence of 1000 Langmuirs of H2S decreased the BH from 0.76 (+- 0.02) eV to 0.40 (+- 0.03) eV. This Schottky barrier lowering can be used to obtain monocrystalline non-alloyed ohmic contacts with low specific contact resistances.

Svensson and Andersson [22] found Ga cluster growth on (001) MBE GaAs with c(2 X 8) surface reconstruction. The Ga interlayer led to an increase in BH and the authors concluded that this should permit BH tuning (0.75-1 eV) while still retaining Al as the contact material for device purposes.

EFFECT OF SURFACE STRUCTURE ON BH

The (001) surface of GaAs has a number of reconstructions with widely varying structure and composition. However, detailed studies of Al interfaces with this surface by Svensson et al [11] and Barret [14] find no dependence of BH on reconstruction in contrast to earlier findings by Cho and Dernier [15] and later work by Wang [17] who measured Al barrier height for clean (100) GaAs and (100) AlAs surfaces as a function of initial As surface coverage, which is associated with different reconstructions. BH decreased with increasing As coverage on the AlAs.

OTHER WORK

Lindau et al [16] studied the electronic structure of the (110) GaAs/Al interface and found that Fermi level pinning is complete at less than 0.2 monolayers. They propose that defect states at, or close to the interface are responsible for the BH. Photoemission core level data suggest a replacement interaction between Al and Ga in the topmost GaAs surface layer.

Srivastava and Arora [18] heat treated near ideal GaAs/Al diodes at 523 deg K and reported an increase in the difference between the I-V and C-V measured values of BH, as well as a 2 or 3 fold reduction in the Richardson constant.

Other investigations of the GaAs/Al barrier include refs [19,20].

REFERENCES

[1] W.E.Spicer, I.Lindau, P.Skeath, C.Y.Su, P.Chye [Phys. Rev. Lett. (USA) vol.44 (1980) p.420]

[2] L.J.Brillson, R.Z.Bachrach, R.S.Bauer, J.McMenamin [Phys. Rev. Lett. (USA) vol.42 (1979) p.397]

[3] J.R.Waldrop [J. Vac. Sci. & Technol. B (USA) vol.2 no.3

(July-Sept 1984) p.445-8]
[4] J.R.Waldrop [Appl. Phys. Lett. (USA) vol.44 no.10 (May 1984) p.1002-4]
[5] S.W.Eglash, M.D.Williams, P.H.Mahowald, N.Newman, I.Lindau, W.E.Spicer [J. Vac. Sci. & Technol. B (USA) vol.2 no.3 (July-Sept 1984) p.481-5]
[6] S.W.Eglash, S.Pan, D.Mo, W.E.Spicer, D.M.Collins [Jpn. J. Appl. Phys. Suppl (Japan) vol.22 suppl.22-1 (1983) p.431-5]
[7] G.P.Schwartz, G.J.Gualtieri [Appl. Phys. Lett. (USA) vol.42 no.3 (1 Feb 1983) p.265-7]
[8] G.P.Schwartz, G.J.Gualtieri [J. Vac. Sci. & Technol. (USA) vol.20 no.3 (Mar 1982) p.674-9]
[9] H.Sakaki, Y.Sekiguchi, D.C.Sun, M.Taniguchi, A.Tanaka [Jpn. J. Appl. Phys. (Japan) vol.20 no.2 (Feb 1981) p.L107-10]
[10] T.Oda, T.Sugano, [Trans. Inst. Electron. and Commun. Eng. Jpn. Sect.E (Japan) vol.E59 (1976) p.56]
[11] S.Svensson, G.Landgren, T.G.Andersson [J. Appl. Phys. (USA) vol.54 no.8 (Aug 1983) p.4474-81]
[12] O.Aina, K.P.Pande [J. Appl. Phys. (USA) vol.56 no.6 (15 Sept 1984) p.1717-21]
[13] J.Massies, J.Chaplart, M.Laviron, N.T.Linh [Appl. Phys. Lett. (USA) vol.38 no.9 (1 May 1981) p.693-5]
[14] C.Barret [J. Vac. Sci. & Technol. B (USA) vol.1 no.3 (July-Sept 1983) p.819-24]
[15] A.Y.Cho, P.D.Dernier [J. Appl. Phys. (USA) vol.49 no.6 (June 1978) p.3328-32]
[16] I.Lindau, P.R.Skeath, C.Y.Su, W.E.Spicer [Surf. Sci. (Netherlands) vol.99 no.1 (Sept 1980) p.192-201]
[17] W.I.Wang [J. Vac. Sci. & Technol. B (USA) vol.1 no.3 (July-Sept 1983) p.574-80]
[18] A.K.Srivastava, B.M.Arora [Solid State Electron. (GB) vol.24 no.11 (Nov 1981) p.1049-52]
[19] P.Skeath, I.Lindau, C.Y.Su, P.W.Chye, W.E.Spicer [J. Vac. Sci. & Technol. (USA) vol.17 (1980) p.511]
[20] R.Ludeke, L.L.Chang, L.Esaki [Appl. Phys. Lett. (USA) vol.23 (1973) p.201]
[21] N.Newman et al [J.Vac.Sci.& Technol.A (USA) vol.3 no.3 (May/June 1985) p.996-1001]
[22] S.P.Svensson, T.G.Andersson [J. Vac. Sci. & Technol. B (USA) vol.3 no.2 (Mar/Apr 1985) p.760-1]

17.8 BARRIER HEIGHT AT THE GaAs/Au INTERFACE

by EMIS Group

November 1985
EMIS Datareview RN=15402

The GaAs/Au Schottky barrier is used as an ohmic or rectifying contact in a wide variety of semiconductor devices, e.g. solar cells and field effect transistors (FETs). Research has been motivated largely by the need to realise the maximum possible barrier height (BH) and to achieve stability during heat treatment. For a short discussion of some general considerations (e.g. the mechanism by which Schottky barriers are formed) see ref [22].

There is considerable chemical mixing and interdiffusion at the GaAs/Au interface (see, for example ref. [18]), rendering it sensitive to the thermal processing needed for device applications. The BH is reduced considerably by heat treatment [1-4], and the associated changes in chemical structure of the interface have been studied by photoemission spectroscopy [5,6], secondary ion mass spectrometry [7], Auger electron spectrometry [2], Rutherford backscattering spectrometry [8,9] and other surface techniques.

BH AT CLEAN INTERFACES

Barrier heights at clean oxide-free Au (and 13 other) interfaces with (100) n-GaAs and p-GaAs were measured by Waldrop [10,11]. X-ray photoemission and low energy electron diffraction analysis verified that the GaAs surface was clean and ordered. At least 2000 A of Au was ultra high vacuum evaporated onto this surface at room temperature. The samples were:

n-GaAs: Se doped to 6 X 10*16/cm*3, LEC
p-GaAs: Cd doped to 8 X 10*16/cm*3, Bridgman

and the room temperature BH values were:

I-V method		C-V method
n-type	p-type	n-type
0.89 eV	0.50 eV	0.87 eV

The I-V values incorporate image force correction of +0.04 eV (n-type) and +0.03 eV (p-type). Note that at a doping level of 6 X 10*16/cm*3 there may be noticeable thermionic field emission which would reduce the BH for the n-type samples.

For other data on BH in practical Schottky barrier device structures, see refs. [13-19] (quoted in ref. [12]).

EFFECT OF INTERFACIAL OXYGEN

GaAs/Au Schottky barriers used in device manufacture normally have at least one monolayer (ML) of oxygen on top of the GaAs. Lu et al [20] showed that oxidation before Au deposition makes the barrier more abrupt by inhibiting Au intermixing with the GaAs. They studied a (110) GaAs/Au interface in which up to 10 MLs of oxygen were applied to the clean cleaved semiconductor surface, using UV and X-ray photoemission spectroscopy. Any coverage exceeding a few tenths of a ML enhanced the abruptness of the barrier. Meirhaeghe et al [21] studied the effect of an interfacial oxide layer on an Au interface with n-GaAs platelets oriented perpendicularly to the [111] direction and found that an oxygen thickness of 3.5 nm gave a definite maximum BH of around 1.20 eV in the BH vs thickness curve (1.5-4.5 nm).

ANNEALING EFFECTS ON BH

Newman et al [3] carried out a detailed study of the effects of annealing Ag,Al and Au on clean, cleaved (110) n-GaAs:Si with a doping level of 10 X 17/cm*3. The diodes were fabricated by slow controlled evaporation of clean metal overlayers to a thickness of 1000 A, and annealed for 10 minutes in nitrogen at various temperatures:

Annealing temperature deg C	Barrier Height (eV) I-V (+-0.02)	Barrier Height (eV) C-V (+-0.05)	Ideality Factor (+-0.02)
Room temp.	0.92	1.00	1.05
150	0.91	0.99	1.06
220	0.85	0.92	1.07
290	0.80	0.89	1.06
430	0.80	0.88	1.06
495	0.80	-	-

The measurements were made at room temperature after annealing. The I-V and C-V data do not include corrections for the image potential or the Gummel-Scharfetter factor respectively. The authors used photoemission spectroscopy to correlate annealing-induced microscopic changes in the electronic and chemical structure of the interface with the electrical measurements, and attributed the decrease in BH to a change in the interface Fermi level position. An Au-Ga rich layer formed at the interface during anneals at 200 to 300 deg C, and leakage currents dominated the I-V characteristics in devices annealed above the Au-Ga eutectic temperature [4].

REFERENCES

[1] A.K.Sinha, J.M.Poate [Appl. Phys. Lett. (USA) vol.23 (1973) p.666]

[2] G.Y.Robinson [J. Vac. Sci. & Technol (USA) vol.13 no.14 (July/Aug 1976) p.884-7]

[3] N.Newman et al [J. Vac. Sci. & Technol. A (USA) vol.3 no.3 (May/June 1985) p.996-1001]

[4] N.Newman, W.G.Petro, T.Kendelewicz, S.H.Pan, S.J.Eglash, W.E.Spicer [J. Appl. Phys. (USA) vol.57 no.4 (15 Feb 1985) p.1247-51]

[5] W.G.Petro et al [J. Vac. Sci. & Technol. (USA) vol.21 (1982) p.585]

[6] W.G.Petro, I.A.Babalola, T.Kendelewicz, I.Lindau, W.E.Spicer,

[J. Vac. Sci. & Technol. A (USA) vol.1 (1983) p.1181]
[7] H.B.Kim, G.G.Sweeney, T.M.S.Heng [Inst. Phys. Conf. Ser. (GB) vol.24 (1975) p.307]
[8] S.Leung, L.K.Wong, D.D.L.Chung, A.G.Milnes [J. Vac. Sci. & Technol. (USA) (1983) p.462]
[9] E.Kinsbron, P.K.Gallagher, A.T.English [Solid State Electron (GB) vol.22 (1979) p.517]
[10] J.R.Waldrop [J. Vac. Sci. & Technol. B (USA) vol.2 no.3 (July-Sept 1984) p.445-8]
[11] J.R.Waldrop [Appl. Phys. Lett. (USA) vol.44 no.10 (May 1984) p.1002-4]
[12] P.Skeath, C.Y.Su, I.Hino, I.Lindau, W.E.Spicer [Appl. Phys. Lett. (USA) vol.39 no.4 (15 Aug 1981) p.349-51]
[13] W.E.Spicer, P.W.Chye, P.R.Skeath, C.Y.Su, I.Lindau [J. Vac. Sci. & Technol. (USA) vol.16 no.5 (Sept-Oct 1979) p.1422-33]
[14] W.E.Spicer, P.W.Chye, P.R.Skeath, C.Y.Su, I.Lindau [Inst. Phys. Conf. Ser. (GB) no.50 (1979) p.216]
[15] W.E.Spicer, I.Lindau, P.R.Skeath, C.Y.Su, P.W.Chye [Phys. Rev. Lett. (USA) vol.44 (1980) p.420]
[16] W.E.Spicer, I.Lindau, P.R.Skeath, C.Y.Su [J. Vac. Sci. & Technol (USA) vol.17 (1980) p.1019]
[17] P.R.Skeath, I.Lindau, P.W.Chye, C.Y.Su, W.E.Spicer [J. Vac. Sci. & Technol. (USA) vol.16 (1979) p.1143]
[18] P.W.Chye, I.Lindau, P.Pianetta, C.M.Garner, C.Y.Sy, W.E.Spicer [Phys. Rev. B (USA) vol.18 no.10 (1978) p.5545-59]
[19] E.H.Rhoderick [Metal-Semiconductor Contacts (Clarendon, Oxford, 1978)]
[20] Z.M.Lu, W.G.Petro, P.H.Mahowald, M.Oshima, I.Lindau, W.E.Spicer [J. Vac. Sci. & Technol. B (USA) vol.1 no.3 (July-Sept 1983) p.598-601]
[21] R.L. van Meirhaeghe, E.S.Vespurten, F.Cardon, W.P.Gomes [2nd EC Photovoltaic Solar Energy Conf. (Berlin, West Germany, 23-6 April 1979) p.946-51]
[22] EMIS Group [EMIS Datareview RN=15271 (Sept 1985) `Barrier height at the GaAs/Al interface´]

17.9 BARRIER HEIGHT AT THE GaAs/Pt INTERFACE

by EMIS Group

November 1985
EMIS Datareview RN=15409

The necessity for reliable, reproducible MESFET (metal semiconductor FET) gates has prompted considerable research into the effect of annealing on the electrical properties of GaAs/metal Schottky barriers. In the case of Pt there is evidence (see below) that this significantly affects the barrier height (BH) as determined by the I-V and C-V methods.

Fontaine et al [1] obtain a photoresponse value of 0.90 eV for the BH both before and after annealing at 350-500 deg C for up to 20 hours. Clean damage-free, oxygen-free n-GaAs oriented in the [100] direction-family and having a carrier density of 1-2 X 10*17/cm*3 was used, and onto this Pt was deposited to a thickness of 1000 A. PtGa and PtAs2 were found at the interface. However, Fontaine et al [1], Murarka [2] and Sinha and Poate [3] all observed an increase in the forward bias I-V value of BH after similar annealing. Also Fontaine et al [1] found that the C-V determined BH was reduced. Fontaine et al [1] note that these annealing changes (e.g. from 0.65 to 0.81 eV for I-V and from 1.11 to 0.98 eV for C-V [1]) are associated with the measurement methods. The series resistance, generation-combination and leakage, non-thermionic current mechanisms (e.g. tunnelling), and non-uniform space charges have less effect on the photothreshold BH than on the I-V and C-V BHs.

Pt significantly dissolves gaseous hydrogen. Aspnes et al [4] found that although this reversibly lowered the work function by almost 1 eV the I-V, C-V and photoresponse BH values were hardly affected, whereas for other Pt-group metals (Ru,Ir) the BHs were considerably changed. Air-exposed samples were used and the authors note that air exposure of Pt-group metals (Ru,Rh,Pd,Os,Ir,Pt) can cause oxygen absorption at the surface and grain boundaries. The observed effects could, they suggest, represent differences between oxygen-containing metals and oxygen-free metals with hydrogen at the interstitial grain boundaries and the surface.

For an experimental study of majority and minority carriers in 2 nm diameter GaAs/Pt Schottky barriers see ref [5]. For a deep level transient spectroscopy (DLTS) investigation of interface states and their reduction by sintering see ref [6]

REFERENCES

[1] C.Fontaine, T.Okumura, K.N.Tu [J. Appl. Phys. (USA) vol.54 no.3 (March 1983) p.1404-12]
[2] S.P.Murarka [SOlid-State Electron (GB) vol.17 no.9 (Sept 1974) p.985-91]
[3] A.K.Sinha, J.M.Poate [Appl. Phys. Lett (USA) vol.23 (1973) p.666]
[4] D.E.Aspnes, A.Heller [J. Vac. Sci. & Technol. B (USA) vol.1 no.3 (July-Sept 1983) p.602-7]
[5] E.Y.Chan, H.C.Card, E.S.Yang, A.R.Kerr, R.J.Mattauch [IEEE Trans.

Electron Devices (USA) vol.ED-26 no.3 (March 1979) p.214-9]
[6] T.Onuma, T.Uenoyama, H.Yugita [Appl. Phys. Lett. (USA) vol.44 no.1 (1 Jan 1984) p.80-1]

17.10 BARRIER HEIGHT AT THE GaAs/W INTERFACE

by EMIS Group

November 1985
EMIS Datareview RN=15408

An abrupt, almost ideal (111) n-GaAs/W interface was produced by Batev et al [1] using the reaction between tungsten hexachloride and hydrogen. The carrier concentration was about 10*16/cm*3 and the ideality of the Schottky diode characteristics was 1.02 (+- 0.01). The room temperature barrier height (BH) was measured as 0.81 (+- 0.01) eV by both I-V and C-V methods. Waldrop and co-workers [2,3] obtained a BH of 0.9 eV by X-ray photoelectron response (XPS) measurements on a non-abrupt (100) n-GaAs/W interface with a similar carrier concentration. The XPS BH is obtained by measuring the interface band bending with respect to the Fermi level and is therefore not sensitive to the interfacial compounds, oxides or annealing factors which affect the Schottky diode characteristics.

Contrary to earlier expectations W and the other refractory metals (Ta,Re,Ir,Mo) are not chemically inert in contact with GaAs. According to Waldrop et al [2,4] their reactivity is about the same as for Fe,Cr and Ti, and more than for Sn,Ag,Au and Al. Mg is considerably more reactive than W.

The effects of annealing oxide-free and oxide-containing interfaces at up to 650 deg C were investigated by Waldrop [3] who ultra high vacuum (UHV) deposited 1000 A or more of W onto (100) n-GaAs wafers by electron evaporation of a thin W wire. The wafers had been cleaned in a sulphuric acid solution and had a carrier density of 5 X 10*16/cm*3. A native oxide layer (Ga2O3 + As2O3) 10 A thick was removed from some samples by heating to 550 deg C to leave an ordered contamination-free surface prior to W deposition. Results for the two kinds of interface are reproduced below:

A. OXIDE-FREE INTERFACE (n-GaAs /Ga+WAs2/ W)

30 minutes anneal temperature deg C	ideality factor	BH (I-V) eV	BH (C-V) eV
initial	1.26	0.66	0.98
225	1.22	0.68	0.99
300	1.12	0.72	0.95
350	1.13	0.70	0.96
450	1.12	0.72	0.97
550	1.19	0.76	1.30
600	1.28	0.76	1.48
650	1.47	0.67	1.26

B. OXIDE-CONTAINING INTERFACE (n-GaAs /Ga2O3+WO3+WAs2/ W)

initial	-	-	1.34
225	1.23	0.68	1.25
300	1.13	0.71	1.16
350	1.11	0.69	1.09
450	1.10	0.65	0.91
550	1.10	0.69	0.95
600	1.12	0.72	1.06
650	1.41	0.71	1.11

Waldrop [op. cit.] determined the true BH (XPS) as 0.90 eV and concluded that because BH (I-V) is some 0.25 eV lower than the true BH and because the ideality is high it is likely that high interface recombination occurs during current transport. The high BH (C-V) values follow from the dielectric properties of the interface.

Ion implantation is a common method of producing source and drain regions in metal semiconductor FETs (MESFETs). High anneal temperatures are needed to activate the implanted impurities. This motivated Matsumoto et al [5] to study the thermal stability of the interface between Se+ implanted (300 keV, 6 X 10*12/cm*2) GaAs and W for 10 minute anneals at 800,850 and 950 deg C. Tungsten patterns (1000 A thick, 130 microns in diameter) were produced by electron beam evaporation and lift off. The ideality stayed at 1.1-1.2 and the BH (I-V) was 0.73 eV for all three anneal temperatures. He+ backscattering analysis revealed interdiffusion between the GaAs and W at 900 deg C, which nevertheless caused no deterioration in the Schottky diode characteristics.

REFERENCES
----

[1] P.M.Batev, M.D.Ivanovitch, E.I.Kafediiska, S.S.Simenov [Phys. Status Solidi A (Germany) vol.45 no.2 (1978) p.671-5]

[2] J.R.Waldrop, S.P.Kowalczyk, R.W.Grant [J. Vac. Sci. & Technol. (USA) vol.21 no.2 (July/Aug 1982) p.607-10]

[3] J.R.Waldrop [Appl. Phys. Lett. (USA) vol.42 no.4 (15 Aug 1982) p.350-2]

[4] J.R.Waldrop, R.W.Grant [Appl. Phys. Lett. (USA) vol.34 (1979) p.630]

[5] K.Matsumoto, N.Hashizume, H.Tanoue, T.Kanayama [Jpn. J. Appl. Phys. (Japan) vol.21 no.6 (June 1982) p.L393-5]

17.11 BAND BENDING AT GaAs/DIELECTRIC INTERFACES

by H.H.Wieder

March 1985
EMIS Datareview RN=14805

The electrical properties of n- and p-type GaAs/insulator interfaces have been determined, primarily, by means of capacitance versus voltage (C-V) measurements made on metal-insulator-semiconductor (MIS) structures.

The insulators were made either by the electrochemical anodization or plasma anodization of semiconductor surfaces or by chemical vapour phase deposition of synthetic insulating layers on the same semiconductors [1,2].

The position of the equilibrium surface Fermi level, E(F)*, in a voltage, V(g) = 0, applied to an n-type MIS structure, is shown below:

```
Conduction band minimum (CBM) ---:---
                                 :
                               0.8 eV
                                 :
                                 :
                         E(F)* ---:---
                                 :
                               0.6 eV
                                 :
   Valence band maximum (VBM) ---:---
```

This coincides with Fermi level pinning by interface states whose density exceeds 10*13/cm*2/eV*1 [3,4].

The surface potential, phi(s), excursion is limited to about 0.45 eV within the lower half of the fundamental bandgap and the lowest interface state density, in this zone, is about 2 X 10*12/cm*2/eV.

Fermi level pinning at such GaAs/insulator interfaces is qualitatively consistent with that measured by means of X-ray photoemission spectroscopic (XPS) methods on ultra high vacuum-cleaved ´free´ (110) oriented GaAs surfaces on which oxygen has been chemisorbed: it is also consistent with the barrier height of corresponding metal/GaAs interfaces [5].

Fermi level pinning is attributed to extrinsic surface states associated with defects produced by the chemisorption process [6].

REFERENCES
----

[1] [First Special Issue on Semiconducting III-V Compound MIS Structures, Thin Solid Films (Switzerland) vol.56 (1979) C.W.Wilmsen and H.H.Wieder, eds.]
[2] [Second Special Issue on Semiconducting III-V Compound MIS Structures, Thin Solid Films (Switzerland) vol.103 (1983)

C.W. Wilmsen and H.H.Wieder, eds.]
[3] L.Meiners [J. Vac. Sci. & Technol. (USA) vol.15 (1978) p.1402]
[4] T.Sawada, H.Hasagawa [Electron Lett. (GB) vol.12 (1976) p.471]
[5] W.E.Spicer, I.Lindau, P.Skeath, C.Y.Su, P.W.Chye [Phys. Rev. Lett. (USA) vol.44 (1980) p.420]
[6] H.H.Wieder [Surf. Sci. (Netherlands) vol.132 (1983) p.390]

17.12 CONDUCTION BAND AND VALENCE BAND OFFSET AT THE GaAs/AlGaAs HETEROSTRUCTURE INTERFACE

by L.Eaves

January 1986
EMIS Datareview RN=15497

During the last year or so, a series of different experimental techniques and analysis methods have been employed to determine the conduction (DELTA Ec) and valence (DELTA Ev) band offset energies at the GaAs/(AlGa)As heterostructure interface. The offsets, sometimes referred to as the conduction and valence band discontinuities, determine the way in which the energy gap difference (DELTA Eg) of the two materials is distributed between the conduction and valence bands. They are critical parameters for many optical and electronic devices. The offset ratio

$$R = \frac{\text{DELTA Ec}}{\text{DELTA Ev}}$$

which is quoted in many of these papers applies to structures in which the (AlGa)As is direct gap, that is with Al mole fraction x less than or equal to around 0.40-0.45. At higher values of x, the X-conduction band minima cross the GAMMA minimum. The variation of DELTA Ev is a continuous function of x.

Most new measurements give R in the range 60:40 to 65.35. This is significantly different from the earlier and widely quoted value of 85:15, obtained several years ago from optical absorption measurements on multiple quantum well structures [1,2] and supported by other work [3,4].

Recent C-V profiling measurements on both nn (0.15 below x below 0.3) and pp (x = 0.3) junctions [5] have indicated offsets DELTA Ec = 0.77x eV and DELTA Ev = 0.47x eV. These are supported by work [6] on nn junction [x below 0.42] which give DELTA Ec = 0.83x eV. Internal photoemission experiments [7] also indicate a ratio R = 60:40 for x up to about 0.4 as do recent optical spectroscopy measurements on parabolic, rectangular and asymmetric double rectangular quantum wells [8,9]. Several authors [10,11,12,13] have measured the offsets by determining the activation energy for thermionic current flow over the barrier. These measurements also give a value of R = 60:40 to 65:35. Of especial interest is the recent work of Batey and Wright [12] which measures DELTA Ev = 0.55x eV over the complete alloy range. This last value is somewhat larger than that, DELTA Ev = 0.45 +- 0.05 eV (x = 1), given by a charge transfer analysis of pp GaAs/AlAs heterojunctions [14, see also 15]. Some recent optical data [16] giving R = 75:25 (+- 0.05) runs counter to this range of new values, though this value has been re-appraised recently to 65:35 by the authors [17]. New photoconductivity measurements on multi-quantum well structures have put R in the range 57:43 to 66:34 [18].

Assessment of this mass of new data clearly indicates that R is significantly closer to 60:40 than the older accepted value of 85:15.

It is worth noting that the values of DELTA Ec + DELTA Ev obtained from the C-V profiling of Watanabe et al [5] add up to DELTA Eg = 1.24x eV in satisfying agreement with the widely quoted value of DELTA Eg = 1.25x eV for x up to 0.45 [19]. However, the larger quoted values of DELTA Ec [6] and DELTA Ev [13] add up to a value for DELTA Eg which is fairly consistent with a recent paper which reports DELTA Eg = 1.45x eV [20]. Clearly the offset problem is not completely settled yet. In the reviewer's opinion, optical measurements on quantum structures are likely to be less accurate in giving DELTA Ec and DELTA Ev than C-V and thermionic current measurements since the quantum well energies are relatively insensitive to the height of the well and involve the complexity of the valence band structure in the description of the bound hole states. However, the consensus of the above data indicates that DELTA Ev = (0.51 +- 0.04)x eV for x = 0 to 1 inclusive and DELTA Ec = (0.80 +- 0.03)x eV for x up to about 0.45. Above this value of x, the X minima falls below the GAMMA minimum and DELTA Ec decreases, falling to 0.2 eV at x = 1.

REFERENCES

[1] R.Dingle, W.Wiegmann, C.H.Henry [Phys. Rev. Lett. (USA) vol.33 (1974) p.827]

[2] R.Dingle [in "Festkorperprobleme/Advances in Solid State Physics", Ed. by H.J.Queisser (Vieweg, Braunschweig, 1975) vol.15 p.21-48]

[3] R.People, K.W.Wecht, K.Alavi, A.Y.Cho [Appl. Phys. Lett. (USA) vol.43 (1983) p.118]

[4] A.C.Gossard, W.Brown, C.L.Allyn, W.Wiegmann [J. Vac. Sci. Technol. (USA) vol.20 (1982) p.694]

[5] M.O.Watanabe, J.Yoshida, M.Mashita, T.Nakanisi, A.Hojo [J. Appl. Phys. (USA) vol.57 (1985) p.5340]

[6] H.Okumura, S.Misawa, S.Yoshida, S.Gonda [Appl. Phys. Lett. (USA) vol.46 (1985) p.377]

[7] M.Heilblum, M.I.Nathan, M.Eizenberg [Appl. Phys. Lett. (USA) vol.47 (1985) p.503]

[8] R.C.Miller, A.C.Gossard, D.A.Klenimann, O.Munteanu [Phys. Rev. B (USA) vol.29 (1984) p.3740]. See also R.C.Miller, D.Al.Kleinmann, A.C.Gossard [Phys. Rev. B (USA) vol.29 (1984) p.7085]

[9] M.H.Meynadier, C.Delalande, G.Bastard, M.Voos, F.Alexandre, J.L.Lievin [Phys. Rev.B (USA) vol.31 (1985) p.5539]

[10] T.W.Hickmott, P.M.Solomon, R.Fischer, H.Morkoc [J. Appl. Phys. (USA) vol.57 (1985) p.2844]

[11] J.Batey, S.L.Wright, D.J.DiMaria [J. Appl. Phys. (USA) vol.57 (1985) p.484]

[12] J.Batey, S.L.Wright, D.J.DiMaria, T.N.Theis [J. Vac. Sci. & Technol. B (USA) vol.3 (1985) p.653.] See also J.Batey, S.L.Wright [to be published in J. Appl. Phys. (USA) (Jan 1986)]

[13] D.Arnold, A.Ketterson, T.Henderson, J.Klem, H.Morkoc [Appl. Phys. Lett. (USA) vol.45 (1984) p.1237]. See also D.Arnold, A.Ketterson, T.Henderson, J.Klem, H.Morkoc [J. Appl. Phys. (USA) vol.57 (1985) p.2280]

[14] W.I.Wang, F.Stern, [J. Vac. Sci.& Technol.B (USA) vol.3 (1983) p.1280]

[15] W.I.Wang, E.E.Mendez, F.Stern [Appl. Phys. Lett. (USA) vol.45 (1984) p.639]

[16] P.Dawson, G.Duggan, H.I.Ralph, K.Woodbridge ["Superlattices and Microstructures" vol.1 (1985) p.231] See also G.Duggan [J. Vac.

Sci Technol. B (USA) vol.3 (1985) p.1224]
[17] G.Duggan, H.I.Ralph, K.Moore [Phys. Rev. B (USA) vol.32 (1985) p.8395]
[18] D.C.Rogers, R.J.Nicholas [J. Phys. C (GB) vol.18 (1985) p.L891]
[19] H.C.Casey, M.B.Panish ["Heterostructure Lasers: Part A; Fundamental Principles" (Academic, New York 1978)]
[20] N.C.Miller, S.Zemon, G.P.Werber, W.Powazinik [J. Appl. Phys. (USA) vol.57 (1985) p.512]

CHAPTER 18

ALUMINIUM GALLIUM ARSENIDE: MISCELLANEOUS PROPERTIES

18.1 Defect energy levels in AlGaAs

18.2 DX defect centres in AlGaAs

18.3 Photoluminescence spectra of AlGaAs

18.4 Electron mobilities in AlGaAs

18.5 Carrier concentrations in LPE AlGaAs

18.6 Carrier concentrations in MOCVD AlGaAs

18.7 Carrier concentrations in MBE AlGaAs

18.8 Etching rates of reactive ion and reactive ion-beam etched AlGaAs

18.1 DEFECT ENERGY LEVELS IN AlGaAs

by A.R.Peaker

August 1985
EMIS Datareview RN=15606

INTRODUCTION

In ternary compounds the properties of defects often vary with alloy composition. In Al(x)Ga(1-x)As most work has been done on material with x = 0.25-0.3 but for a few commonly occurring states the characteristics have been studied over a range of compositions. In general, the deep level concentration are higher than in GaAs, indeed for compositions with x greater than 0.25 a large proportion of the dopants S, Se, Te, Si, Ge and Sn appear as deep centres with apparent DLTS activation energies from 190 to 560 meV. These so called DX centres are the source of the persistent photoconductivity effect and are reviewed separately [1]. It is not clear whether the higher concentrations of other deep states in GaAlAs are simply due to higher reactivity of aluminium or to a fundamental thermodynamic property of the ternaries. There is, however, no doubt that the presence of oxygen or water vapour in the growth environment has the effect of dramatically increasing the deep state concentration in AlGaAs in contrast to GaAs where the effects are negligible [2]. Some states observed in AlGaAs can be related to similar or identical states in GaAs. The most important of these is EL2 as it is usually the trap present in highest concentration in bulk, VPE and MOVPE GaAs and high quality MOVPE AlGaAs with low aluminium content (x less than 0.2).

In the data that follows unless otherwise stated the energies are the experimentally observed slopes of the Arrhenius plots with T*2 correction. The intercept cross section values are derived from the observed emission rate by dividing the prefactor of the exponential term by v X Nc where v is the thermal velocity and Nc the density of states. There is some uncertainty regarding the most appropriate value of this product in AlGaAs especially in the region around x = 0.4. Unfortunately few authors quote the values actually used and so it is not known whether the substantial variations reported represent different experimental observations or discrepancies in the analysis.

EL2 IN AlGaAs

The concentration of the deep state EL2 in MOVPE AlGaAs is comparable with that in MOVPE GaAs for normal growth conditions [3]. This is typically between 5 X 10*13 and 5 X 10*15/cm*3. The concentration is essentially independent of aluminium content [4] although as in GaAs it is found to increase with increasing V/III ratio in the gas stream [5]. EL2 is not seen in as-grown MBE or LPE GaAs but has been reported in MBE AlGaAs [6].

There are major discrepancies in the literature regarding the properties of EL2 in AlGaAs. Wagner et al [4] in high carrier concentration material (n = 10*17 to 10*18/cm*3) observed an apparent activation energy of 0.84 eV independent of aluminium content and quote a single value of intercept cross section of 2.4 X 10*13 cm*-2. Matsumoto et al [7] have examined the EL2 level in low doped material and also observed that neither the EL2 concentration nor the activation energy changes with x but the intercept cross section does.

Aluminium content x	0	0.11	0.22	0.3
Activation energy meV	830	830	820	820
Intercept cross sectn cm*2	1.5 10*-13	5 10*-15	4.4 10*-16	4.4 10*-16

The results for x = 0.3 were published in a later paper by the same group [5]. There was no intentional doping and growth was at 750 deg C with a V/III ratio of 20.

Johnson et al [3] have also examined EL2 in selenium doped MOVPE AlGaAs with a range of aluminium content (x = 0-0.33) grown at 750-825 deg C. In contrast to Matsumoto et al [7] they see a quite dramatic change of activation energy:

Aluminium content x	0	0.05	0.14	0.25	0.33
Activation energy meV	830	850	930	950	1220

The doping was quite low (n = 4 X 10*16/cm*3) as was the trap concentration (3 X 10*14/cm*3). It is quite inconceivable that such a disparity could be due to experimental techniques and so the only explanation is that the studies were really of different states. There is indeed evidence in GaAs that EL2 may have a number of variants with slightly different properties. Gatos and Lagowski [8] have identified two such states which have been called EL2 and EL0. Although the emission properties are very similar, the directly measured (real) capture cross section of EL0 is four times that of EL2. It is possible that the structural differences between these states are manifest as differences in emission in AlGaAs and this is the reason for the discrepancies in the experimental observations.

OTHER ELECTRON TRAPS

The DX centre reviewed separately [1] is present in AlGaAs layers with x greater than 0.2. In MOVPE material the EL2 state is nearly always seen. Other electron states appear much less reproducibly. Perhaps the most common have been associated with the presence of oxygen or water vapour during growth. Wallis [2] intentionally introduced oxygen during low pressure (300 mbar) MOVPE. Three traps were associated with oxgyen but only one was fully characterised. It had an activation energy of 410 meV at x = 0.05. Its concentration was more than 10*15/cm*3 (greater than EL2) with only very small oxygen partial pressures (10*-5 mbar). The concentration increased with increasing x. Similar behaviour of what appears to be the same state was observed by Casey et al [9] in MBE material and several others in MOVPE.

Sakamoto [10] observed the trap in material produced from atmospheric pressure MOVPE and also attributed it to oxygen contamination. He observed that the concentration of another trap at 270 meV was the same as the 410 meV state over two orders of magnitude and regarded them as different charge states of the same centre. Similar observations were also made by Andre [11]. In material in the range x = 0.25 to 0.4 the

DLTS peak of the deeper oxygen related state is very close to the DX centre (which is usually present in higher concentrations) and so is not easy to identify in this composition range. Both Wallis and Sakamoto also looked for the state in oxygen contaminated GaAs, its absence indicates that both aluminium and oxygen are necessary for its creation. The table summarises observations of the two states.

These states are very significant for two reasons. Firstly, they can be present in high concentrations and are acceptor like. This can result in n-type material becoming semi-insulating. Secondly, it appears they can act as powerful non-radiative recombination centres; a good negative correlation is observed between the photoluminescence efficiency and the trap concentration [10,15].

Composition x	Activation Energies in meV and name given		Approx concn cm*-3	Ref
0.05	290	410	10*15	[2]
0.06		410		[12]
0.1		460	10*16	[2]
0.11-0.13	270(A)	410(B)	10*14-10*16	[10]
0.15	300	500	10*15	[11]
0.2		550	10*16	[2]
0.25	240(EM1)	410(EM3)	10*16	[13]
0.3	250(E1)	450(E3)	10*15	[5]
0.5		640		[9]

Another state which appears to be important in the recombination kinetics has an activation energy of around 660 meV. Tsai considers it to be related to LED efficiency [12]. It is generally present in concentrations in the 10*15/cm*3 range [5] and has a large hole cross section (greater than 10*-15/cm*2) but a quite small electron cross section (6 X 10*-18) as measured at 330 deg K for x = 0.25 [13]. It would therefore only be expected to be important as a recombination centre in n-type material.

It is possible that this state has also been observed in MBE material as ME6 (710 meV at x = 0.2) [14]. There is a good inverse correlation between the trap concentration and the room temperature photoluminescence intensity. The trap increases in concentration with increasing V/III flux ratio but decrease dramatically with increasing substrate temperature [14].

This is consistent with the observation that the `quality´ of MBE AlGaAs improves with increasing growth temperature. However, unlike the GaAs case it appears to make little difference whether dimeric or tetrameric arsenic is used for growth. Chen et al [16] growing AlGaAs with x = 0.3 observe a reduction of a factor of 2 in trap concentration with the dimer (in contrast to the tetramer) at low temperature (610-660 deg C) but no difference at the more normal growth temperature of 700 deg C [16]. In a more comprehensive study, Mooney et al [17] covering the range x = 0.04 to 0.36 observed that the effect of the arsenic species is negligible compared to the effects of temperature and V/III flux ratio.

Work on deep states created by electron damage has been undertaken by several groups [18-21]. The results agree well and so only the most comprehensive study is quoted here [21].

Composition x	Corrected activation energy meV E1	E2	E3
0	40	180	330
0.14	120	200	490
0.25	170	340	680
0.47	300	430	850
Introduction rate (cm*-1)	2.1	1.88	0.73

The introduction rate is measured at a dose of 10*15e/cm*2 for x = 0.25. The results are interesting in that, while E1 and E2 change proportionally in energy with the gap as the composition is changed, E3 remains fixed in relation to the valence band. Lang [19] assigns E3 as the gallium vacancy.

HOLE TRAPS

Measurements on hole traps are reported mainly in MOVPE material with some work on LPE layers. In all cases in the first table below the aluminium content is in the range x = 0.25-0.3.

Growth technique and dopant	Apparent activation energy meV	Measurement temp (K)	Measured hole cross section cm*2	Measured electron cross section cm*2	
LPE(Ge)	180		-	-	[22]
MOVPE (S and u)	200	150	10*-15	1.2 X 10*-18	HM [13]
MOVPE (S and u)	270	160	1.5 X 10*-15	3 X 10*-17	HM2[13]
LPE (Sn)	240	145	above 5 X 10*-15	2 X 10*-19	Ly [22]
MOVPE (S and u)	440	230	3.5 X 10*-15	10*-19	HM3[13]
MOVPE (u)	740		-		H1 [5,24]
MOVPE (S and u)	770	330	2.5 X 10*-15	2 X 10*19	HM4[13]
MOVPE (u)	860		-	-	H2 [24]

Arrhenius plots are given in the references; u signifies that there was no intentional dopant.

Lang [18] examined four states which are seen in LPE GaAs, they are the A and B centres and defects associated with Cu and Fe. The apparent activation energy as a function of composition is given in the table. These values are derived from graphical information in the original paper.

Aluminium content x	Apparent activation energy A(HB5)	Cu(HB5)	Fe(HB3)	B(HB2)
0	400	440	520	710
0.1	410	-	560	753
0.15	-	490	580	-

0.2	-	520	610	-
0.22	458	-	-	820
0.28	-	-	-	855

The HB notation is that used for the states in GaAs by Mitonneau [25]. All four defects maintain their relative position in the gap i.e. the energy level changes proportionally with the gap over the measured range.

In MOVPE the defects HM3, HM4, H1 and H2 are reported in the 10*14 - 10*15/cm*3 concentration range while HM1 and HM2 are in the 2 X 10*16/cm*3 range. In LPE the copper and iron related states are only seen in material intentionally doped with those elements. Lang gives no concentrations for the A and B states in AlGaAs. However, the 180 meV state is reported at a concentration of 6 X 10*12 and the 480 meV at 2 X 10*16/cm*3.

An extrapolation of the characteristics of the A centre (HB5) seen in LPE agrees approximately with the activation energy and peak position of HM3 seen by Wu et al [13] in MOVPE. An iron related state (HB3) is commonly seen in MOVPE GaAs but no state with similar characteristics to the Fe state in LPE AlGaAs has been reported in MOVPE.

REFERENCES

[1] A.R.Peaker [EMIS Datareview RN=15602 (Jan 1986) "DX defect centres in AlGaAs"]

[2] R.H.Wallis [Inst. Phys. Conf. Series vol.56 (1981) p.73-82]

[3] N.M.Johnson, R.D.Burnham, D.Fehete, R.D.Yingling ["Defects in Semiconductors", Eds Narayam & Tan (North Holland, Amsterdam 1981) p.481]

[4] E.E.Wagner, D.E.Mars, C.Hom, G.B.Stringfellow [J. Appl. Phys. (USA) vol.51 no.10 (1980) p.5434-7]

[5] P.K.Bhattacharya, T.Matsumoto, S.Subramanian [J. Cryst. Growth (Netherlands) vol.68 no.1 (1984) p.301-4]

[6] K.Hikosaka, T.Mimura, S.Hiyamizu ["Gallium Arsenide and Related Compounds", 1981, Inst. Phys. Conf. Ser. vol.63 p.233-238]

[7] T.Matsumoto, P.K.Bhattacharya, M.J.Ludowise [Appl. Phys. Lett. (USA) vol.41 no.7 (1982) p.662-4]

[8] H.C.Gatos, J.Lagowski [Mat. Res. Soc. Symp. vol.46 (1985) p.153-67]

[9] H.C.Casey, A.Y.Cho, D.V.Lang, E.H.Bicollian, P.W.Foy [J. Appl. Phys. (USA) vol.50 (1979) p.3484]

[10] M.Sakamoto, T.Okada, Y.Mosi [J. Appl. Phys. (USA) vol.58 no.1 (1985) p.337-40]

[11] J.P.Andre, M.Boulou, A.Mircea-Roussel [J. Cryst. Growth (Netherlands) vol.55 (1981) p.192]

[12] M.J.Tsai, M.M.Tashima, B.L.Twu, R.L.Moon [Inst. Phys. Conf. Series vol.65 (1982) p.85]

[13] S.Nantsuka, K.Yamanaka, M.M.Lara, M.Ishii [Jpn. J. Appl. Phys. (Japan) vol.23 (1984) p.L112-4]

[14] K.Yamanaka, S.Nanitsuka, M.Mannoh, S.T.Yussa, Y.Nomura, M.Mihara, M.Ishii [J. Vac. Sci. & Technol. B (USA) vol.2 no.2 (1984) p.229-32]

[15] P.K.Bhattacharya, S.Subramanian, M.J.Ludowise [J. Appl. Phys. (USA) vol.55 no.10 (1984) p.3664-8]

[16] R.T.Chen, N.H.Sheng, D.L.Miller [J. Vac. Sci. & Technol. B (USA) vol.3 no.2 (1985) p.652]

[17] P.M.Mooney, R.Fischer, H.Markoc [J. Appl. Phys. (USA) vol.57 no.6 (1985) p.1928-31]

[18] D.V.Lang, R.A.Logan, L.C.Kimmerling [Proc. 13th Int. Conf. on Semiconductors, Rome,1976, Ed F.C.Fumi (North Holland, Amsterdam) p.615-8]

[19] D.V.Lang, R.A.Logan, L.C.Kimmerling [Phys. Rev. B (USA) vol.15 no.10 p.4874-82]

[20] A.F.C.Kravchenko, V.Ya Prints [Sov. Phys. Semicond. (USA) vol.12 no.8 (1978) p.p.952-3]

[21] S.Loualiche, G.Cuillot, A.Nouailhat, J.Bourgoin [Phys. Rev. B (USA) vol.26 no.12 p.7090-2]

[22] S.S.Li, C.Y.Lin, S.M.Bedasir, J.A.Hutchly [J. Electron. Mat. (USA) vol.11 no.2 (1982) p.273-87]

[23] T.Uji, T.Suzuki, T.Kamejima [Appl. Phys. Lett. (USA) vol.36 no.8 (1980) p.655-7]

[24] S.Subramanian, P.K.Bhattacharya, M.J.Ludowise [Solid State Electron. (GB) vol.27 no.6 (1984) p.537-41]

[25] A.Mitonneau, G.M.Martin, A.Mircea [Electron. Lett. (GB) vol.13 no.22 (1977) p.666-7]

18.2 DX DEFECT CENTRES IN AlGaAs

by A.R.Peaker

November 1985
EMIS Datareview RN=15602

In Al(x)Ga(1-x)As with x greater than 0.22 a significant concentration of the Group IV and Group VI dopants appear as deep donors. This behaviour was first reported by Springthorpe [1] in Te doped LPE layers but is now known to occur with S, Te, Se, Si, Ge and Sn impurities in LPE [2], MOVPE [3] and MBE [4] material. These deep donors exhibit a phenomenon known as persistent photoconductivity. Essentially, when the sample is cooled down the deep donors retain their charge and quite predictably do not contribute to the conductivity below about 100 deg K. However, on photoexcitation they release an electron. The probability of recapture is then small and the electron remains free to contribute to the conductivity. At temperatures below 80 deg K this effect is very pronounced; the photoconductivity remaining for hours or even days. This is significant because for aluminium content where x is greater than 0.27 the deep donors are present in a higher concentration than the shallow donors [5]. This has dramatic device implications. It has been shown quite definitively that the DX centre is responsible for pinch off voltage shift in modulation doped FETs [6] and is probably responsible for self pulsation of lasers [7].

For compositions with small aluminium content (where x is less than 0.2) the donors are shallow as predicted from a simple hydrogenic model (about 6 meV). For higher aluminium content the activation energy of the donors as measured by the Hall technique is greater (50-150 meV). This is because an increasing proportion of the dopant occurs as a DX centre. However, DLTS measurements yield a much larger activation energy, 430 meV in the case of silicon in Al(0.4)Ga(0.6)As [2]. This arises because the DLTS technique gives an apparent activation energy which include the activation energy of the capture cross section. The cross section activation energy has been measured to be 330 meV for the silicon DX centre. The difference corresponds to the Hall (or so-called equilibrium) energy. The threshold photon energy required in the same material to free an electron which then contributes to the persistent photoconductivity is 1 eV [2]. This was interpreted by Lang and Logan [2] as a Stokes shift resulting from a massive lattice relaxation [8]. This is the basis of the Lang model for the DX centre which postulates the state as consisting of a donor complexed with X where X is possibly an arsenic vacancy [9]. There are many observations which support the model but alternative explanations of the persistent photoconductivity have been suggested. None of the models explain all the observed phenomena.

A selection of the measured parameters are listed below but it is necessary to add some cautionary notes. The properties of the DX centres have been investigated in most detail in the composition range where x varies from 0.2 to 0.4. In this region the concentration of the centre varies from zero to contributing most, if not all, of the donor concentration. It is also the region in which the band structure of GaAlAs changes dramatically [10] the GAMMA, X and L minima crossing as the composition changes and being sufficiently close in energy for each to have a siginficant electron occupancy. In DLTS the usual analysis of data assumes that the deep states have a low concentration compared to the shallow donors; applying this analysis in the DX case results in large errors (possibly a factor of 5) in concentration determinations and cross section measurements. If the material is highly doped additional errors appear in the activation energy determination. In the Hall measurement errors of up to a factor of 2 are possible in activation energy and somewhat less in concentrations using a simple analysis. In both the DLTS and Hall measurements a more comprehensive treatment is possible. In the published DX studies few authors have stated explicitly whether the simplified analysis or a more detailed treatment has been used.

Schubert and Ploog [5] have studied GaAlAs for a range of x from 0 to 0.4 in Si doped MBE material using Hall measurements analysed by a comprehensive Fermi-Dirac method. They interpret the system as a very shallow donor interacting with the GAMMA minima and a deep donor of 140 (+-10) meV related to the X minima. The energies are independent of aluminium composition but the concentration of the deep centre changes dramatically. At the doping levels they used (greater than 10*17/cm*3), for x less than 0.2, all the donors were shallow; at x = 0.27 there were equal concentrations of deep and shallow donors, while at x = 0.4 all the donors were deep. A similar dependence of concentration of the DX centre has been observed by Tachikawa et al [11]. Chand et al [12] have studied MBE Si doped AlGaAs over the entire compositional range (x = 0 to 1). Again, they identify a shallow donor related to the GAMMA minima but regard the deep level as tied to the L band, the energy difference being around 160 meV. The measured value of the ionisation energy for moderate doping in the range of x from 0.22 to 0.40 can be expressed as:

$$Ed = 707x - 146 \text{ meV}$$

Chand [12] has also studied the persistence photoconductivity effect as a function of atomic Si density. For a given composition the persistent carrier density (PCD) is proportional to silicon concentration. As x is changed the PCD increases up to x = 0.32 then falls rapidly. For x greater than 0.43 the PCD carrier density is less than 5% of the atomic silicon density.

The most comprehensive Hall study has been undertaken by Lifshitz et al [13]. They examined Sn doped LPE GaAlAs over the compositional range x = 0.15 to 0.7 and modulated the position of the GAMMA, L and X minima with hydrostatic pressure. The results at atmospheric pressure show a change in apparent ionisation energy from 6 to 101 meV over the compositional range with a peak at 130 meV at a composition near x = 0.45. This behaviour is explained as being due to a 6 meV level tied to the GAMMA minima, a 150 meV level attached to the L minima and a 101 meV level tied to the X minima, a view supported by the high pressure measurement. In addition to the compositional effects a dependence of activation energy on doping level is seen [14], the energy increasing at lower doping levels.

DLTS measurements show that the apparent activation energy has a clear dependence on the donor species.

TABLE 1: DLTS ACTIVATION ENERGIES (meV) FOR DX CENTRE ELECTRONS

x	Growth	S	Se	Te	Si	Ge	Sn	Ref
0.3	LPE						190 and 320	[11]
0.32	MOVPE				430	330	210	[15]
0.4	LPE		280	280	430		190	[2]
0.43	MOVPE	280	280	280				[15]
0.5	LPE		290	320	560		360	[11]
0.7	LPE						240 and 270	[11]

Capture cross sections of certain levels have been shown to activate with temperature. At x = 0.4 the activation energy is 180 meV for Se and Te; for Si it is 330 meV [2].

The photoionisation energies also depend strongly on donor species. The threshold for excitation with x = 0.4 at 50 deg K is 600 meV for Te, 800 meV for Sn and 1000 meV for Si [2]. The tellurium photoionisation spectrum has been investigated in detail [9]. The threshold energy is not very temperature dependent and the spectra are indistinguisable over the aluminium range x = 0.37 to 0.6. However, a marked shift to lower threshold energies becomes apparent as x changes from 0.37 to 0.27.

MBE GaAlAs is doped almost exclusively with silicon although some work has used tin doping. The DLTS spectra of the DX centre in GaAlAs:Si grown by MBE appears to be slightly different to LPE and MOCVD in that two distinct but similar peaks are present [3], both peaks exhibit the characteristics associated with DX centres. The energies attributed to these peaks in the literature vary but as the peaks overlap deconvolution is necessary and this rather than any physical difference may well be the source of the discrepancies. However, a few authors report only a single peak (as in LPE) the double peak being seen only in MBE material in the range of x from 0.25 to 0.3.

TABLE 2: DX ACTIVATION ENERGIES (meV) IN Si DOPED MBE

x	lower temp peak	higher temp peak	ref
0.26	420	400	[16]
0.3	400	470	[17]
0.3	440	570	[3]
0.34	440	-	[18]
0.4	-	400	[16,19]

Zhan et al [3] measured the electron capture cross section of the states associated with the two peaks of x = 0.3 material and found them to have identical values of 8.3 X 10*-22 cm*2 at 205 deg K but the cross sections have slightly different activation energies; 330 meV for the lower temperature peak and 370 meV for the other. Zhan has made appropriate corrections for the high trap concentrations.

A number of measurements of the DX centre have also been done using modulation doped FETs [6,17,20,21] and quantum well structures [22]. This work is not reviewed here despite its immediate relevance to devices as a number of additional complications result from the mode of charge separation in the heterostructure making interpretation of the deep state properties even more problematic. In particular Mooney et al [23] have pointed out that Debye tail effects are very large in such cases, introducing particular difficulties for DLTS and capture measurements.

Logan´s model for the centre was of a simple substitutional donor paired with another defect [8]. This model is quite consistent with most of the observed features. There is no doubt that the intentionally added Group IV or Group VI impurity is involved, the concentration dependence and variation with donor species make this indisputable; what is uncertain is the nature (or even existence) of the defect it is complexed with. As the state and its concentraiton dependence is essentially the same in LPE and MOVPE (and very similar in MBE) the possibility of X being a chemical impurity can be eliminated. Lang et al [9] proposed that the X defect was an arsenic vacancy. This is consistent with symmetry studies using ballistic phonons [24] in which it is observed that the concentration of the DX centre decreased near climb induced dislocation networks but not near glide induced networks [25]. Very importantly such a structure would be expected to produce a state with a shallow binding energy (100 meV) but with a large (750 meV) Franck-Condon energy (Stokes shift). Lang et al [9] have fitted the experimental data for the tellurium DX centre to such a model and found it totally consistent.

However, after Lang´s publication detailed results on MOVPE and MBE material became available and these cast some doubt on the arsenic vacancy assignment for X. This is essentially because one might reasonably expect a much smaller arsenic vacancy concentration in an arsenic rich growth environment than in gallium rich LPE. The complexing behaviour is also somewhat bizarre. The results imply that at x less than 0.2 all the donors are isolated while for x greater than 0.4 virtually all are complexes. Van Vechten [26] dismisses such hand waving thermodynamics because it does not take account of Fermi level effects which he considers may be dominant.

However, Saxena [27] made a very important observation which seems to have been overlooked in the present controversy. He noted that at low temperatures (less than 100 deg K) the Hall mobility increased when the DX centre was photoexcited. At 20 deg K the increase was a factor of two. This appears to conflict with the DX model as at these temperatures ionised impurities are a major source of scattering. Even with large lattice relaxation it seems highly unlikely that an increase in the number of ionised centres could actually result in an increase in mobility.

Lifshitz et al [13] in their Hall effect studies subjected Al(0.15)Ga(0.85)As to hydrostatic pressure and observed a tellurium donor ionisation behaviour similar to the DX centre. The DX centre is not seen in this composition at low pressures. This work was extended using DLTS by Mizuta [27] to GaAs:Si and GaAs:Sn. The application of hydrostatic pressure to GaAs on low aluminium content alloys shows a configuration of GAMMA, L and X minima very similar to the higher aluminium content AlGaAs. This is particularly important for the GaAs case as it allows a simulation of the band structure Al(0.3)Ga(0.7)As without the aluminium. At 30 kbar a behaviour of the silicon donor in GaAs similar to that of the DX centre was observed. It would be a very remarkable coincidence for the defect reactions in GaAs to be identical to those in GaAlAs at exactly the pressure where the band structures were the same. This seems to confirm earlier suggestions that the DX centre behaviour is a function of the band structure rather than the defect itself.

It has therefore been suggested that the DX centre is a simple substitutional donor [5,27] and that the persistent photoconductivity is due to interaction of the deep donor with the conduction band X minima. This does not explain the large Stokes shift as even in the extremely unlikely event of this occurring for a simple substitutional donor it would not occur on both the Group III and Group V sublattices. This is a prerequisite for any model as similar results are observed for both Group IV and Group VI donors.

This discussion of the physical nature of the defect may seem somewhat esoteric as the phenomenological effects are extensively documented. However, the DX centre concentration is so large as to dominate device behaviour for x greater than 0.25. In AlGaAs quantum wells and some heterojunction devices the way in which the DX defect interacts with the band structure will be a central issue in the device physics and there is no doubt that a considerable amount of work is still needed to resolve this problem.

REFERENCES

[1] A.J.Springthorpe, F.D.King, A.Becke [J. Electron. Mat. (USA) vol.4 (1975) p.101]

[2] D.V.Lang, R.A.Logan [Inst. Phys. Conf. Ser. vol.43 (1979) p.433-6]

[3] B.L.Zhan, K.Ploog, E.Gomelin, X.Q.Zheng, M.Schulz [Appl. Phys. A (Germany) vol.28 (1982) p.223-7]

[4] J.J.Yang, C.A.Mandy, W.I.Simpson [Appl. Phys. Lett. (USA) vol.40 no.3 (1982) p.244-6]

[5] E.F.Schubert, K.Ploog [Phys. Rev. B (USA) vol.30 no.12 (1984) p.7021-9]

[6] P.M.Mooney, P.M.Solomon, T.N.Theis ["Gallium Arsenide & Related Compounds", 1984, Inst. Phys. Conf. Ser. vol.74 p.617-22]

[7] J.L.Merz, J.P.Van der Zeil, R.A.Logan [Phys. Rev. B (USA) vol.20 (1979) p.654]

[8] D.V.Lang, R.A.Logan [Phys. Rev. Lett. (USA) vol.39 no.10 (1977) p.635-39]

[9] D.V.Lang, R.A.Logan, M.Janos [Phys. Rev. B (USA) vol.19 no.2 (1979) p.1015-30]

[10] H.J.Lee, L.Y.Juravel, J.C.Wooley, A.J.Springthorpe [Phys. Rev. B (USA) vol.21 no.2 (1980) p.659-69]

[11] M.Tachikawa, M.Mizuta, H.Kukimoto [Jpn. J. Appl. Phys.(Japan) vol.23 no.12 (1984) p.1594-7]

[12] N.Chand, T.Henderson, J.Klem, W.T.Masselink, R.Fischer, Yia-Chung Chang, H.Movkoc [Phys. Rev. B (USA) vol.50 no.8 (1984) p.4481-92]

[13] N.Lifshitz, A.Jayaraman, R.A.Logan, H.C.Card [Phys. Rev. B. (USA) vol.21 no.2 (1980) p.670-8]

[14] H.Kunzel, K.Ploog, K.Wunstel, B.L.Zhon [J. Electron. Mat. (USA) vol.13 (1984) p.281]

[15] O.Kumagai, H.Kawai, Y.Mori, K.Kaneko [Appl. Phys. Lett. (USA) vol.45 no.12 (1984) p.1322-3]

[16] K.Hikosaka, T.Mimura, S.Hiyamizu [Proc. of Gallium Arsenide Conf. Osio 1981 Inst. Phys. Conf. Ser. 1982 p.233]

[17] S.Subramanian, A.Schuller, T.R.Arthur [J. Appl. Phys. (USA) vol.58 no.2 (1985) p.845-51]

[18] M.O.Watanabe, K.Morizuka, M.Mashita, Y.Ashizawa, Y.Zahta [Jpn. J. Appl. Phys. (Japan) vol.23 no.2 (1984) p.L103-5]

[19] K.Yamanaka, S.Naritsuka, M.Mannoh, T.Yuasa, Y.Namura, M.Mihara, M.Ishii [J. Vac. Sci. & Technol. B (USA) vol.2 no.2 (1984) p.229-32]

[20] D.M.Collins, D.E.Mars, B.Fischer, C.Kocot [J. Appl. Phys. (USA) vol.54 no.2 (1983) p.857-61]

[21] M.Takikawa, M.Uzeki [Jpn. J. Appl. Phys. (Japan) vol.24 no.3 (1985) p.303-10]

[22] T.C.L.G.Sollner, H.Q.Le, C.A.Conea, W.D.Goodhue [Appl. Phys. Lett. (USA) vol.47 no.1 (1985) p.36-8]

[23] P.M.Mooney, N.S.Caswell, P.M.Solomon, S.L.Wright [Proc. Mat. Res. Symp., USA, vol.46 (1985) p.403-8]

[24] V.Naranyanamurti, R.A.Logan, M.A.Chin [Phys. Rev. Lett. (USA) vol.43 (1979) p.1536]

[25] D.V.Lang, P.M.Petroff, R.A.Logan, W.D.Johnston [Phys. Rev. Lett. (USA) vol.42 no.20 (1979) p.1353-6]

[26] J.A.Van Vechten [Proc. Mat. Res. Symp. vol.46 (1985) p.83-103]

[27] M.Mizuta, M.Tachikawa, H.Kukimoto, S.Minomura [Jpn. J. Appl. Phys. (Japan) vol.24 no.2 (1985) p.L143-6]

[28] A.K.Saxena [Solid State Electron. (GB) vol.25 no.2 (1982) p.127-31]

18.3 PHOTOLUMINESCENCE SPECTRA OF AlGaAs

by EMIS Group

March 1985
EMIS Datareview RN=15048

Photoluminescence (PL) spectra of Al(x)Ga(1-x)As are used in the measurement and diagnosis of Al content, carrier concentrations, trap densities and impurities.

A calibration curve giving peak photon energy in terms of x at room temperature is obtained in [1]:

Ep = 1.42 + 1.45 x - 0.25 x*2 eV

This gives a least squares fit with a statistical error of +- 1.6% for x up to 0.45. The samples were LPE grown and x was measured by electron microprobe using an Al-Cu standard. All samples had a free electron concentration below 2 X 10*17/cm*3.

Ref [20] reports PL bound exciton peaks vs x at 75 deg K for LPE samples Si doped to an atomic fraction of about 3 X 10*-5:

x	0	0.07	0.14	0.24	0.32	0.40
peak (eV)	1.513	1.585	1.683	1.779	1.910	1.999

PL half width (PLHW) gives a rough indication of carrier concentration in multilayered (non-quantum) structures, since for x below 0.35 PLHW is largely determined by carrier concentration and is almost independent of x. Some typical data for high quality LPE and VPE samples at room temperature [2]:

Carrier Concn (/cm*3)	Approx PLHW (meV)
n = 10*17	37
n = 10*18	60
n = 10*19	200
p = 10*17	43
p = 10*18	60
p = 10*19	85

The lower rate of PLHW increase with carrier concentration for p-type samples arises because of the larger effective mass of holes.

Ref [3] examines deep traps in high quality MOCVD AlGaAs for x up to 0.4. Total trap density Nt is tentativley related to the edge luminescence Ipl (arb. units) by Ipl = 2.96 (Nt/n)*-0.53. PL and other methods reveal 5 dominant electron traps with thermal ionisation energies (Et) from 0.25 to 0.82 eV and hole traps with Et´s of 0.74 and 0.87 eV. Four of the traps appear to be due to Al-O bonds forming part of the microstructure and are absent when the crystal growth ambient is kept relatively free of O2 and H2O. See also [4] for a PL investigation of O2 and H2O contamination in MOCVD AlGaAs.

O2 and H2O effects on PL spectra of MOCVD AlGaAs doped with Se to 1.5 X 10*18/cm*3 and deposited at 700 deg C are given in [5]. Some typical data are

Wavelength nm	Approx. PL intensity (arbitrary units)	Contamination
640	near zero	no O2 or H2O
693	1 (peak)	
755	near zero	
670	zero	0.1 ppm O2
690	near zero (peak)	
715	zero	
650	near zero	12 ppm H2O
720	0.6 (peak)	
770	near zero	

Emission peak energies in MOCVD AlGaAs:Zn (p = 10*18/cm*3) increase monotonically with temperature and nearly linearly with the vapour phase V/III ratio [6].

Various Ep data are summarised in [6]:

Dopant and Growth Method	Growth Temp. deg C	x	Ep (eV)	Ref
Zn MOCVD	700	0-0.46	1.65	[7]
Zn LPE	775-800	0-0.75	1.6-1.7	[8]
Zn MOCVD	800	0.15-0.62	1.5-1.7	[6]
Ge LPE	785	0.40	1.55	[9]
Ge LPE	790-800	0.40	1.55	[10]
Ge LPE	855	0-0.041	1.4-1.6	[11]

[12] investigates the effect of changes in ME3 and ME6 trap concentrations due to growth and alloy composition changes for MBE AlGaAs:Si. ME3 does not affect PL intensity but ME6 reduces the near band-edge PL intensity at room temperature by a factor of 100 as the ME6 density rises from 10*14 to 10*16/cm*3 (x up to 0.38).

OMVPE AlGaAs shows a strong dependence of PL spectra on substrate for given growth parameters (700 deg C, 76 Torr, V/III = 40, growth rate = 600 A/min) [13]. Data on Al(0..43)Ga(0.57)As (unintentionally doped, Zn doped and Si doped) deposited on various substrates (LEC undoped, n+ Bridgeman, LPE GaAs:Si and LPE AlGaAs:Si) indicate Si redistribution into the OMVPE layers is the dominant effect. High quality surface morphologies are obtained on films deposited on AlGaAs substrates exposed to air when a brief in situ vapour etch is used prior to growth.

[14] observes a low temperature (77 deg K) PL peak 100 meV below the band-to-band transition peak for MOCVD n-AlGaAs but only at growth temperatures below 740 deg C.

An MOCVD solar cell structure is examined in [15]. Strong PL intensity is found for x in the 0.28-0.35 range with degradation above x = 0.38 attributed to deep level densities as well as direct/indirect band edge crossover effects.

Other works of interest include: a quantitative model for the linewidth of the excitonic and band-to-acceptor recombinations in MBE AlGaAs [16]; luminescence line shape analysis in an e-h plasma [17]; PL of Mg and Be implanted AlGaAs [18]; and laser induced luminesence of Al(0.22)Ga(0.78)A:Ge [19].

REFERENCES

[1] N.C.Miller, S.Zemon, G.P.Wheeler, W.Powazinik [J. Appl. Phys. (USA) vol.57 no.2 (15 Jan 1985) p.512-5]

[2] G.B.Stringfellow, H.T.Hall Jr [J. Electron. Mater. (USA) vol.8 no.3 (1979) p.201-26]

[3] P.K.Bhattacharya, T.Matsumoto, S.Subramanian [J. Cryst. Growth (Netherlands) vol.68 no.1 (Sept I 1984) p.301-4]

[4] P.K.Bhattacharya, S.Subramanina, M.J.Ludowise [J. Appl. Phys. (USA) vol.55 no.10 (15 May 1984) p.3664-8]

[5] H.Terao, H.Sunakawa [J. Cryst. Growth (Netherlands) vol.68 no.1 (Sept I 1984) p.157-62]

[6] M.Sakamoto, T.Okada, Y.Mori, K.Kaneko [J. Appl. Phys. (USA) vol.55 no.10 (15 May 1984) p.3613-6]

[7] Z.Y.Xu, V.G.Kreismains, C.L.Tang [J. Appl. Phys. (USA) vol.54 (1983) p.4536]

[8] K.Sugiyama, T.Kawakami [Jpn. J. Appl. Phys. (Japan) vol.10 (1971) p.1007]

[9] V.Swaminathan, N.E.Scrumaker, J.L.Zilko, W.R.Wagner, C.A.Parsons [J. Appl. Phys. (USA) vol.52 (1981) p.412]

[10] M.Ishi, A.Ito [Jpn. J. Appl. Phys. (Japan) vol.17 (1978) p.515]

[11] K.Kaneko, M.Ayabe [J. Appl. Phys. vol.51 (1981) p.6337]

[12] K.Yamanaka et al [J. Vac. Sci. & Technol. (USA) vol.B2 no.2 (Apr-June 1984) p.229-32]

[13] J.R.Shealy, C.F.Schaus, L.F.Eastman [J. Cryst. Growth (Netherlands) vol.68 no.1 (Sept I 1984) p.305-10]

[14] M.Sakamoto, T.Okada, H.Kawai, Y.Mori, K.Kaneko [Jpn. J. Appl. Phys. (Japan) vol.23 no.7 (July 1984) p.L461-3]

[15] M.J.Ludowise, W.T.Dietze [J. Appl. Phys. vol.55 no.12 (15 June 1984) p.4318-21]

[16] E.F.Schubert, E.O.Gobel, Y.Horikoshi, K.Ploog, H.J.Queisser [Phys. Rev. B (USA) vol.30 no.2 (15 July 1984) p.813-20]

[17] A.Selloni, S.Modesti, M.Capizzi [Phys. Rev. B (USA) vol.30 no.2 (15 July 1984) p.821-31]

[18] M.Yokota, S.Ashigaki, Y.Makita, T.Kanayama, H.Tanoue, I.Takayasu [Nucl. Instrum. & Methods (Netherlands) vol.209/210 (1983) p.711-8]

[19] R.P.Salathe, H.H.Gilgen [Mater. Res. Soc. Symp. Proc (USA) vol.17 (1981)]

[20] V.Swaminathan, M.D.Sturge, J.L.Zilko [J. Appl. Phys. (USA) vol.52 no.10 (Oct 1981) p.6306-11]

18.4 ELECTRON MOBILITIES IN AlGaAs

by EMIS Group

April 1985
EMIS Datareview RN=15056

The mobility of electrons (ME) in Al(x)Ga(1-x)As is affected by temperature (T), x, doping, deep electron traps and crystal growth parameters, not all of which are under the control of the crystal grower. The highest mobilities are observed for x below 0.35, where transport in the GAMMA valley predominates. The state of the art precludes general conclusions on the variation of ME with the various factors and this Datareview confines itself to a brief illustrative survey of some recent data.

Bhattacharya et al [1] carried out a detailed Hall and high field study of metalorganic CVD (MOCVD) AlGaAs for x up to 0.6 and for T in the range 20-600 deg K. They conclude that at room temperature space charge scattering is the main influence on ME for x up to 0.3, while intervalley scattering is dominant for x in the 0.3-0.5 range. There is a minimum at x = 0.45 due to a combination of intervalley, space charge and alloy scattering. For x above 0.6 electrons travel in the X valley where their effective mass is high and mobility low.

Data on the variation of Hall ME with x in MBE AlGaAs:Si were obtained by Salmon et al [2]. The substrate was (100) oriented and held at a substrate temperature (Ts) of 650 deg C. The doping level was 10*18/cm*3. The following data were obtained at room temperature.

x	Approx. ME cm*2/V.s
0.2	950
0.3	250
0.4	220
0.6	15
0.8	15

The variation with T was investigated for a free electron concentration of less than 1 X 10*13 / cm*3 at 50 deg K. ME increased with T at constant slope between 4 and 200 deg K. E.g. for x = 0.33

T deg K	Approx. ME cm*2/V.s
50	6
100	62
300	250

The fact that ME increased with T indicates that no 2 dimensional electron gas (2DEG) was formed, since Collins et al [10] report that the mobility of electrons in a 2DEG falls by an order of magnitude as T rises from 4 to 200 deg K.Salmon et al prevented a 2DEG by isolating the measured layer from the GaAs/AlGAs interface with an unintentionally doped AlGaAs layer 0.2 microns thick.

Kunzel et al [8] also studied MBE AlGaAs:Si on a (100) substrate. The effect of deep trap levels and Ts on ME is shown in the following table:

Ts deg C	net donor concn(/cm*3) at 77 deg K	deep trap to net donor concn ratio	thermal activation energy (eV) deep trap	shallow level	Hall ME (300 K) cm*2/V.s
675	9.0 X 10*15	42	0.410	0.018	250
640	2.3 X 10*15	80	0.400	0.047	90
620	2.5 X 10*14	580	0.372	0.075	30
600	-	-	0.342	-	SI

The V/III beam flux ratio is also important in achieving high mobility MBE layers. Nomura et al [3] report an optimum V/III ratio of 2.3 at Ts=780 deg C in growing AlGaAs:Si on a (001) substrate:

x	Ts deg C	V/III	n /cm*3	Hall ME at room temp. cm*2/V.s
0.22	750	1.1	1.4 X 10*15	3870
0.21	750	1.4	1.4 X 10*15	1820
0.19	750	1.9	highly resistive	
0.22	780	2.3	8.8 X 10*15	3520
0.23	780	2.8	9.4 X 10*15	3030
0.21	780	3.8	10 X 10*15	3010

Stringfellow [4] experimented on AlGaAs produced by various LPE and VPE techniques. The samples had similar ME vs T relationships with maxima at around 150-200 deg K or lower. E.g. for x = 0.13 LPE samples having a free electron concn (n) of 3.5 X 10*15/cm*3 have a maximum ME of about 3000 cm*2/V.s at about 150 deg K, rising from 2300 cm*2/V.s at 75 deg K and falling back to similar level at 300 deg K. The mobility values decreased as x increased between 0.03 and 0.35. Ref [4] also shows clearly that ME varies very little with n = 10*15 to 10*19/cm*3, either at 77 deg K or at room temperature.

In certain situations a 2DEG can form leading to orders of magnitude increases in ME. Matsumoto et al [5], for example, obtained room temperature mobilities of nearly 3000 cm*2/V.s in an organometallic VPE undoped n-Al(0.4)Ga(0.6)As layer 5-10 microns thick on a SI (100) GaAs:Cr substrate. When the substrate is removed ME drops to about 230 cm*2/V.s. The anomalously high ME arises because a 2DEG is formed by electron transfer from the epitaxial layer to the substrate. This effect is not observed for comparable LPE layers, probably because the interface is more diffused due to substrate dissolution when brought into contact with the melt.

ME can be resolved into various components, one of which is proportional to 1 / SQRT T in highly compensated AlGaAs. Stringfellow et al [6,7] deduce that this component is due to scattering by the C acceptor in unintentionally doped samples with C as the major shallow acceptor.

Finally, for a theoretical analysis of electron scattering effect on mobility see ref [9]; and for a review of electrical transport properties of AlGaAs see ref [11].

REFERENCES

[1] P.K.Bhattacharya, U.Das, M.J.Ludowise [Phys. Rev. B (USA) vol.29 no.12 (15 June 1984) p.6623-31]

[2] L.G.Salmon, I.J.D´Haenens [J. Vac. Sci. & Technol. (USA) vol.B2 no.2 (Apr-June 1984) p.197-200]

[3] Y.Nomura et al [J. Electrochem. Soc. (USA) vol.131 no.11 (Nov 1984) p.2630-3]

[4] G.B.Stringfellow [J. Appl. Phys. (USA) vol.50 no.6 (June 1979) p.4178-83]

[5] T.Matsumoto, P.K.Bhattacharya, J.Darmawan, M.J.Ludowise [Appl Phys. Lett. (USA) vol.41 no.11 (1 Dec 1982) p.1075-7]

[6] G.B.Stringfellow, H.Kunzel [J. Appl. Phys. (USA) vol.51 no.6 (June 1980) p.3254-61]

[7] G.B.Stringfellow [Appl. Phys. Lett. (USA) vol.36 no.7 (1 Apr 1980) p.540-2]

[8] H.Kunzel, K.Ploog, K.Wunstel, B.L Zhou [J. Electron. Mater. (USA) vol.13 no.2 (March 1984) p.281-308]

[9] D.L.Rode, P.A.Fedders [J. Appl. Phys. (USA) vol.54 no.11 (Nov 1983) p.6425-31]

[10] P.M.Collins, D.E.Mars, B.Fischer, C.Kocot [J. Appl. Phys. (USA) vol.54 (1983) p.857]

[11] A.K.Saxena [J. Inst. Electron. & Telecommun. Eng. (India) vol.29 no.3 (1983) p.97-107]

18.5 CARRIER CONCENTRATIONS IN LPE AlGaAs

by D.Lee

January 1985
EMIS Datareview RN=15609

Most material has been grown at 700-850 deg C with cooling rates 0.2-0.5 deg C/minute under gas ambients of helium, nitrogen or hydrogen. Ga melts are generally baked extensively in a hydrogen atmosphere before growth to reduce contamination [1]. Exposure of Cr doped SI GaAs substrates to hydrogen at about 800 deg C usually forms a thin p-type layer with shallow acceptor concentration about 10*17 /cm*3 [1] which must be removed by solution etching prior to LPE growth [1,2].

The dominant background n-type impurity has been reported to be sulphur [3-5] (segregation coefficient greater than 100 [4]) the level of which can be reduced by a factor of 5 by the addition of 0.9 ppm of oxygen or 11% hydrogen to the gas ambient (He) [4,5].

Background n-type carrier densities between 5 X 10*14 and 5 X 10*15 /cm*3 have been reported [3,6,7] with compensation ratios (Nd+Na)/(Nd-Na) as low as 2 for Al fractions x = 0 - 0.2 [3]. From mobility fits Nd was found to increase with x (5 X 10*14 - 1.6 X 10*15 /cm*3) for growth runs from new melts whereas growth from once used melts gave Nd = (4 - 5) X 10*14 /cm*3 independent of x, suggesting some gettering of impurities [3]. A p-type background of low 10*16 /cm*3 has also been reported (at x = 0.4) [4].

The majority of doping studies have been on p-type material, with Ge being the most widely studied dopant [2,7-9]. Doping densities in the range mid 10*16 to mid 10*18 /cm*3 have been reported (with low compensation in the range 7 X 10*16 to 2 X 10*17 /cm*3) [7]. Ge solubility has been reported to decrease with x [8] but Springthorpe et al [2] suggest total Ge incorporation independent of composition. However Ea increases rapidly with x (up to 67 meV at x = 0.3 and greater than 100 meV at x = 0.6) which reduces free carrier density, particularly at high values of x [2,9].

Zn and Be have smaller Ea [10,11] (e.g. about 30 meV for Zn at x = 0.3 [11]) but these dopants diffuse readily into the GaAs substrate [11,12] to form a p-type layer which can dominate the electrical properties. Diffusion depths of 0.7 micron at 700 deg C and 8 microns at 900 deg C increasing with x have been reported for Be [12]. Hole concentrations of more than 10*18 /cm*3 have been reported for Be doped material with x up to 0.9 [10]. Doping in the range 10*17 to 10*18 /cm*3 (for x up to 0.65) has been reported using Mg, which has much smaller Ea (about 13 meV at x = 0.3). Whilst some diffusion was observed, particularly at high temperatures, the effect could be contained by use of undoped GaAlAs buffer layers [13]. Si is amphoteric in LPE, due to the Ga rich growth conditions, p-type material is obtained for T less than 800 deg C but is heavily self compensated [5].

N-type doping is reported using Se [14], Te [2,14,15] and Sn [6,8]. Carrier densities of 5 X 10*16 - 2 X 10*18 /cm*3 have been reported for Te [2,15] and 5 X 10*16 - 2.5 X 10*17 /cm*3 for Sn (0.7-5 mole % Sn, x = 0.24) [8]. Free carrier concentration is independent of x at low values of x but for x between about 0.2 and 0.4 the carrier concentration is observed to decrease dramatically and the observed donor ionization energy to increase to a maximum of about 150 meV [2]. At the same time a persistent increase in carrier density is observed on illumination at temperatures below about 100 deg K. Typical values for Te doped Ga(0.64)Al(0.36)As are 1.5 X 10*17 /cm*3 at 300 deg K, 1.7 X 10*16 /cm*3 at 60 deg K in the dark and 3.2 X 10*17 /cm*3 after illumination at 60 deg K [14]. Similar phenomena are observed for all donors in LPE and also in MOCVD and MBE grown material and are due to the effects of deep levels (DX centres [15,16]).

The carrier density is reported to increase linearly with liquid phase dopant concentration, for all n- and p-type dopants [2,8,13].

REFERENCES

[1] J.J.Hsieh [Handbook on Semiconductors vol.3 (S.P.Keller ed, North Holland 1980) p.415-97]

[2] A.J.Springthorpe, F.D.King, A.Becke [J. Electron. Mater. (USA) vol.4 no.1 (1975) p.101-18]

[3] A.Chandra, L.F.Eastman [J. Electrochem. Soc. (USA) vol.127 no.1 (1980) p.211-16]

[4] P.J.Anthony, J.L.Zilko, V.Swaminathan, N.E.Schumaker, W.R.Wagner, J.C.Norberg [Appl. Phys. Lett. (USA) vol.38 no.6 (1981) p.434-6]

[5] V.Swaminathan, M.D Sturge, J.L.Zilko, N.E.Schumaker, W.R.Wagner, C.A.Gaw [J. Electrochem. Soc. (USA) vol.129 no.7 (1982) p.1563-6]

[6] A.K.Saxena [Solid State Electron. (GB) vol.25 no.2 (1982) p.127-31]

[7] A.W.Nelson, P.N.Robson [J. Appl. Phys. (USA) vol.54 no.7 (1983) p.3965-72]

[8] K.K.Shih, G.D.Pettit [J. Electron. Mat. (USA) vol.3 no.2 (1974) p.391-408]

[9] Y.P.Yuan, K.Mohammed, J.L.Merz [J. Appl. Phys. (USA) vol.57 no.8 (1985) p.2896-9]

[10] S.Fujita, S.M.Bedair, M.A.Littlejohn, J.R.Hauser [J. Appl. Phys. (USA) vol.51 no.10 (1980) p.5438-44]

[11] K.Masu, M.Konagai, K.Takahashi [J. Appl. Phys. (USA) vol.51 no.2 (1980) p.1060-4]

[12] K.Masu, M.Konagai, K.Takahashi [J. Appl. Phys. (USA) vol.54 no.3 (1983) p.1574-8]

[13] S.Mukai, Y.Makita, S.Gonda [J. Appl. Phys. (USA) vol.50 no.3 (1979) p.1304-7]

[14] R.J.Nelson [Appl. Phys. Lett. (USA) vol.31 no.5 (1977) p.351-3]

[15] D.V.Lang, R.A.Logan, M.Jaros [Phys. Rev. B (USA) vol.19 no.2 (1979) p.1015-30]

[16] E.F.Schubert, K.Ploog [Phys. Rev. B (USA) vol.30 no.12 (1984) p.7021-9]

18.6 CARRIER CONCENTRATIONS IN MOCVD AlGaAs

by D.Lee

January 1986
EMIS Datareview RN=15610

In the majority of cases growth is reported from trimethyl gallium (TMG), trimethyl aluminium (TMA) and arsine (AsH3). The background carrier density is sensitive to V/III [the ratio of Group V (i.e. As) to Group III (i.e. Ga+Al) species in the reactor], the growth temperature and the purity of the reagents.

At low V/III undoped material is p-type. The hole density decreases with increasing V/III until at some critical value the material changes to n-type whereafter the electron density increases with V/III [1]. For V/III ratios around the critical value there can be an extensive range where semi-insulating material is obtained. The critical V/III ratio increases with Al fraction (x) and decreases with increasing growth temperature [2,3]. At high values of V/III the electron density is reported to increase linearly with As partial pressure and x, and at low V/III the hole density increases exponentially with x [3].

The dominant shallow acceptor impurity is generally carbon [e.g 4] and the background level has been reported to increase with increasing aluminium content [5]. Carbon acceptor concentrations as high as 10*18 cm*-3, decreasing with increasing V/III have been reported in GaAlAs grown from TMG and TMA. On growing from the triethyl reagents (TEG, TEA) at low pressure the carbon background was reduced to 10*16 cm*-3 independent of V/III [6].

MOCVD GaAlAs is extremely sensitive to trace quantities of oxygen or water in the reagents or reactor. Systematic studies have demonstrated that oxygen reduces the free electron density and luminescence efficiency whereas water has little direct effect on n but serves to reduce the aluminium content and growth rate of the layer [7]. It has been suggested that TMA reacts with oxygen to form volatile dimethylaluminium methoxide whereas with water yields aluminium hydroxide which deposits on the walls of the reactor [7].

Various techniques have been reported to reduce the effects of residual oxygen and water such as passing reagents through molecular sieve [8] or liquid InGaAl alloy bubblers [9], by the use of heated graphite baffles [8,9] or preheated reaction zones at the inlet of reactor [5] and elevated growth temperatures [8]. Residual oxygen and moisture in the reactor itself has been reduced by the growth of aluminium containing buffer/getter layers prior to the layer of interest [11]. Typical improvement due to oxygen/water gettering is an increase in free electron density of GaAlAs (grown at 800 deg C, V/III = 40) from 1 X 10*16 - 5 X 10*16 cm*-3 using a Ga/In/Al bubbler at room temperature, implying an oxygen related deep acceptor concentration of about 5 X 10*16 cm*-3 [9].

It has been reported that when growth is optimized to reduce oxygen and water, the electrical quality is improved as the growth temperature is reduced to 650 deg C [5]. However, in general, optimum material is produced at high growth temperatures (750 - 800 deg C) and high V/III ratios (relative to GaAs). Reported background carrier densities vary widely (typically around 10*16 cm*-3) but good quality undoped material with n down to low 10*15 cm*-3 has been reported [9,12].

N-type GaAlAs has been grown by doping with Si (using silane [13,14] or disilane [15]) and Se (using H2Se [1,2,16,17]). Doping ranges were about 10*16 cm*-3 to low or mid 10*18 cm*-3 in each case. The doping efficiency of silane has been reported to be about a factor of 10 greater than that of H2Se [14]. Typical dopant mole fractions are 5 X 10*-8 - 5 X 10*-7 cm*-3 for H2Se. In both cases the doping efficiency decreases with increasing temperature [14,15]. Kuech et al [15] claim higher efficiency and doping density independent of growth temperature using disilane. In general the measured doping density increased linearly with dopant mole fraction. However Bottka at al [13] observed a square law dependence on silane mole fraction and similar deviation from a linear dependence was observed at low temperature with small mole fractions of H2Se [14,17].

The room temperature carrier concentration derived from C-V measurements on Si doped GaAlAs is reported to be nearly independent of alloy composition and in close agreement with the Si concentration from SIMS [15]. However Hall measurements on both Si and Se doped material give carrier concentrations strongly dependent upon x [15,18,19]. For low aluminium fractions the Hall carrier concentration is nearly independent of x and in good agreement with C-V measurements [15], for but x greater than 0.25 the carrier concentration decreases sharply to a minimum and the observed donor ionization energy increases to a maximum at about x = 0.47. At the same time a persistent increase in carrier concentration is observed on illumination at low temperatures [15]. Similar phenomena are observed for all donors in LPE and MBE grown material and are due to the effects of deep levels (DX centres [20,21]) (further information is available in another EMIS datareview [22]).

P-type doping has been reported using diethyl Zn (DEZ) [1,16,17,23] and Be (DEB) [13]. Doping densities in the range low 10*16 cm*-3 - to high 10*18 cm*-3 have been achieved for DEZ mole fractions 10*-6 - 2 X 10*-4 cm*-3. The doping efficiency decreases with increasing x [1] but hole densities upto 10*18 cm*-3 have been reported for x = 0.8 [23]. The doping efficiency of DEB is much higher, similar doping densities being achieved for DEB mole fractions of 3 X 10*-9 - 4 X 10*-8 cm*-3 [13]. Doping density generally increases approximately linearly with DEZ mole fraction but Bottka et al [13] report a square law dependence for DEB.

Oxygen and hexacarbonyl Cr have both been used to produce semi-insulating GaAlAs [24].

REFERENCES
----

[1] T.Nakanisi [J. Cryst. Growth (Netherlands) vol.68 no.1 (1984) p.282-94]

[2] J.P.Andre, M.Boulou, A.Mircrea-Rousse [J. Cryst. Growth (Netherlands) vol.55 (1981) p.192-7]

[3] Y.Mori, M.Ikeda, H.Sato, K.Kanoko, N.Watanabe ["Gallium Arsenide

and Related Compounds", Oiso, (1981) (Inst. Phys. Conf. Ser. no.63) p.95-100]

[4] G.B Stringfellow, R.Linnebach [J. Appl. Phys. (USA) vol.51 no.4 (1980) p.2212-7]

[5] T.F.Keuch, E.Veuhoff, D.L.Woolford, J.A.Bradley ["Gallium Arsenide and Related Compounds", Biarritz (1984) (Inst. Phys. Conf. Ser. no.74) p.181-6]

[6] N.Kobayashi, T.Fukui [Electron. Lett. (GB) vol.20 no.21 (1984) p.887-8]

[7] H.Tarao, H.Sunkawa [J. Cryst. Growth (Netherlands) vol. 68 no.1 (1984) p.157-162]

[8] M.J.Tsai, M.M.Tashima, R.L.Moon [J. Electron. Mater.(USA) vol.13 no.12 (1984) p.437-46]

[9] J.R.Shealy, V.G.Kreismanis, D.K.Wagner, J.M.Woodall [Appl. Phys. Lett. (USA) vol.42 no.1 (1983) p.83-5]

[10] G.B.Stringfellow, G.Hom [Appl. Phys. Lett. (USA) vol.34 no.11 (1979) p.794-6]

[11] S.D.Hersee, M.A. di Forte-Poisson, M.Baldy, J.P.Duchemin [J. Cryst. Growth (Netherlands) vol.55 (1981) p.53-7]

[12] M.Erman, P.Frijlink, J.B.Theeton, C.Alibert, S.Gaillard ["Gallium Arsenide and Related Compounds", Biarritz, 1984 (Inst. Phys. Conf. Ser.no.74) p.327-32]

[13] N.Bottka, R.S.Sillmon, W.F.Tseng [J. Cryst. Growth (Netherlands) vol.68 no.1 (1984) p.54-9]

[14] J.Hallais, J.P.Andre, A.Mircrea-Roussel, J.Varon, M.C Boissy, A.T.Vink [J. Electron. Mater. (USA) vol.10 no.4 (1981) p.665-82]

[15] T.F.Kuech, E.Veuhoff, B.S Meyerson [J. Cryst. Growth (Netherlands) vol.68 no.1 (1984) p.48-53]

[16] S.Nakatsuka, Y.Ono, T.Kajimura, M.Nakamura [15th Conf. Solid State Devices & Materials, Tokyo (1983) extended Abstracts p.297-300]

[17] V.Swaminathan, J.L.Zilko, S.F.Nygren [Mater. Lett. (Netherlands) vol.2 no.4B (1984) p.308-12]

[18] J.J.Yang, W.I.Simpson, L.A.Moudy ["Gallium Arsenide and Related Compounds", Oiso ,1981 (Inst. Phys. Conf. Ser no.63) p.107-112]

[19] J.J.Yang, L.A.Moudy, W.I.Simpson [Appl. Phys. Lett (USA) vol.40 no.3 (1982) p.244-6]

[20] D.V.Lang, R.A.Logan, M.Jaros [Phys. Rev. B (USA) vol.19 no.2 (1979) p.1015-30]

[21] E.F.Schubert, K.Ploog [Phys. Rev. B (USA) vol.30 no.12 (1984) p.7021-9]

[22] A.R.Peaker [EMIS Datareview RN=15602 (Nov 1985) "DX defect centres in AlGaAs]

[23] V.Aebi, C.B.Cooper III, R.L.Moon, R.R Saxena [J. Cryst. Growth (Netherlands) vol.55 (1981) p.517-25]

[24] J.P.Andre, C.Schiller, A.Mittoneau, A.Briere, J.Y.Aupied ["Gallium Arsenide and Related Compounds", Albuquerque 1982 (Inst. Phys. Conf Ser no.65) p.107-12]

18.7 CARRIER CONCENTRATIONS IN MBE AlGaAs

by D.Lee

January 1986
EMIS Datareview RN=15611

Most material has been grown with V/III flux ratios between about 2 and 5 and substrate temperatures between 600 and 720 deg C. Best quality material is generally reported to be obtained at about 700 C but Heiblum et al [1] report improved quality GaAlAs at 600 deg C on reduction of the growth rate from 1.4 to 0.14 microns per hour (for Al ractions x in the range 0.3-0.43). Kunzel et al [2] report good quality material (x = 0.33) only between 630 and 680 deg C with semi-insulating material obtained below 600 deg C. Intentially undoped GaAlAs is reported to be p-type [1,2] with published background carrier densities of 5 X 10*14 cm*-3 (x = 0.43) [1] and below 10*15 cm*-3 [2]. The principal background acceptor species is carbon [2,3].

The majority of published work has been concentrated on GaAlAs doped with silicon from elemental Si cells typically heated to 950-1200 deg C. Si doped GaAlAs is n-type over the full composition range [2] with a typical doping range of mid 10*16 to mid 10*18 cm*-3 [4]. Nearly all Si is electrically active up to 2 X 10*10 cm*-3, with precipitation occurring at higher concentrations [2,4,5]. Both the Si and free electron concentration are reported to increase proportionally with Si vapour pressure up to about (2-4) X 10*18 cm*-3 [4-6]. The Si concentration is reported to decrease with increasing substrate temperature above about 650 deg C (possibly due to a decrease in the effective V/III ratio) [4,5].

Chen et al [7] studied the effect of As vapour species on the electrical properties of Si doped GaAlAs (x=0.3) and observed an increase in doping efficiency and a factor of two reduction in total trap concentration with the use of As2 rather than As4 for substrate temperatures of 610 and 660 deg C. At 700 deg C similar doping efficiencies were observed with both As2 and As4.

Kondo et al [8] studied the effect of hydrogen during MBE growth of Si doped GaAlAs (x=0.2, 680 deg C). Whilst the photoluminescence intensity was increased by a factor of 7 for hydrogen pressures of 5 X 10*-8 to 5 X 10*-7 torr the effect on the carrier concentration was negligible (an increase from 7 X 10*17 to 9 X 10*17 cm*-3).

Briones et al [9] investigated Si doping using silane and report carrier concentrations betweeh about 10*17 and 10*18 cm*-3 increasing as the square root of the silane to Group III flux ratio. The same workers also studied the effect of doping with H2S and report carrier densities between 10*17 and 10*18 cm*-3 (x = 0.14 & 0.43) but very small doping efficiencies (10*-5 - 10*-4) [10].

Tin has also been used as a dopant in MBE GaAlAs [6,11]. Carrier densities in the range 2 X 10*16 to 2 X 10*18 cm*-3 were reported for substrate temperatures in the range 580-630 deg C and x between 0.17 - 3.75. Mobilities comparable to LPE grown GaAlAs (with n = 10*17 to 10*18 cm-3) were obtained in material grown at 630 deg C [11].

Many authors have reported the variation of carrier concentration with alloy composition in n-type GaAlAs [2,6,9-16]. For low aluminium fractions donors are shallow and the carrier concentration is independent of x. As x increases from about 0.2 to 0.4 an increasing fraction of donors become deep (DX centres [13,18]), although the total number of donors is reported to remain constant [14,15]. The carrier concentration derived from 300 deg K Hall measurements is seen to decrease sharply with increasing x and the observed donor ionization energy to increase to a maximum value of around 250 meV [6,11,14,17]. For higher values of x Ed is reported to decrease and n to increase smoothly towards AlAs. In the range of x where deep levels are dominant a persistent increase in carrier concentration is observed on illumination at temperatures below about 100 deg K [13,16,17]. Room temperature C-V measurements give the total donor concentration (i.e. shallow donors plus DX centres) [15] which can be more than a factor of 10 greater than the carrier concentration derived from 300 deg K Hall measurements [19]. C-V measurements at 77 deg K, however, give only the shallow donor concentration (due to the small thermal ionization rate of the DX centre at 77 deg K [15]. Ishikawa et al [16] report the concentration of DX centres to be independent of MBE growth conditions (Si doped material). Similar effects are observed for all shallow donor impurities in MBE and also LPE & MOCVD grown GaAlAs. DX centres are discussed in more detail in another Datareview [20].

The most common p-type dopant in GaAlAs is Be [21-23]. Hole concentrations in the range mid 10*16 to 3 X 10*19 cm*-3 have been achieved [22], with recent layers reported to be generally uncompensated [23]. The carrier concentration is reported to be practically independent of substrate temperature, V/III ratio and also of x up to 0.33 for p above 10*18 cm*-3 although p decreased slightly with increasing x at higher doping levels [22]. More recently, Chand et al [23] report the binding energy of Be to increase from 15 to 41 meV as x increased from zero to 0.7. Chand et al [23] also observed a persistent reduction in hole concentration together with an increase in mobility on illumination at 77 deg K for material with x between 0.2 - 0.4 independent of Be concentration. Ilegems [22] investigated the effects of annealing on p-n junctions and concluded that no appreciable diffusion of Be occurs at temperatures up to 800 deg C.

Zinc has a very low sticking coefficient in GaAlAs growth (about 10*-7) but Takahashi and co-workers [24,25] have demonstrated that an effective sticking coefficient of up to 0.03 (independent of x) can be achieved by use of medium energy (0.2-1.5 keV) Zn+ ions. Hole densities in the range 10*16 to 10*19 cm*=3 are reported, the carrier concentration being controllable by varying the ion current [25].

Whilst hole concentrations up to 10*19 cm*-3 have been reported using Mg this dopant has a very small sticking coefficient which is also a strong function of x [26].

REFERENCES
----

[1] M.Heiblum, E.E.Mendez, L.Osterling [J.Appl.Phys. (USA) vol.54 no.12), (1983) p.6982-8]

[2] H.Kunzel, K.Ploog, K.Wunstel, B.L.Zhou [J.Electron.Mat. (USA) vol.13 no.2 (1984) p.281-308]

[3] G.Wicks, W.I.Wang, C.E.C.Wood, L.F.Eastman, L.Rathbun [J.Appl.Phys. (USA) vol.52 no.9 (1981) p.5792-6]

[4] R.Fischer, T.J.Drummond, R.E.Thorne, W.G.Lyons, H.Morkoc

[Thin Solid Films (Switzerland) vol.99 (1983) p.391-7]
[5] T.J.Drummond, W.G.Lyons, R.Fischer, R.E.Thorne, H.Morkoc [J.Vac.Sci. & Technol. (USA) vol.21 no.4 (1982) p.957-60]
[6] T.Ishibashi, S.Tarucha, H.Okamoto [Jpn. J.Appl.Phys. (Japan) vol.21 no.8 (1982) p.L476-8]
[7] R.T.Chen, N.H.Sheng, D.L.Miller [J.Vac.Sci. & Technol.B (USA) vol.3 no.2 (1985) p.652]
[8] K.Kondo, S.Muto, K.Nanbu, T.Ishikawa, S.Hayamizu, H.Hashimoto [Jpn.J.Appl.Phys. (Japan) vol.22 no.2 (1983) p.L121-3]
[9] F.Briones, D.Golmayo, L.Gonzalez, J.L.de Miguel [J.Vac.Sci. & Technol.B (USA) vol.3 no.2 (1985) p.568-71]
[10] F.Briones, D.Golmayo, L.Gonzalez, J.L. de Miguel [Appl.Phys.A vol.36 no.3 (1985) p.147-51]
[11] H.Morkoc, A.Y.Cho, C.Radice [J.Appl.Phys. (USA) vol.51 no.9 (1980) p.4882-4]
[12] T.Ishakawa, J.Saito, S.Sasa, S.Hiyamizu [Jpn.J.Appl.Phys. (Japan) vol.21 no.11 (1982) p.L675-6]
[13] E.F.Schubert, K.Ploog [Phys.Rev.B (USA) vol.30 no.12 (1984) p.7021-9]
[14] M.O.Watanabe, H.Maeda [Jpn.J Appl.Phys. (Japan) vol.23 no.9 (1984) p.L734-6]
[15] M.O.Watanabe, K.Morizuka, M.Mashita, Y.Ashizawa, Y.Zohta [Jpn. J Appl.Phys. (Japan) vol.23 no.2 (1984) p.L103-5]
[16] T.Ishikawa, K.Kondo, S.Hiyamazu, A.Shibatomi [Jpn. J Appl.Phys. (Japan) vol.24 no.6 (1985) p.L408-10]
[17] N.Chand et al [Phys.Rev.B (USA) vol.30 (1984) p.4481-92]
[18] D.V Lang, R.A Logan, M.Jaros [Phys.Rev.B (USA) vol.19 no.2 (1979) p.1015-30]
[19] M.Heiblum [J.Vac.Sci.& Technol. (USA) vol.3 no.3 (1985) p.820-2]
[20] A.R.Peaker [EMIS Datareview RN=15602 (Nov 1985) "DX defect centres in AlGaAs"]
[21] K.Ploog [in "Crystals" vol.3: III-V Semiconductors (Springer Verlag, 1980) p.73-162]
[22] M.Ilegems [J.Appl.Phys.(USA) vol.48 no.3 (1977) p.1278-87]
[23] N.Chand et al [J. Vac. Sci. & Technol.B (USA) vol.3 no.2 (1985) p.644-8]
[24] N.Matsunaga, T Suzuki, K.Takahashi [J.Appl.Phys. (USA) vol.49 no.12 (1978) p.5710-5]
[25] T.Suzuki, M.Konagai, K.Takahashi [Thin Solid Films (Switzerland) vol.60 (1979) p.85-9]
[26] A.Y Cho, J.R.Arthur [Solid State Chem. vol.10 no.3 (1975) p.157-91]

18.8 ETCHING RATES OF REACTIVE ION AND REACTIVE ION-BEAM ETCHED AlGaAs

by S.P.Speakman

December 1985
EMIS Datareview RN=15495

This Datareview discusses the etch rate of AlGaAs as a function of Al content and discusses present selectivity over GaAs, using dry etching processes [1].

It would be potentially useful to have available a dry etching process which gives a high selectivity between GaAs and AlGaAs, for the fabrication of structures such as GaAs waveguides embedded within an AlGaAs layer. The selectivity of such a process depends upon the etching technique chosen and the proportion and type of chemically active ions used. To date only RIE (reactive ion etching) [2-6] and RIBE (reactive ion beam etching) [7] techniques have yielded any data appertaining to Al(1-x)Ga(x)As etching as a function of the Al content and selectivity over GaAs, (no data on selectivity over masking materials are included here). Since selectivity is an important consideration in device fabrication it seems surprising that this topic has not attracted more attention.

RIE has been used with the following gas mixtures:-

Gas Mixture	GaAs:AlGaAs Etch Ratio	Ga Content x	Etch Rate micron/min.	Growth Type
CCl2F2/He	200 : 1	0.7	---	---
Cl2/Ar	4 : 1	0.67	0.6	LPE
BCl3	1 : 1	0.45	0.15	MOCVD
BCl3/Cl2	1.5 : 1	0.45	0.4	MOCVD
CCl4/Cl2	1 : 1	----	---	---
CCl2F2/Ar	1.08 : 1	----	0.14	MBE

From the above data it is not viable to establish an etch rate of AlGaAs as a function of Al content due to the different gas chemistries involved. Moreover, some of the AlGaAs layers are grown via LPE, MOCVD and MBE, introducing a further complication. However, it is encouraging to note that in all cases RIE has been shown to produce vertical walled profiles with clean, smooth surfaces, as well as exhibiting good run to run reproducibility.

For the Cl2-Ar gas mixture [3], it has been observed that the AlGaAs etch-rate decreases as the Al content increases for double heterostructure etching, although no data were presented.

At the time of writing the reviewer is aware of only one paper on RIBE etching of AlGaAs [7]. Here, a purpose-built etching system was designed using only a Cl2 etching atmosphere. The GaAs to AlGaAs etching ratio was shown to be 1:1. In this study it was further shown that the equal rate of etching remains constant for variations of Al content from 0.15 to 0.30. Equal rate etching was also independent of ion-extraction voltage in the range 400-500 V. As with the RIE technique, smooth, clean, nearly vertical walled profiles were observed.

It is most apparent from the data collated to date that no observations of AlGaAs dry etching quicker than GaAs have been reported, irrespective of Al content. It will be interesting to see how magnetron ion etching and laser-enhanced reactive ion etching can influence the GaAs:AlGaAs etch selectivity.

REFERENCES

[1] G.F.Doughty [Semiconductor Symposium (Birmingham, 1984) p.63]

[2] H.Tamura, H.Kurihara [Jpn. J. Appl. Phys. (Japan) vol.23 no.9 (1984) p.L731-3]

[3] H.Yamada, H.Ito, H.Inaba [J. Vac. Sci. & Technol. B (USA) vol.3 (1985) p.884-8]

[4] H.Hikosaka et al [Jpn. J. Appl. Phys. (Japan) vol.20 (1981) p.L847-50]

[5] J.E.Curran, D.J.McCulloch [Vacuum (GB) (extended abstracts) vol.35 no.10-11 (1985) p.504-6]

[6] H.Nagasaka, H.Okano, N.Motegi [Proc. Symp. on Dry Processing (Institute of Electrical Engineers of Japan, Tokyo, 1982) p.79]

[7] K.Asakawa, S.Sugata [Jpn. J. Appl. Phys. (Japan) vol.22 no.10 (1983) p.L653-5]

SUBJECT INDEX

ablation, 16.5
absorption edge, 7.2, 9.2
activation efficiency, 3.1, 3.2
adsorption
  Al, 17.2
AlGaAs
  carrier concentrations, 18.5, 18.6, 18.7
  defect energy levels, 18.1, 18.2
  direct-indirect transition, 18.2
  doping efficiency, 18.6, 18.7
  DX centres, 18.2
  EL2, 18.2
  electron mobility, 18.4
  epitaxial, 18.5, 18.6, 18.7
  etching rates, 18.8
  growth rate, 18.6, 18.7
  persistent photoconductivity, 18.2
  photoluminescence intensity, 18.7
  photoluminescence spectra, 18.3
alloying, 17.4
amorphization, 16.1, 16.3
amorphous GaAs, 7.4
amphoteric impurities, 2.2, 3.1, 3.3, 12.9
annealing
  laser, 3.1, 3.2, 17.4
  thermal, 2.2, 3.1, 3.2, 5.10, 12.2, 13.9, 13.10, 17.1, 17.3, 17.4, 17.8
antisite defects, 12.1, 12.3, 13.3, 13.4, 13.5, 13.7
anti-Stokes emission, 12.7
Arrhenius plots, 13.6
asymmetric dimer model, 15.1
atomic arrangement, 15.1

bakeout, 5.3, 12.4
band gap
  direct-indirect transition, 12.8, 18.2
  free electron concentration effects, 8.9
  pressure dependence, 5.12, 7.2, 12.8
  temperature dependence, 7.1, 8.5
bulk GaAs
  defect energy levels, 13.4
  electron lifetime, 5.14
  electron mobility, 5.1, 5.2
  hole diffusion length, 6.6
  hole lifetime, 6.7
  hole mobility, 6.1, 6.2
  IR absorption bands, 10.1
  photoconductivity spectra, 11.1
  photoluminescence spectra, 12.1, 12.2, 12.3
bulk modulus, 1.4
  adiabatic, 1.4
  isothermal, 1.4
Burstein shift, 8.9

capture cross sections, 13.6
carrier concentration profiles, 2.1, 3.1, 3.2
carrier concentrations, 2.1, 3.1, 3.2, 3.3, 8.9
  AlGaAs, 18.5, 18.6, 18.7
carrier ionization rates
  doping dependence, 4.2, 4.3
  electric field dependence, 4.3
  orientation dependence, 4.1
  temperature dependence, 4.1
chemical shift, 11.2, 11.4, 12.10
chemisorption, 17.2, 17.7, 17.11
compliance, 1.6
conduction band degeneracy, 8.9
conduction band effective mass, 8.9
conductivity, 2.1
  temperature dependence, 2.1
conductivity-type conversion, 2.1, 2.2, 3.1, 3.2, 12.6, 13.10, 14.2
contact ageing, 17.4
contact resistivity, 6.1, 17.4

dead space, 4.2, 4.3
Debye temperature, 1.7, 1.8
deep levels: see defect energy levels, EL2
defect densities, 13.1, 13.2
defect energy levels (see also EL2), 13.5
  AlGaAs, 18.1, 18.2
  electron irradiation, 13.10
  interface states, 17.9
  IR absorption, 10.1, 13.3
  IR imaging, 13.1
  LPE, 13.6
  MBE, 13.9
  melt-grown, 13.4
  MOCVD, 13.7
  MOVPE, 13.8
  nomenclature, 13.6, 13.7, 13.8, 13.9
  photoconductivity spectra, 11.1, 11.2, 11.3, 11.4, 11.5, 11.6, 11.7
  photoluminescence spectra, 12.1, 12.2, 12.3, 12.4, 12.5, 12.6, 12.7, 12.8, 12.9, 12.10
  SI LEC, 13.1, 13.2
  VPE, 13.7
defect mapping, 13.1
defect production rate
  electron irradiation, 13.10
defect visualisation, 16.5
deformation potential, 7.3
density, 1.1
  pressure dependence, 6.4
  stoichiometry, 1.1
  temperature dependence 1.1
dielectric constant
  free carrier effects, 8.9
  infrared, 8.1, 8.2
  static, 8.1, 8.2
  temperature dependence, 8.2
diffusion, 3.1, 3.2

diffusion coefficients of impurities
  concentration dependence, 14.3, 14.4, 14.5
  Cr, 14.1
  doping dependence, 14.5
  Mn, 14.2
  S, 14.6
  Si, 14.4
  Sn, 14.5
  stoichiometry, 14.3
  temperature dependence, 14.1, 14.2, 14.3, 14.4, 14.5, 14.6
  vapour pressure dependence, 14.2, 14.3, 14.4, 14.6
  Zn, 14.3
diffusion profiles, 14.1, 14.2, 14.3, 14.4, 14.5, 14.6
dipole moments, 8.1
dislocation density, 2.1, 13.1, 13.2
dislocations, 12.1, 13.1, 13.3
dispersion relations, 8.3
distribution coefficient, 13.6
dopant identification, 12.8
doping efficiency, 3.1, 3.2
  AlGaAs, 18.6, 18.7
Drude model, 8.9

EL2, 2.1, 11.1, 12.1, 13.1, 13.2, 13.3, 13.4, 13.7, 13.9
  AlGaAs, 18.2
elastic properties, 1.4, 1.5, 1.6
electrical activity, 2.2, 5.10
electroabsorption, 9.2
electron diffusion coefficient, 5.13
electron diffusion length, 5.13
electron effective mass, 5.11
  pressure dependence, 5.12
electron irradiation, 13.10
electron lifetime, 5.14
electron mobility
  AlGaAs, 18.4
  bulk, 5.1, 5.2
  carrier concentration dependence, 5.3
  doping dependence, 5.6, 5.9, 5.10
  ion-implanted, 5.10
  LPE, 5.3, 5.11
  MBE, 5.7, 5.8, 5.9
  orientation dependence, 5.6
  pressure dependence, 5.11, 6.4
  temperature dependence, 5.2, 5.5, 5.8
  VPE, 5.4, 5.5, 5.6
electro-optical coefficients, 9.1
  temperature dependence, 9.1
energy band offsets at heterostructure interface, 17.12
etching rates, 16.1, 16.2, 16.3, 16.4, 16.5
  AlGaAs, 18.8
etch pits, 16.1, 16.2, 17.3

etched selectivity, 18.8
extinction coefficient, 8.1, 8.3, 8.5, 8.6, 8.7,
  8.8

facets, 15.1
Fermi level, 12.1
Fermi level pinning, 17.7, 17.11
FET, 2.2, 5.1, 12.5, 18.2
Fick's law, 14.1, 14.2, 14.3
Franz-Keldysh effect, 9.2
free-carrier absorption, 8.9

Gruneisen parameter, 1.7

Hall factor, 6.1
heterostructure interface, 17.12
hole diffusion coefficient, 6.6
hole diffusion length, 6.6
hole effective mass, 6.4
  pressure dependence, 6.5
hole lifetime, 6.7
hole mobility
  doped GaAs, 6.2
  ion implanted GaAs, 6.2
  pressure dependence, 6.4, 6.5
  temperature dependence, 6.2, 6.3
  undoped GaAs, 6.1
hopping conduction, 2.2

interface band bending, 17.11
interface degradation, 12.4
interface state density, 17.11
interface structure
  Ag/GaAs, 17.1
  Al/GaAs, 17.2
  Au/GaAs, 17.3
  Au-Ge/GaAs, 17.4
  Cu/GaAs, 17.5
interfacial oxides, 17.8, 17.10
ion-beam-assisted etching, 16.4
ion beam milling, 16.3
ion implantation, 2.2, 3.1, 3.2, 5.10, 6.2
  dual implantation, 3.1, 3.2
IR absorption, 8.3, 10.1, 10.2, 13.1, 13.2
IR absorption coefficient, 10.1
IR imaging of deep levels, 13.1
isoconcentration technique, 14.3

laser-assisted etching, 16.5
laser pulse initiated diffusion, 5.6
laser self-defocussing, 10.2
lattice dilation, 1.7, 7.1
  impurities, 1.3
lattice disorder, 2.2
lattice parameter, 1.2, 1.3
  doping dependence, 1.3
  facet effect, 1.2
  stoichiometry, 1.2
  surface damage effects, 1.2
  temperature dependence, 1.2

lattice scattering, 6.1
lattice vibrations, 8.3
line defects, 13.2
localised vibrational modes, 10.1
LPE GaAs
  carrier concentrations, 3.3
  carrier ionization rates, 4.1, 4.3
  defect energy levels, 13.6
  electron diffusion length, 5.13
  electron effective mass, 5.12
  electron lifetime, 5.14
  electron mobility, 5.3, 5.11
  hole diffusion length, 6.6
  hole mobility, 6.4
  photoconductivity spectra, 11.3
  photoluminescence spectra, 12.4
LSI, 12.2

MBE GaAs
  carrier concentrations, 3.3
  defect energy levels, 13.9
  electron mobility, 5.7, 5.8, 5.9
  interface structures, 17.1, 17.2
  photoconductivity spectra, 11.6
  photoluminescence spectra, 12.7, 12.8
  Schottky barrier heights, 17.6, 17.7
  surface structure, 15.1
melting point, 1.7, 1.8
MESFET, 2.2
metallization, 17.1, 17.3, 17.4
MIS structures, 17.11
minority carriers, 5.12, 5.14, 6.6, 6.7
MOCVD GaAs
  band gap, 7.2
  carrier concentrations, 3.3
  defect energy levels, 13.7
  electron diffusion length, 5.13
  electron mobility, 5.4, 5.6
  hole lifetime, 6.7
  photoconductivity spectra, 11.5
  photoluminescence spectra, 12.6
MOVPE GaAs
  defect energy levels, 13.8
  electron mobility, 5.5
  photoluminescence spectra, 12.7

native defects, 13.2, 13.3, 13.4
neutron transmutation doping, 11.7, 12.9

optical absorption coefficient, 8.1, 8.5, 8.7, 8.8, 8.9
optical functions, 8.1, 8.2, 8.3, 8.4, 8.5, 8.6, 8.7, 8.8
  extrinsic effects, 8.9
optical gap (amorphous), 7.4
  composition dependence, 7.4
  pressure dependence, 7.4
oscillator energies, 8.4
oscillator strengths, 8.4

out-diffusion, 2.1, 3.2, 14.1, 14.2, 17.3
oxidation, 15.2, 15.3
oxidation rates, 15.3
oxide layers, 8.6, 15.3, 16.2
  structure, 15.2

pattern formation, 16.1, 16.3, 16.5
permittivity, 8.1, 8.3, 8.4
phase modulators, 9.1
photoconductivity, 11.1
photoconductivity quenching, 11.1
photoconductivity spectra, 11.2
  bulk GaAs, 11.1
  epitaxial GaAs, 11.3, 11.4, 11.5, 11.6
  neutron transmutation-doped, 11.7
  shallow donors, 11.2, 11.3, 11.4, 11.5, 11.6, 11.7
photoluminescence decay, 12.7
photoluminescence intensity, 12.1, 12.2, 12.4, 12.6, 12.7, 12.8, 12.10
  AlGaAs, 18.7
photoluminescence quenching, 12.8
photoluminescence spectra
  AlGaAs, 18.3
  bulk GaAs, 12.1, 12.2, 12.3, 12.10
  doped GaAs, 12.2, 12.8, 12.9, 12.10
  epitaxial GaAs, 12.4, 12.5, 12.6, 12.7, 12.9, 12.10
  growth rate dependence, 12.7
  ion implanted GaAs, 12.8, 12.9
  KP lines, 12.7
  pressure dependence, 12.8, 12.10
  shallow donors, 12.10
  temperature dependence, 12.7
plasma etching, 16.2
point defects, 13.2, 13.3, 15.1

reactive ion-beam etching, 16.4
reactive ion etching, 16.4
reactivity of metals, 17.10
recombination, 5.13, 5.14, 6.7, 12.6, 13.6, 13.10
reflectance, 8.3, 8.7, 8.9
refractive index, 8.1, 8.3, 8.4, 8.5, 8.6, 8.8, 10.2
residual stress, 12.2
resistivity, 2.1, 2.2
  magnetic field dependence, 2.1
  stoichiometry, 2.1
reststrahlen region, 8.3
Richardson constant, 17.7

Schottky barrier heights
  Ag/GaAs interface, 17.6
  Al/GaAs interface, 17.7
  annealing effects, 17.8, 17.9, 17.10
  Au/GaAs interface, 17.8
  interfacial layer effects, 17.7, 17.8, 17.10
  Pt/GaAs interface, 17.9
  surface structure effects, 17.6, 17.7
  W/GaAs interface, 17.10

Schottky barriers, 17.1, 17.3, 17.5, 17.6
Schottky diodes, 17.6, 17.7, 17.10
semi-insulating LEC
  defect densities, 13.1, 13.2
  native defects, 13.3
  photoluminescence spectra, 12.1, 12.2
  resistivity, 2.1, 2.2
shallow acceptors, 12.5
sheet carrier concentration, 5.6, 5.10
sheet resistivity, 2.2
  annealing effects, 2.2
  ion dose effects, 2.2
  temperature dependence, 2.2
  thermal stability, 2.2
solar cells, 18.5
specific heat, 1.8
sputter etching, 16.3
stiffness constants, 1.5
  adiabatic, 1.5
  isothermal, 1.5
stoichiometry, 1.1, 1.2, 12.3
superlattices, 12.6, 17.12
surface layers, 8.6, 8.7, 12.2
surface relaxation, 15.1, 17.2
surface roughness, 16.2, 16.4
surface structure, 15.1
  electronic structure, 15.1
  orientation, 15.1
  stoichiometry, 15.1

thermal conductivity, 1.9
  doping dependence, 1.9
thermal diffusivity, 1.10
thermal expansion coefficient, 1.7
trap nomenclature, 13.6, 13.7, 13.8, 13.9
traps, 13.4, 13.6, 13.7, 13.9, 13.10, 18.3
two-dimensional electron gas, 18.4
two-photon IR absorption coefficient, 10.2

vias, 16.5
volume defects, 13.2
VPE GaAs
  carrier concentrations, 3.3
  defect energy levels, 13.7
  electron mobility, 5.4, 5.5, 5.6
  photoconductivity spectra, 11.4
  photoluminescence spectra, 12.5

wafer uniformity, 12.2
wet etching, 16.1, 16.5
work function, 15.1, 17.9